INTRODUCTION TO PLANT DISEASES: Identification and Management

Second Edition

INTRODUCTION TO PLANT DISEASES: Identification and Management

Second Edition

George B. Lucas
C. Lee Campbell
Leon T. Lucas

An AVI Book
Published by Van Nostrand Reinhold
New York

An AVI Book
AVI is an imprint of Van Nostrand Reinhold

Library of Congress Catalog Card Number 91-39434
ISBN 0-442-00578-4

Manufactured in the United States of America

Published by Van Nostrand Reinhold
115 Fifth Avenue
New York, NY 10003

Chapman and Hall
2-6 Boundary Row
London, SE 1 8HN

Thomas Nelson Australia
102 Dodds Street
South Melbourne 3205
Victoria, Australia

Nelson Canada
1120 Birchmount Road
Scarborough, Ontario M1K 5G4, Canada

16 15 14 13 12 11 10 9 8 7 6 5 4 3 2 1

Library of Congress Cataloging-in-Publication Data

Lucas, George Blanchard.
Introduction to plant diseases : identification and management / by George B. Lucas, C. Lee Campbell, Leon T. Lucas.—2nd ed.
p. cm.
"An AVI book."
Includes bibliographical references and index.
ISBN 0-442-00578-4
1. Plant diseases. 2. Phytopathogenic microorganisms—Control. 3. Plant diseases—Diagnosis. 4. Phytopathogenic microorganisms—Identification. I. Campbell, C. L. (C. Lee.) II. Lucas, L. T. III. Title.
SB731.L83 1992
632′.3—dc20

91-39434
CIP

To
Robert Aycock
Friend, Scientist, Scholar, Mentor

Contents

Preface to the Second Edition xiii

1 Agriculture, Plant Diseases, and Human Affairs 1

The Study of Plant Diseases 3
Plant Disease Losses 4
Effects of Modern Agriculture on Plant Diseases 5
What Is a Plant Disease? 6
Recognizing Plant Diseases 6
Management of Plant Pests 7

2 Causes of Plant Diseases 9

Infectious Plant Diseases 9
Disease Complexes 12
Noninfectious Plant Diseases 12
Names of Pathogens and Diseases 14

3 History of Plant Pathology 15

In Ancient Times 15
From 1600 to 1800 16
The Golden Age of Biology, 1840 to 1900 17
The Twentieth Century 18

4 Development of Plant Diseases **20**

The Occurrence of Plant Disease 20
The Disease Cycle 21
Environmental Effects on Disease Development 25
Epidemiology of Plant Diseases 27

5 Integrated Pest Management **30**

Evolution of Pest Management Practices 30
Low Input Sustainable Agriculture (LISA) 33
The IPM Concept 34
Economic Thresholds 35
Sources of Information 35

6 Management of Plant Diseases **38**

Avoidance of the Pathogen by Exclusion and Evasion 39
Eradication of Pathogens 39
Reduction of Inoculum 41
Adjustment of the Environment 52
Biological Control 54
Disease-Resistant Cultivars 58
Certified Seed and Seed Storage 62
Summary 66

7 Chemical Management of Plant Diseases **67**

How Pesticides Work 68
Names of Pest Management Chemicals 69
Pesticide Safety 69
Pesticide Formulations 72
Soil Treatment 75
Seed Treatment 81
Postharvest Treatment 81
Treatment of Growing Plants 82
Pesticide Sprayers and Dusters 83
Spray Equipment Parts 87
Pesticide Application Guidelines 95
Cleaning and Storing Equipment 97
Calibration of Equipment 98
Forecasting Plant Diseases as an Aid to Pesticide Scheduling 105
Protecting the Environment 107
Summary 111

8 Biotechnology **112**

Vegetative Propagation 115
Regeneration of Plants from Cells 118

Transplanting DNA from One Cell to Another 121
Citizen Concerns and the Need for Biotechnology 125

9 Nematodes 127

General Nature and Importance 127
Size, Distribution, and Anatomy 129
Feeding Behavior and Plant Injury 130
Life History, Reproduction, and Dispersal 132
Enemies of Nematodes 133
Environmental Effects on Nematode Survival 134
Nematodes and Other Diseases 135
Management 136

10 Diseases Caused by Nematodes 139

Root-Knot Nematodes 140
Cyst Nematodes 145
Lesion Nematodes 147
Stem and Bulb Nematodes 150
Stubby-Root Nematodes 151

11 Fungi 154

General Nature and Importance 154
Growth, Reproduction, and Classification 157
Distribution and Dispersal 160

12 Diseases Caused by Soilborne Fungi 162

Phytophthora Root Rots 163
Damping-off 170
Rhizoctonia Rots 173
Fusarium Root Rots, Yellows, and Wilts 178
Verticillium Wilts 181
Southern Blight 183
Crown and Stem Rots and Watery Soft Rots 186
Cylindrocladium Black Rot 189

13 Diseases Caused by Airborne Fungi 192

Rusts 192
Smuts 201
Downy Mildews 204
Powdery Mildews 212
Ergot 215
Blast of Rice 217
Gray Mold or Botrytis Blight 219
Cercospora Leaf Spots 221

Helminthosporium Leaf Spots 225
Alternaria Leaf Spots 230
Anthracnose 233
Apple Scab 236
Rhizopus Soft Rot 240

14 Fungal Diseases of Shade and Forest Trees and Decay in Wood 243

Chestnut Blight 244
Dutch Elm Disease 245
Oak Wilt 248
Annosus Root and Butt Rot 249
Fusiform Rust 250
Dogweed Anthracnose 253
Little Leaf of Pine 254
Wood Decay in Buildings 255

15 Bacteria 259

Size and Structure 260
Identifying Bacteria 260
Growth and Reproduction 263
Distribution and Dispersal 264
Managing Plant-Pathogenic Bacteria 264

16 Diseases Caused by Bacteria and Mycoplasmas 266

Bacterial Wilt 267
Fire Blight of Apple and Pear 269
Soft Rots 272
Bacterial Wilt of Cucumbers 274
Bacterial Canker of Tomato 274
Bacterial Spot of Peach and Plum 276
Crown Gall 278
Mycoplasma Diseases 280

17 Viruses 283

Discovering Plant Viruses 283
Naming of Viruses 285
Symptoms of Virus Diseases 286
Properties and Morphology 287
Identification and Detection 288
Transmission 289
Management 290

18 Diseases Caused by Viruses **291**

Tobacco Mosaic 292
Vein Banding or Streak 297
Tobacco Vein Mottle 299
Tobacco Etch 299
Tomato Spotted Wilt 299
Cucumber Mosaic 302
Barley Yellow Dwarf 304
Soilborne Wheat Mosaic 305
Maize Dwarf Mosaic and Maize Chlorotic Dwarf 306
Viroid Diseases 308

19 Diseases Caused by Parasitic Plants **309**

Broomrapes 310
Witchweed 311
Mistletoes 312
Dodder 314

20 Abiotic Agents **316**

Deficiencies and Toxicities 316
Weather Extremes 318
Abnormal Water Conditions 320
Phytotoxic Chemicals 321
Air Pollution 324

Glossary **333**

Suggested Readings **347**

Index **351**

Preface to the Second Edition

Every year we see a remarkable increase in scientific knowledge. We are learning more each day about the world around us, about the numerous biological organisms of the biosphere, about the physical and chemical processes that shaped and continue to change our planet. The cataloging, retrieval, dissemination, and use of this new information along with the continued development of new computer technology provide some of the most challenging problems in science as we enter the Information Age.

With the explosion of knowledge in science, it is especially important that students in introductory courses learn not only the basic material of a subject, but also about the newest developments in that subject. With this goal in mind, we have prepared a second edition of *Introduction to Plant Diseases: Identification and Management.* We prepared this edition with the same general purpose that we had for the first edition—to provide practical, up-to-date information that helps in the successful management of diseases on food, fiber, and landscape plants for students who do not have a strong background in the biological sciences. We included new information on (1) the precise identification of diseases and the pathogens that cause them, (2) the development of epidemics of plant diseases, (3) the application of biotechnology in plant pathology, (4) the use of alternative methods of crop production and disease management that help protect the environment, and (5) diseases that have become more important since the first edition was published.

We are indebted to our colleagues and students for their criticisms, suggestions, and ideas about what an introductory textbook should be. We thank William L. Klarman and Larry F. Grand for their recognition of the importance of preparing a textbook such as ours. We are also grateful to the members of the Department of Plant

Pathology and to the faculty and staff of the Cooperative Extension Service, the Agricultural and Life Sciences Research Service, and the Academic Affairs Division of the College of Agriculture and Life Sciences at North Carolina State University who allowed us to use their photographs and other materials.

Our special thanks to Joyce Denmark for her dedication and tolerance as she worked her magic at the word processor to transform handwritten drafts and revisions into readable prose. We also especially want to thank W. Marvin Williams and Charles R. Harper for the preparation and printing of the new figures, Scot C. Nelson and George R. Hess for assistance with the computer-generated graphics, and William E. Brintnall, Jr. for assistance in assembling and preparing the manuscript for submission. Also, our thanks to the editors and staff of Van Nostrand Reinhold for their assistance and enthusiasm.

Finally, we are indebted to our families, especially Vernelle, Karen, and Joy, whose love has sustained us.

Raleigh, North Carolina

GEORGE B. LUCAS
C. LEE CAMPBELL
LEON T. LUCAS

INTRODUCTION TO PLANT DISEASES: Identification and Management

Second Edition

1

Agriculture, Plant Diseases, and Human Affairs

The history of mankind is the story of a hungry creature in search of food.

—Hendrick Van Loon

At any given time, on any given day, there is enough food to feed all the people on earth for only about three weeks. Even that small amount of food is unequally distributed, and famine stalks much of the world. One-third of the earth's population awakes hungry each morning and goes to bed hungry at night. Hunger is a way of life in many developing countries. As stated in a 1980 report to President Carter: "For hundreds of millions of the desperately poor, the outlook for food and other necessities of life will be no better by the year 2000. For many it will be worse. . . . Hunger and disease will claim more babies and young children and more of those surviving will be mentally handicapped by childhood malnutrition." A 1989 survey of leaders in the population field resulted in 100% of the respondents saying they expect today's population of 5 billion to be more than 10 billion in the next century, which will place such enormous demands on the biosphere, (e.g., arable soils, water supplies, and food) that in much of the world, human existence may be threatened. This dismal and alarming forecast underscores two of the most important problems facing humankind in the twentieth and twenty-first centuries—overpopulation and hunger.

Another problem we face is sustaining sufficient agricultural production to feed a hungry world while still preserving a clean, safe, and healthy environment for the very people we are trying to feed. Pollution of water and natural areas by pesticides and other agrichemicals is of growing concern. Pollution of the air around us by automobile exhausts and industrial processes is increasing. Loss of precious soils due to improper management and erosion threatens to limit production of vital food crops in many areas of the world. As we approach the twenty-first century, our success and survival as a species on this planet depends on our ability to preserve and protect our natural resources—yet we must continue to place increasing demands on these resources if we are to feed a hungry world.

What has all this to do with plant diseases? Well, humans depend upon plants for their very existence. Only green plants and some blue-green algae can convert solar energy into food. Therefore, humans and other animals exist on earth as guests of the plant kingdom. Three-fourths of the total world food supply is drawn from the grasses, and all human civilization rests on the cultivation of cereal grains. Although more than 3000 species of edible plants have been used for food throughout our history, today only about 30 plant species make up 90% of our food supply. In fact, most of the human food supply worldwide is derived from the following 20 crops:

- Banana
- Barley
- Cassava
- Citrus
- Coconut
- Corn (maize)
- Oats
- Peanut
- Pineapple
- Potato
- Pulses (beans, peas)
- Rice
- Rye
- Sorghum
- Soybean
- Sugar beet
- Sugarcane
- Sweet potato
- Wheat
- Yam

In addition to the food we eat, plants provide food for the animals we use for meat, for work, and for pleasure. Lumber from the forests provides shelter and furniture; plant fibers furnish clothing, fabrics, rope, and paper. Other plants supply medicines. Many beverages and drugs—including beer, cocoa, coffee, tea, whiskey, wine, quinine, opium, and tobacco—are derived from plants. All fossil fuels such as oil and coal, on which our current high standard of living depends, originated from plants or from animals that lived on plant substances. Plants beautify our surroundings, purify the air and water, stabilize soil, prevent floods and erosion, and protect our natural resources. Flowers, fields, and forests help satisfy our longings for beauty, fulfill our aesthetic cravings, and provide relaxation and recreation.

Plants and animals coexist. Plants provide food for animals, either directly or indirectly, and in turn use the carbon dioxide released into the atmosphere during the respiration of animals, burning of fuels, and processes of decay. Humans have learned to cultivate certain plant species for use as food, shelter, or clothing. These relatively few species have been selected from wild plants and improved to produce larger yields of superior quality. As a result of improvements in farm management, pest control, herbicides, and irrigation and fertilization methods since 1950, U.S. farmers have produced food in abundance for Americans year after year, with some left over for export.

However, plants do get sick. They have diseases. Plant diseases cause losses in yield, and, at times, losses are so great that famine results. Famine then leads to human suffering and diseases. People who are undernourished and weak from hunger are easy prey to cholera, pneumonia, stomach disorders, and other infectious diseases and parasites. Famine and epidemics ride side by side. People huddled together in unsanitary, crowded conditions are easy prey to unrest and panic. As Seneca said, "A hungry people listens not to reason nor cares for justice, nor is bent by any prayers."

Throughout human history, reductions in crop yields from diseases, pests, or bad weather have had profound effects on the whole human race and on specific, local populations. Despite the improved plant cultivars (cv) and more efficient cultural techniques that now are available to grow crops, many of the world's people still do not have an adequate diet, and hunger is their constant companion. Among the many plant diseases that have influenced world history are the following:

- Bacterial wilt
- Chestnut blight
- Coffee rust
- Downy mildew
- Ergot of rye
- Phytophthora root rot
- Potato late blight
- Rice blast
- Root knot
- Smuts of cereals
- Rusts of wheat
- Tobacco mosaic

THE STUDY OF PLANT DISEASES

Plant pathologists study the causes and development of plant diseases and develop practical management methods for those diseases. The study of plant diseases is known as *plant pathology* or *phytopathology.* These words are derived from the Greek words *phyton,* meaning plant, *pathos,* meaning suffering, and *logos,* meaning reason.

By understanding the causes of plant diseases, learning to recognize them, and developing ways to reduce crop losses, plant pathologists help to increase food and fiber production and to protect the food supply. In this way, they combat hunger, reduce suffering, and make the world a better, safer place for all people.

Plant pathologists usually are concerned with disease in many plants rather than with disease of a single plant. Occasionally, diseases on a single plant such as a large shade tree or ornamental shrub will receive special attention, whereas the loss of a single plant in a large wheat field usually would not be noticed. A medical doctor, in contrast, usually deals with disease on an individual basis.

The first step in a disease management program is to identify the disease. It is essential to diagnose the problem correctly before selecting and using a pest management practice. Sometimes it is easy to isolate a disease-causing organism, but it may take years to identify the real cause of the trouble. Some farmers have had enough experience with plant pests and are familiar enough with disease symptoms that they can correctly diagnose plant troubles. However, when some unknown problem occurs, the grower should seek professional help.

Many different areas of science and agriculture are involved in the study of plant diseases. Plant pathology began as an offshoot of *botany,* the study of plants. To understand diseases and develop management methods, one must study the physiology, anatomy, taxonomy, and growth of plants. The following sciences also contribute to the understanding of the causes of plant diseases:

Agronomy. The science of field crop production and soil management.

Bacteriology. The study of bacteria.

Biochemistry. The chemistry of plant and animal life.

Biotechnology. The collection of processes and techniques, many at the molecular and cellular levels, that involve understanding and modification of biological systems, including genetically engineered microorganisms, plants, and animals.

Botany. The study of plants and their structure, function, classification, and ecology.

Chemistry. The study of the composition, properties, and structure of substances.

Ecology. The study of the relations between organisms and their environment.

Edaphology. The study of soils.

Engineering. The study of machines and engines; also, the practical application of scientific principles.

Entomology. The study of insects.

Forestry. The science of planting and taking care of forests.

Genetics and plant breeding. The study of sexual reproduction and inheritance in plants.

Horticulture. The science of cultivating flowers, fruits, vegetables, or ornamental plants.

Meteorology. The study of weather and climate.

Mycology. The study of fungi.

Nematology. The study of nematodes.

Physics. The study of matter and energy and their interaction.

Statistics. The description of data using mathematical techniques to evaluate the uncertainty of inductive inferences.

Virology. The study of viruses.

Weed science. The study of weeds and their control.

Plant pathologists use centrifuges, complicated chemical equipment, computers, electron microscopes, greenhouses, light microscopes, photography, spectrophotometers, sterile growth chambers, and many biochemical and physiological techniques to gain information leading to the development of more efficient plant disease management methods.

PLANT DISEASE LOSSES

Although there are at least 50,000 diseases of economic plants and new diseases are discovered every year, it is difficult to accurately assess losses from disease. Worldwide, nematodes alone cause staggering losses of $100 billion annually. Annual losses to plant diseases in the United States average about 15% of the total agricultural production, or more than $15 billion.

Plant diseases vary widely in the damage they cause. *Annihilating* diseases may wipe out a crop completely; for example, chestnut blight annihilated the American chestnut throughout most of its range in the United States. *Devastating* diseases, such as southern corn leaf blight, may be severe for a few years then gradually subside. *Limiting* diseases, including many root diseases, build up to the point where it is not economical to grow a crop in a particular field. Then there are *debilitating* diseases that occur almost every year and weaken crops, resulting in depressed yields and poor quality. No crop ever produces its full yield. Many bad things can happen to a crop between sowing the seed and using the harvest.

Plant diseases do their damage in different ways. One major effect of disease in plants is reduced plant vigor, resulting in decreased yields. Another major impact is reduced quality; for example, wheat rust infection reduces the size of grain kernels and lowers carbohydrate content; diseased apples are discolored and unsightly; decayed wood is weak or unusable. Some plant pathogens contaminate or infest soil to such an extent that a desirable field cannot be used for growing a principal crop unless the soil is fumigated or crop rotations are established.

Disease management expenses constitute one of the major losses associated with plant disease. Some diseases, such as apple scab and brown patch of turf, are often kept in check only by repeated applications of pesticides. This requires time, equipment, and labor, all of which add to the cost of production. Misuse or misapplication of pesticides results in environmental pollution, injury, and/or death to many living things, including humans. The cost of pesticide misuse is difficult to measure.

Some plant pathogens form *toxins* in stored grain, peanuts, hay, or other products. If such moldy products are fed to animals or eaten by humans, they may become sick or die. These costs are difficult to estimate accurately.

Additional information on the effects of plant disease on human affairs is given in Chapter 3 and in the discussions of individual diseases.

EFFECTS OF MODERN AGRICULTURE ON PLANT DISEASES

The great plains of the United States and Canada are known as the breadbasket of the world, and elsewhere throughout the United States farmers produce a wide array of crops upon which our society depends. The fortuitous combination of suitable soil and climate in large areas of the country, skilled growers and farmers, and researchers and educators dedicated to obtaining and disseminating new knowledge had led to very high levels of agricultural production. However, the struggle against plant diseases never ends. The primary mission of agricultural research centers is to develop production practices and new disease-resistant cultivars that reduce the adverse effects of pests and unfavorable environmental conditions on plant growth while maintaining the integrity of the environment.

Despite the development and widespread use of high-yielding cultivars of improved quality and of new plant protection technologies, the possibility of severe damage from disease and pests has not been eliminated. Indeed, some farm practices actually increase the severity of plant disease. Inefficient land management, sanitation, and agronomic practices may favor the buildup and survival of plant pathogens or predispose crop plants to increased damage from disease.

Failure to destroy crop residues or weed hosts that may harbor disease agents, improper land preparation and erosion prevention, use of infested seed, improper use of fertilizers and irrigation, growing the same cultivar in the same field year after year (*monoculture*), failure to control weeds and insects—all these may contribute to disease increase. Introducing a highly susceptible crop into a region occupied by an aggressive pathogen or permitting a virulent pathogen to invade an area where a susceptible crop is grown may result in epidemics and huge crop losses. Plant breeders may develop a cultivar, resistant to a major disease, that turns out to be susceptible to a minor disease; the minor disease may soon become a major one.

Failure to maintain *genetic diversity* in crop plants also can aggravate disease problems. For example, when a desirable high-yielding, disease-resistant cultivar or several closely related cultivars resistant to only one strain of a pathogen are introduced over a wide area, new strains of the pathogen may arise in these "pure" stands of genetically similar plants. An epidemic occurs, huge losses result, and after a few years the "new" desirable cultivar is no longer usable—the risk of disease is too great.

WHAT IS A PLANT DISEASE?

A plant disease may be defined as any disturbance that prevents the normal development of a plant and reduces its economic or aesthetic value. A disease interferes with the normal function of some part of the plant, resulting in lower yields or reduced quality. For the farmer, this leads to a decrease in income; for the consumer, reduced food supplies of lower quality and higher prices; and for the homeowner and turf manager, lower aesthetic value, less beauty, and higher maintenance costs.

Diseases are caused by living agents (microorganisms and parasitic plants) and nonliving agents (environmental factors, faulty nutrition, and chemical substances). The microorganisms and parasitic plants are called *pathogens* or *biotic agents*. The nonliving environmental and nutritional factors and chemical substances are called *abiotic agents*. In some diseases both pathogens and abiotic agents are involved.

A plant disease develops over a period of time, whereas an injury that may be caused by an insect feeding on a plant or damage from equipment or machinery occurs immediately or over a very short period of time. Usually a disease has been developing for several days before any symptoms on the plant are visible.

RECOGNIZING PLANT DISEASES

The visible reactions of a plant to disease are called *symptoms*. These include wilting, stunting, yellowing, death, and abnormal growth of part or all of a plant. A diseased plant is recognized by comparing it to a healthy plant. However, although plants rarely are free of all diseases during their life span, they may appear healthy. Therefore, to determine if a plant has a disease, we must know as much as possible about the growth, development, culture, and appearance of normal, healthy plants. Agronomy, horticulture, and forestry are sciences that deal with the growth and culture of plants.

Accurate diagnosis of a plant disease depends on the identification of the pathogen or causal agent. Microscopic examination or isolation of the pathogen is often necessary for diseases caused by microorganisms. The identification of the causal agent of a disease that has not been described previously or one that is not well known must meet four criteria, called Koch's postulates:

1. The pathogen or abiotic agent must be consistently associated with the disease.
2. The pathogen must be isolated in pure culture or an abiotic agent identified.
3. When the pathogen or abiotic agent is applied to healthy plants, the same symptoms develop as on the diseased plants originally examined.
4. For diseases caused by biotic agents, the same pathogen (described in step 2) must be isolated from the diseased plants that were inoculated with the isolated pathogen (step 3).

When all four criteria are met, the proof of pathogenicity for that organism or abiotic agent is established. Modifications to step 2 must be made for pathogens that do not grow in culture, such as viruses. For many virus diseases, the virus can be purified from plant sap. The pure virus then is used in step 3 to fulfill the proof of pathogenicity.

MANAGEMENT OF PLANT PESTS

Plant diseases cause never-ending problems. Sound management of plant diseases requires knowledge and judgment because the more we know about the pathogens, the more chance we have of producing a healthy crop. Research is needed to gather information and knowledge about plant pests, their capabilities, and resources, about crop plants and plant resistance, and about the environment in which plant diseases occur. The object is to provide crop plants with every possible advantage against pathogens, to manage diseases at the least possible cost in terms of dollars and cost to the environment), and to minimize losses.

Successful disease management costs money. Therefore, growers must know how to expend their resources wisely. If they do not, costs and losses may be intolerably high. Crop yield may be lost, and then the money spent for disease management is wasted. Every dollar spent in pest management should return three to four dollars to growers.

Continuous pest management is basic to efficient production. Plant pests are extremely resourceful and constantly challenge the judgment, skills, and resources of the farmer. Hence, growers must know not only the current capabilities of the pathogen at any given time and place, but also their own ability to apply corrective measures. Farmers must make suitable preparations in advance of the outbreaks of disease that are bound to come. Sometimes the pathogens can infect plants without warning; plant resistance breaks down, and epidemics occur.

A successful strategy to manage plant diseases involves two main objectives: to reduce the initial amount of the pathogen present and to reduce the rate of pathogen increase and spread. Suitable management systems must be developed. Machinery and resources must be available in adequate amounts at the correct time and place to implement management strategies. Management practices must be safe, efficient, and effective. Growers need to know what each management practice is for, how and when it is to be employed, and how effective it is. They should use chemicals wisely and avoid contaminating or polluting the environment.

Overall disease management practices may be complex, especially where more than one plant pest is involved. Successful management may require the use of various combinations of measures that assist and strengthen each other. Sometimes diseases occur unexpectedly, and large yield losses occur. Such occurrences require a reevaluation of the present management strategy and the development and employment of new methods designed to prevent recurrent disasters. Successful management of plant diseases must not be neglected if a hungry world is to be fed.

2

Causes of Plant Diseases

INFECTIOUS PLANT DISEASES

Infectious plant diseases are caused by living (*biotic*) agents, or pathogens. These pathogens can be spread from an infected plant or plant debris to a healthy plant. *Microorganisms* that cause plant diseases include *nematodes, fungi, bacteria,* and *mycoplasmas.* We also classify *viruses* and *viroids* as biotic agents because they must have living cells for reproduction and are composed of nucleic acid and protein. Some higher plants that produce seeds are parasitic on other plants and are considered to be pathogens. Each type of plant pathogen is discussed briefly in this chapter. More detailed information is given in the chapters dealing with the individual diseases. The relative sizes of various types of pathogens are illustrated in Fig. 2.1.

Nematodes

Nematodes are small eel-like worms, many of which are too small to be seen with the unaided eye. Nematodes feed on plants by means of a miniature hypodermic-like structure called a *stylet,* which is used to suck liquid nutrients out of plant cells. Nematodes reproduce by laying eggs that hatch into young nematodes called *larvae* or juveniles. The larvae develop through a series of four molts, in which the outer skin is shed, until they become adults. Some nematodes can complete their life cycle in less than 30 days.

The most important nematode pathogens injure plant roots, but different types of nematodes can feed on almost any part of a plant. Some nematodes feed mainly on the outside of plant tissues (*ectoparasites*), whereas others wriggle inside the tissues

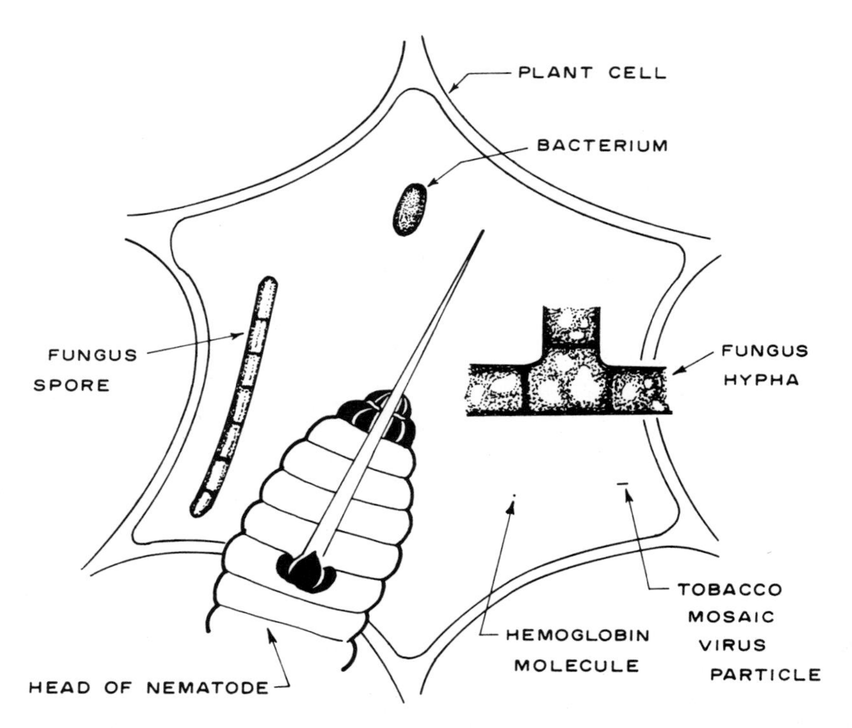

FIGURE 2.1. Relative sizes of plant pathogens. The actual sizes (averages) are: diameter of plant cell =15 μm (micrometers); diameter of nematode head = 4 μm; diameter of fungus hypha = 2 μm; length of bacterium = 1 μm; length of tobacco mosaic virus particle = 0.3 μm = 300 nm (nanometers); diameter of hemoglobin molecule = 5 nm. *Adapted from G. N. Agrios.*

and feed from within (*endoparasites*). Nematodes cause damage to plants by injuring cells, removing cell contents, or changing normal plant growth processes. Chemicals that are used to kill nematodes are called *nematicides.*

Fungi

Fungi are small threadlike organisms composed of tiny filaments called *hyphae,* which are composed of strands of simple cells so small they cannot be seen without a microscope. The tangled mass of intertwined hyphae is collectively called *mycelium.* Fungi may develop into colonies such as those often seen on bread or vegetables that have been stored improperly or for a long time. Some fungi grow inside plant tissues or organic debris and cannot be seen unless magnified; others develop into large structures such as mushrooms or toadstools.

Most fungi produce microscopic "seedlike" spores, which are often spread from one plant to another by wind, water, or people. The spores land on a suitable substrate (plant) and germinate. The hyphae may penetrate the plant surface directly, through natural openings or through wounds. Fungi damage plant tissues by producing toxins, enzymes, or growth-regulating substances that alter or destroy plant tissues. Many different types of chemicals, called *fungicides,* have been developed to kill fungi that cause plant diseases.

Bacteria

Bacteria are small, single-celled organisms that possess rigid cell walls. Bacterial cells must be magnified about 400 times to be seen through a light microscope. Bacteria reproduce by a process called binary *fission* in which one cell divides into two cells. Under suitable conditions, with plenty of available nutrients, some bacteria can divide about every 30 min. If every bacterial cell divided into two cells every 30 min, a total of 8,388,608 cells would be produced after 12 hr from a single cell.

The rapid multiplication of bacteria and the production of toxins and enzymes that destroy plant tissues contribute to the damage caused by bacteria. Only rod-shaped bacteria cause plant diseases. They enter plants through wounds or natural openings. Some bacteria produce growth-regulating substances that cause the surrounding tissues to grow abnormally. Antibiotics called *bactericides* have been used to manage some plant diseases caused by bacteria.

Mycoplasmas

Mycoplasmas, first shown to cause plant diseases in the 1960s, are small single-celled organisms that occur inside plant cells. Mycoplasmas do not have rigid cell walls; each is enclosed by a triple-layer cell membrane. These cell membranes can be easily ruptured if the cells are not in a favorable environment.

Many diseases that were earlier thought to be caused by viruses, and were called *yellows* diseases, are caused by mycoplasmas. These organisms are spread from one plant to another by a *vector,* usually an insect such as a leafhopper. Mycoplasmas can also be spread in infected plant parts. Tetracyclines are antibiotic substances that have been used to kill some mycoplasmas in plants.

Viruses and Viroids

Viruses are very small particles, usually less than 300 nm in length, composed of nucleic acid and protein. They can be seen only under an electron microscope at magnifications of 2000 to 3000 times. Viruses that infect plants have several different shapes: some are spherical; some are, comparatively speaking, long rigid rods; some are long flexuous rods; and some are short thick rods. Nucleic acid (usually ribonucleic acid, or RNA) composes the center of the particles and is surrounded by and connected to a protein coat. The RNA in viruses directs certain enzyme systems in plant cells to manufacture components used to produce more virus particles.

Viruses cause diseases in plants by diverting energy and structural components, normally used in plant growth, into reproductive processes for the virus. Viruses are spread from one plant to another by insect or nematode vectors or by humans. Another major means of spread is in infected plant parts. Viroids are smaller than viruses. They act like viruses, but the infectious particle is simply a strand of RNA and contains no protein. Therapeutic chemicals that inactivate plant viruses and viroids are not yet available.

Parasitic Plants

Parasitic plants are higher plants that reproduce by true seed. Most of these parasitic plants have modified rootlike structures that attach to plant tissues to obtain nutrients and water, but do not have root systems that can absorb nutrients form the soil. Parasitic plants weaken the host by using nutrients that normally would be utilized by the host plant. Mistletoes are among the most frequently seen parasitic plants. In some forest trees attacked by mistletoes, the plant tissues may be so disorganized that the wood is weakened, and the branches are deformed.

DISEASE COMPLEXES

Frequently plant diseases, especially those induced by soilborne pathogens, involve more than one pathogen. When this double trouble happens, increased damage to or death of the plant may occur. A plant may be able to resist the attack of one parasite, but when two or more parasites attack the same plant, it may be overwhelmed. Whenever the combined damage of two or more pathogens acting together is much greater than the sum of the damage caused by the pathogens acting alone, we called this a *plant-disease interaction*. For example, tomatoes suffering from root knot may be killed more rapidly if attacked by the Granville-wilt bacterium. The combination of cold injury, bacterial canker, improper pruning, and root-rotting organisms causes a short life for peach trees. Air pollutants predispose ponderosa pine to attack by bark beetles.

Environmental extremes—such as too much or too little water, temperature, or relative humidity (RH)—often affect disease development; nutritional excesses or deficiencies may increase the susceptibility of plants to some diseases. Insects, nematodes, fungi, weeds, and even people often transport pathogens from one area to another. Satisfactory management programs for disease complexes require accurate diagnosis of the problem and the combined use of various practices if the affected crop is to be grown profitably.

NONINFECTIOUS PLANT DISEASES

Noninfectious plant diseases are caused by nonliving (*abiotic*) agents that cannot be transmitted from one plant to another. These agents are usually environmental, nutritional, or chemical factors. Environmental extremes in temperature, moisture, or light often are unfavorable for plant growth. Nutritional deficiencies and excesses cause disease-like problems that must be correctly identified if they are to be prevented or eliminated. Chemically induced plant diseases result from improper soil pH, improper use of fertilizers and pesticides, chemical spills, and air pollution. In recent years air pollutants from automobiles and industrial processes have been shown to damage many plant species, and when these pollutants become part of natural rain, the quality of water in many lakes also can be reduced (this phenomenon is called *acid rain*).

TABLE 2.1. Classification of a host plant, a parasitic plant and several pathogens.

		Causal agent of				
	Old white rose	Stem rust of wheat	Root knot	Bacterial wilt	Citrus stubborn	Leafy mistletoe
Kingdom	Planta	Myceteae	Animalia	Procaryotae	Procaryotae	Planta
Phylum	—	—	Nemata	—	—	—
Division	Magnoliophyta	Eumycota	—	I. Gracilicutes	III. Tenericutes	Magnoliophyta
Subdivision	—	Basidiomycotina	—	—	—	—
Class	Magnoliopside	Urediniomycetes	Secernentea	Scotobacteria	Mollicutes	Magnoliopsida
Order	Rosales	Uredinales	Tylenchida	Pseudomonadales	Spiroplasmatales	Santales
Family	Rosaceae	Pucciniaceae	Heteroderidae	I. Pseudomadaceae	III. Spiroplasmataceae	Viscaceae
Subfamily	—	Pucciniodeae	Meloidogninae	—	—	—
Genus	*Rosa*	*Puccinia*	*Meloidogyne*	*Pseudomonas*	*Spiroplasma*	*Phoradendron*
Species	*alba*	*graminis*	*incognita*	*solanacearum*	*citri*	*villosum*
Subspecies	—	—	*incognita grahami*			
Variety	—	tritici	—			

NAMES OF PATHOGENS AND DISEASES

All living things have common names and scientific names. The common names may vary from country to country and from one language to another. However, in 1753 the Swedish naturalist Linneaus proposed the binomial system of nomenclature in which each type (species) of organism is given a Latin name consisting of two parts. Latin was chosen because it was a widely known language used by scholars in many countries. The first part of the binomial is the *genus* or generic name, and the second is the *species* or specific name. Also, the name, or an abbreviation of the name, of the person who first described the plant or animal often is written following the binomial. Thus, the name for the white rose is *Rosa alba* L., which signifies that Linneaus originally gave the name *Rosa alba* to this particular kind of rose. Species names are underlined in typed material and generally italicized in printed material, as they are in this book.

Once a Latin binomial has been applied to a certain animal or plant, the same name can never legitimately be used to apply to any other species. The scientific name of one of the most widespread root-knot nematodes is *Meloidogyne incognita* (Kofoid and White) Chitwood. Kofoid and White first described this nematode, but Chitwood later revised the genus and gave it a new name that had priority. Fungi, bacteria, and parasitic plants also have Latin binomials. However, the naming of viruses is still being discussed.

In addition to the names for genus and species, living organisms are classified in more aggregated groupings such as families, orders, classes, and kingdoms (Table 2.1). Not all levels of classification are used for all organisms.

Often the scientific names of plants and animals have many syllables and are difficult to pronounce. Despite this disadvantage, the binomial system has proved useful and efficient for more than 200 years.

Pathogens are given common names, such as root-knot nematodes, to use in nonscientific settings in place of the scientific name, that is, *Meloidogyne incognita*. Diseases usually are given a common name that describes some part of the symptoms, or sometimes a portion of the pathogen name is used. Brown patch is a common name descriptive of the brown dead areas in turfgrasses that are caused by the fungus *Rhizoctonia solani* Kühn. Root knot refers to the galls on roots that are caused by the root-knot nematode. An example of a pathogen name being used as part of the common name is Fusarium wilt, a disease of many kinds of plants that is caused by the fungus *Fusarium oxysporum* (Schlecht) Snyder and Hansen. Common names of diseases may differ among different countries or regions of the world; however, the scientific name of a pathogen is the same worldwide.

Common names for viruses usually indicate the host affected or symptoms produced by that virus. For example cucumber mosaic causes a mosaic symptom on cucumber and some other plants. See Chapter 17 (Naming of Viruses section) for a fuller discussion.

3

History of Plant Pathology

IN ANCIENT TIMES

Plant diseases have been recognized for centuries. The ancient Romans worshipped Robigus, the god of wheat, and sacrificed red dogs to him at the Robigalia (a great festival) to save the grain from the destructive red dust. Aristotle, Plato's student, recorded plant diseases as early as 350 B.C., and his colleague Theophrastus observed and speculated about diseases of cereals, legumes, and trees. Evidently, plant diseases were destructive in ancient times and the people lived in fear of famine. Thus, the belief in supernatural causes of plant diseases is not surprising. The idea of Divine punishment for human sins, as evidenced by the occurrence of plant disease and the resulting human suffering, is very old. This belief was encouraged by religious leaders who, because of their own ignorance, held similar beliefs and it also helped preserve the tradition of the infallibility of the Church. People blamed diseases on evil spirits, displeasure of the gods, or unfavorable positions of the stars or moons.

Early biblical writings (for example, the books of Amos, Deuteronomy, and Kings I in the Old Testament) tell of mildews, blasts, and blights. Shakespeare in *King Lear* tells of the foul fiend who walks at night and mildews the white wheat. In the Middle Ages, outbreaks of ergot fungus in Europe poisoned the bread made from infected grain. Convulsions seized those who ate the rotten bread; they went mad and died painful deaths. These unfortunate victims of a plant disease were considered possessed and, at times, accused of engaging in witchcraft.

For 2000 years after the time of Theophrastus, little was added to the knowledge about plant diseases. Their nature and causes were shrouded in superstition and speculation. Most diseases were attributed to supernatural causes such as evil spirits.

From the time of Christ until the Renaissance little that was new was written about plant diseases. The Middle Ages that brought European civilization to a near standstill also enveloped the study of botany and plant pathology. However, increased trade between Europe and the Near East led to the development and rise around A.D. 1000 of a rich middle class of merchants and tradesmen with leisure time available. Many were educated and became doctors, lawyers, or teachers. In their spare time, some became amateur naturalists, studying all forms of life and recording what they saw. Down through the centuries there were a few people in every generation who wanted to find out more about their world. However, throughout the Western world these curious naturalists were hampered by the Christian church, which sought to regulate scientific inquiry, demanded a literal interpretation of the Bible, and preserved the doctrine of the infallibility of church teachings.

Until the Middle Ages and well into the Renaissance, most people could neither read nor write. The few books available were copied by hand, and communications were slow and expensive. New knowledge about plant diseases, or anything else, traveled slowly and to relatively few people. About 1450, Johann Gutenberg invented the printing press, an accomplishment that ranks with the invention of the wheel or the harnessing of fire in human history. With this remarkable machine, multiple copies of books could be made and distributed widely. Education of the masses had begun.

FROM 1600 TO 1800

Two centuries after Gutenberg, in about 1683, Anton von Leeuwenhoek, a Dutch lens grinder, built a crude microscope that enabled him to produce and see magnified images too small to be visible to the naked eye. Leeuwenhoek examined many things: raindrops, saliva, and rotting substances. What he saw delighted and surprised him. He discovered protozoa, bacteria, and other microorganisms. Leeuwenhoek's microscope opened up a whole new world to observation. With these new observations, the idea that diseases were the result of supernatural causes was severely shaken.

The belief that became widely accepted was that the microorganisms associated with disease arose spontaneously from affected tissues. This was known as the doctrine of *spontaneous generation*. It may seem strange to us today to think that microorganisms are the result and not the cause of disease, but from the late 1600s until the mid-1800s, the idea of spontaneous generation was accepted as fact.

With the development of the microscope, a whole new world was opened up to naturalists and amateur scientists. Many excellent descriptions of microorganisms associated with plant diseases were published.

In 1728, Duhamel de Monceau in France described the saffron disease of crocus. He conclusively showed that the disease was caused by the fungus *Rhizoctonia*, and that the fungus was contagious—it could spread among plants and cause epidemics. He also proposed sound management measures for the disease. Unfortunately, his work received almost no attention from other scientists and was lost to plant pathology and the rest of the scientific world. Duhamel was ahead of his time. One can only wonder how much further advanced medical and botanical sciences would be if Duhamel's accurate statement of the germ theory had been widely publicized and accepted in 1728.

In 1743, J. T. Needham first described plant-parasitic nematodes in wheat galls, but a hundred years passed before M. J. Berkeley, in England, observed the root-knot nematode.

P. A. Micheli, an Italian, in 1729 used the microscope to study many fungi. He showed that "seeds" of fungi grew and produced more "seeds." In 1755, Mathieu Tillet, in France, proved experimentally that wheat bunt is contagious and could be prevented by seed treatments. Slowly but surely careful observers were showing that plant diseases did not arise spontaneously but were caused by living things—"germs."

In 1807, Isaac Benedict Prevost from Switzerland proved conclusively that bunt, a disease of wheat, is caused by a fungus and could be controlled by dipping seed in copper sulfate. Prevost pointed out the importance of the environment in the development of the disease. He was also ahead of his time. His great contribution was rejected by the authorities, but his seed treatment became widely used despite lack of approval by the so-called learned men.

THE GOLDEN AGE OF BIOLOGY, 1840 to 1900

A mysterious and terrible *potato murrain* appeared in western Europe and the United States from 1830 to 1850. In 1845, this potato murrain, or late blight, as it is known today, destroyed the Irish potato crop. As a result, a million people died of starvation and malnutrition, and a similar number of Irish citizens emigrated from Ireland to the United States and Canada. This tragedy forced scientists to investigate the disease and find ways to control it.

Heinrich Anton DeBary, a medical doctor and one of the outstanding biologists of his time, proved experimentally that a fungus was the cause of the late blight of potato and had his work accepted by the general scientific community. At the age of twenty-two he published a book presenting overwhelming evidence that fungi are the causes not the results of plant diseases. DeBary also made many other important discoveries about plant diseases. He is called the Father of Plant Pathology not only because of his scientific ability, but also because he was a great researcher, compiler, and educator. He trained more than 60 scientists from many countries who became prominent in the field of plant pathology and mycology.

In the Golden Age of Biology from 1840 to 1900, the foundations of modern biology were laid and much progress was made. In 1858 and 1859, the epic publications of Charles Darwin and G. W. Wallace established the evolutionary theory. In 1861, Louis Pasteur of France demolished the concept of spontaneous generation and showed that microorganisms only arose from other identical microorganisms. Robert Koch showed in 1876 that disease of cattle (anthrax) was caused by a microorganism. He also introduced the poured-plate method for isolating causal organisms from diseased tissue in 1881. Isolation techniques were improved by the development of a culture dish with a lid—the petri dish—by one of Koch's assistants, R. J. Petri. In 1884, Koch provided a series of rules for proving that a microorganism causes a disease, which became known as Koch's postulates (see Chapter 1).

Julius Gotthelf Kühn, a German, published the first plant pathology text in 1858. Another German, Robert Hartig, devoted his life to the study of forest tree diseases and published two books on the subject, in 1874 and 1882. Hartig is widely known as the Father of Forest Pathology. T. J. Burrill, at the University of Illinois, and J. C. Arthur, at Cornell University, showed between 1877 and 1885 that fire blight of apple and pear was caused by a bacterium.

In 1885, Frenchman P. M. A. Millardet discovered Bordeaux mixture for the control of downy mildew of grapes. Erwin F. Smith began his long and productive career with the U.S. Department of Agriculture (USDA) in 1886 and became one of this country's greatest bacteriologists and plant pathologists. Also in the 1880s, Adolph Mayer in Holland demonstrated that tobacco mosaic could be spread by juice from infected plants. Dimitri Iosifovich Ivanovski, a Russian, in 1892 claimed that the entity that caused mosaic was smaller than a bacterium. In 1898, M. W. Beijerinck, working in Holland, demonstrated that the infectious substance that caused mosaic increased in tobacco plants. He was the first to use the word *virus,* from the latin word for poison, in relation to a plant disease. In 1883, H. Hashimoto, a Japanese rice grower, recorded that a leafhopper transmitted the agent (he did not know it was a virus) causing the dwarf disease of rice. M. B. Waite demonstrated in 1931 that honey bees could act as vectors for the bacterium that causes fire blight.

THE TWENTIETH CENTURY

In the twentieth century, plant pathology matured as a science, with startling and rapid discoveries. Departments of plant pathology were established at major universities around the world. Experiment stations, first established in the late 1800s, continued to grow in number, and by 1900 sixty such stations had been established in the United States. Professors such as E. C. Stakman, L. R. Jones, and H. H. Whetzel attracted and trained many students in the United States; pioneer researchers, including E. F. Smith, forged ahead in new areas of plant pathology. Jones in the 1920s restated Prevost's idea that environment was important in the development of plant diseases; this time the idea was accepted. N. A. Cobb, G. Steiner, B. G. Chitwood, and their co-workers expanded nematology.

In 1935, W. M. Stanley won the Nobel Prize for demonstrating that tobacco mosaic virus was a chemical (a protein) that could reproduce itself in living cells. F. C. Bawden and N. W. Pirie showed in 1936 that Stanley's protein, in a more purified liquid crystalline state, was a nucleoprotein, and that a virus was made up of a nucleic acid, such as ribonucleic acid (RNA), and protein together. G. A. Kausche and his colleagues in 1939 first saw virus particles in the amazing electron microscope.

In the early 1940s, scientists formulated the first carbamate fungicides; chemical nematicides came into use in the 1950s. Systemic fungicides were developed and first used in the 1960s, and the *sterol-inhibitor* fungicides were introduced in the 1970s and early 1980s.

Also in the 1940s and 1950s, scientists learned more about the physiology of plant diseases. They learned how pathogens attack and colonize plants and studied antimicrobial compounds produced in plants.

In 1963, J. E. Vanderplank published his book on epidemics and management of plant diseases. His ideas permitted scientists to describe mathematically the factors that contribute to plant disease development.

In 1967, Y. Doi and co-workers in Japan observed a new kind of organism (mycoplasmas) in infected plants; these were transmitted by leafhoppers. In 1971, T. O. Diener discovered viroids, which are smaller than viruses and thus the smallest known infectious agents of plant disease.

In addition to the development of new pesticides and the description of newly found pathogens in the 1960s and 1970s, several new areas of research began to contribute to the understanding of plant diseases. Scientists today are studying the biochemistry of compounds and enzymes involved in the development of plant disease. Other scientists are working on the epidemiology of plant diseases. Through these studies, they are seeking a more complete understanding of pathogen spread and disease development in plant populations over large geographic areas. Biological control of plant diseases is receiving increased attention. In addition, recent advances in biotechnology have enabled plant pathologists to utilize novel sources of disease resistance and to use processes such as plant transformation to shorten the time it takes to develop disease-resistant cultivars. Based on advances in these areas, scientists will be able to provide more efficient management measures for plant diseases.

Today, the flood of published research continues to increase. The storage, retrieval, and use of this information is one of the most important problems in science. New discoveries in computer science, biotechnology, genetics, physiology, and many other sciences, all are having their impact on plant pathology.

A hungry world awaits. As world population grows, the successful management of plant disease is becoming increasingly essential.

4

Development of Plant Diseases

THE OCCURRENCE OF PLANT DISEASE

For a plant disease to occur, a virulent pathogen, a susceptible host plant, and a favorable environment must be present together over a certain period of time. In Fig. 4.1, each circle represents one of these factors, and the amount of disease is indicated by the total area common to all three circles. Each circle or factor of disease development can occur in a favorable or an unfavorable state.

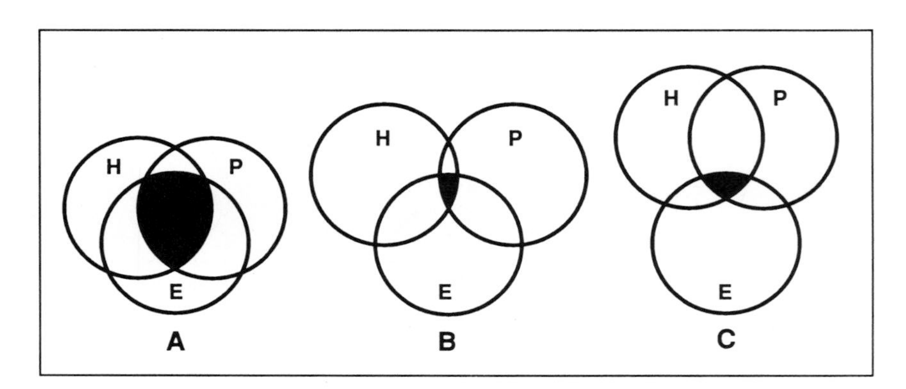

FIGURE 4.1. Circle diagrams with the elements needed for plant disease to occur: H = host; P = pathogen; and E = environment. The amount of black area where the three factors overlap represents the relative amount of disease that develops. Each circle or factor can occur in a favorable or an unfavorable state.

A more susceptible host combined with a more virulent pathogen and a more favorable environment results in more disease than would occur if one or more of the three components were unfavorable. In Fig. 4.1A, all components are favorable, and a large amount of disease occurs (disease intensity is proportional to the darkened area). In Fig. 4.1B, all three components are less favorable (i.e., less susceptible host, less virulent pathogen or lower inoculum density, less favorable environment), and less disease occurs. In Fig. 4.1C, the host and the pathogen are favorable, but the environment is less favorable, and little disease occurs.

This concept helps to explain why most plants are not affected by the many different diseases that exist; that is, most plants are not affected by the many thousands of diseases that can occur because one of the factors for disease occurrence is absent. For example, spores from many different fungi on nearby crops might land on the leaves of an oak tree at the edge of a field, but very few, if any, of these spores cause disease on the oak tree because it is resistant to these fungi; no susceptible host is present, so no disease develops.

THE DISEASE CYCLE

The term *disease cycle* is used to describe the relationship of a pathogen to its host in an environment and the development of a disease over time. A general disease cycle, shown in Fig. 4.2, includes the life cycle of the pathogen and the host. A *life cycle* of a pathogen or a host considers only those steps or stages that relate to the growth and reproduction of the individual organism. The *disease cycle* illustrates the phases of the interaction between the host and the pathogen known as disease.

A knowledge of the disease cycle is essential to the development of effective management methods for a disease. Although the disease cycle is shown only for a single plant, we usually are interested in the development of disease in a population of plants. The study of the survival and spread of pathogens and the development of a disease in a population of plants over time is called *epidemiology.*

Pathogen Survival

Many pathogens can survive in the absence of a susceptible host during conditions unfavorable for disease development. In cold climates pathogens of annual plants may survive from one growing season to the next in plant debris or in seeds or in the soil. In frost-free regions, others survive on growing plants. Many pathogens of perennial plants survive in infected tissues of the host plants or in propagating materials. When conditions are favorable, these pathogens begin to grow and spread to susceptible plants or tissues. Pathogens such as viruses or some rusts can survive in *alternate hosts.*

The survival mechanisms just described apply to pathogens that are reduced to a low level if susceptible plants are not grown in the area for several years. Other pathogens can survive for many years in the soil in the absence of a susceptible host. These pathogens may be able to compete with saprophytic organisms in the soil or to produce resistant structures that do not begin to grow unless susceptible plants are present. It is obvious that successful management methods for these different types of pathogens will vary.

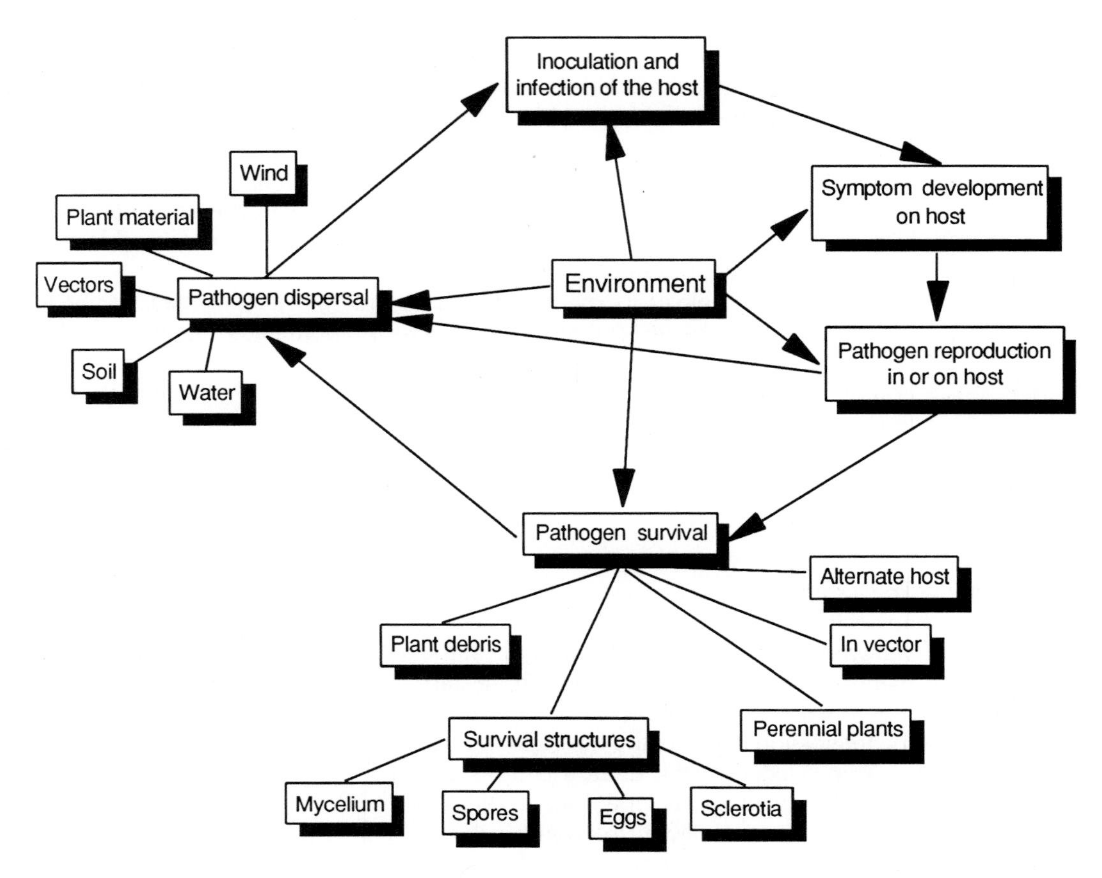

FIGURE 4.2. A general cycle of development for a plant disease.

Pathogen Dispersal

Some pathogens have the ability to spread only a few centimeters, whereas others can be transported for many kilometers. Pathogens are usually spread by wind, water, or soil, in plant parts, or by a vector.

By Wind

Tiny, fragile fungal spores are well adapted to air travel and are blown near and far by the wind. Spores of rust fungi have been detected at altitudes of 10,000 m and are blown for at least 1500 km on turbulent air currents. But spores of other fungi may be carried only a few centimeters by wet wind or spattering rain.

Winds strong enough to blow dust particles have enough power to carry insects, nematode eggs, and billions of invisible fungal spores and bacterial cells capable of causing disease outbreaks to distant places. Some fungi possess specialized fruiting structures that propel the spores into air currents, thus assuring wider distribution.

By Water or Soil

Rainwater transports soil and debris infested with fungi, bacteria, and nematodes along rows, terraces, and ditches and from infested fields into disease-free fields, seed beds, drainage ponds, creeks, or streams. Infested water used for irrigation or transplanting may carry spores to new locations and cause disease outbreaks. Some fungi and bacteria are spread in blowing rain. These types of pathogens must remain moist until they land on a susceptible plant, or they will die. Other pathogens are spread in water as propagating units such as *zoospores* (swimming spores) or in debris that may be washed from place to place. This group includes soil-inhabiting pathogens that can survive in the soil and includes nematodes, fungi, bacteria, and parasitic plants.

On Plant Material

Many pathogens are spread by infected transplants or cuttings. Infected seed is responsible for many disease outbreaks, and the use of contaminated seed has been responsible for introducing plant pathogens into new agricultural areas. Humans are the principal agent for spreading diseased plants to new areas in the course of normal production and marketing practices. During the sorting and grading of fruit, vegetables, and ornamentals for market, discarded culls, leaves, and other infected parts can be a source of pathogens to start disease outbreaks.

By Vectors

Insects, nematodes, fungi, and animals are important agents in transporting pathogens from one place to another. Fungi, bacteria, mycoplasmas, and viruses that cause diseases are spread by insect vectors from diseased to healthy plants (Table 4.1).

TABLE 4.1. Some insect vectors of plant pathogens.

Insect	Pathogen transmitted
Aphids	Cucumber mosaic virus
	Peanut stunt virus
	Potato leaf roll virus
	Potato virus Y
	Sugarcane mosaic virus
	Tobacco mosaic virus
Bark beetles	Chestnut-blight fungus
	Dutch elm disease fungus
	Oak-wilt fungus
Cucumber beetles	Corn-wilt bacterium
Honey bees, wasps, flies	Ergot fungus
	Fire-blight bacterium
Leafhoppers	Beet curly-top virus
	Asters yellows mycoplasma
Plum curculio	Peach brown-rot fungus
Seed corn maggot	Soft-rot bacterium
Thrips	Tomato spotted-wilt virus
White flies	Tobacco leaf-curl virus
	Cotton leaf-curl virus

Some viruses and mycoplasmas reproduce in their insect vectors; this unusual relationship makes their management difficult.

Pathogen Growth on the Host Plant and Appearance of Symptoms

The process whereby a pathogen comes in contact with the host is called *inoculation*. Mineral salts, sugars, and amino acids exuded from seeds, roots, and leaves during normal plant growth may stimulate germination of pathogens in the spermosphere (seeds), rhizosphere (roots), and phyllosphere (leaves). For disease to develop, the pathogen must penetrate the host and invade tissue, in a process called *infection*. After infection, the pathogen begins to grow and establish an intimate relationship with the host. This process is called *colonization* of the host.

Pathogens penetrate host tissues in one of three ways: by direct penetration of the outer layers of plant tissue; by entry into natural openings; or by entry through wounds. Nematodes pierce plant cells with their *stylets* and suck out the cell contents. Spores of many fungi germinate and get inside a plant by growing through natural plant openings such as *stomates* and *hydathodes;* some fungal produce *infection pegs* that penetrate the leaf surface directly; some fungi produce substances called *enzymes* that dissolve plant tissue and allow growth of the mycelium into the plant. Most bacteria gain entry into the plant through natural openings or through wounds. Many viruses are injected into plants by insect vectors while feeding; others penetrate through the microscopic injuries resulting from the wounding or rubbing of plants.

Some pathogens penetrate plant surfaces but do not cause disease because an incompatible (unsuitable) host-pathogen relationship develops. A resistant plant may respond to penetration by undergoing complicated biochemical and physiological changes that kill the pathogen or prevent its spread through the plant. For a disease to develop, a relationship must be established so that a pathogen can continue to grow and colonize host tissue.

As the pathogen continues to grow in the plant, symptoms will begin to appear at a certain time after penetration, depending on the particular pathogen and host plant involved. Visible symptoms are the responses of the plant to disease processes that have been occurring over a period of time. The time between inoculation and the appearance of symptoms is known as the *incubation period*.

Many biochemical and physiological changes occur in the plant before and during the development of disease symptoms. These changes may be due to the production of *toxins* by the pathogen or may result from the disruption of normal metabolic processes in the plant by the growth and reproduction of the pathogen. In fact, most of the major physiological processes in a plant can be disturbed by the invasion of a pathogen. *Photosynthesis* can be reduced dramatically if pathogens grow into and disrupt or kill green leaf and stem tissue. The *respiration rate* of plant tissues invaded by a pathogen may increase as a result of a stimulation of some physiological processes by the pathogen. As disease continues to develop and more and more tissue becomes chlorotic (yellow due to lack of active chlorophyll) or necrotic (dead), the respiration rate will decline. *Transport systems,* which move substances in and out of individual cells and over longer distances in the xylem and phloem, can be damaged physically or disrupted by pathogens or altered by chemicals produced by the pathogens, such as toxins.

Some diseases, including many of the root rots, have only one cycle during a growing season. These are called *monocyclic diseases.* Other diseases, particularly ones that occur in aerial parts of the host, develop secondary, or repeating, disease cycles during a growing season. Recurrent crops of spores, maturing every week or 10 days, provide inoculum to infect nearby healthy plants. These are called *polycyclic diseases.*

The disease cycle is completed when the pathogen reaches the survival stage. Many pathogens produce specialized structures, such as *sclerotia* and *resistant spores,* that enable the pathogen to resist freezing, drying, or other adverse conditions.

ENVIRONMENTAL EFFECTS ON DISEASE DEVELOPMENT

For hundreds of years farmers blamed disease on bad weather; many people still believe bad weather causes disease. Today we know this is not true, but we also know that weather greatly affects the frequency and rate of disease development, the amount of disease, and the damage caused by plant disease epidemics. The reaction of the host and pathogen, independent and in combination, to the many environmental factors determines if and how much disease will develop.

Temperature

Different plants and pathogens thrive at different temperatures. Thus, it is no surprise that there are warm-weather (bacterial wilt) and cool-weather (potato late blight) diseases. Many diseases develop at temperatures optimum for pathogen development but unsuitable for the host. For example, the fungus *Gibberella* (*Fusarium*) *zeae* causes a root rot of both wheat and corn. The fungus does most damage to wheat (a cool-weather crop) at warm soil temperatures, whereas it does most damage to corn (a warm-weather crop) when soil temperatures are low. Temperature also affects growth and sporulation of pathogens. Some downy mildews and rust fungi produce few spores when day temperatures rise above 30°C and remain above 20°C at night.

Moisture

The three major sources of moisture are rain, dew, and irrigation, but fog and high relative humidity (RH) also supply moisture. The occurrence of many diseases in a particular region is correlated with the amount and distribution of rainfall. In some cases, rainfall determines not only disease severity but also whether a disease will occur at all. Many diseases of young tender tissues are favored by high moisture. The spores of many fungal pathogens will not germinate unless enveloped in a film of water, and free water in plant tissues is favorable for the growth of bacteria. On the other hand, hot, dry weather prevents many epidemics from developing. Dry soils inhibit the growth of nematodes and other soilborne pathogens. The amount of moisture in soil determines the amount of free oxygen, and, therefore, controls soil aeration. If soil aeration is not sufficient, plant roots may be stressed and may become more susceptible to disease.

Wind

Wind can spread pathogens and their insect vectors from field to field and over long distances. Wind has a greater effect when accompanied by rain. Windblown rain releases fungal spores and bacteria from infected tissues, blows them to new locations, and deposits them on wet plants where they may start new infections. On the other hand, dry winds may suppress disease development by rapidly drying plant surfaces and decreasing the RH. Wind-blown dust particles may injure tender leaves, stems, and flowers. Plants rubbing against each other during wind storms may be wounded, and thus may be more easily infected.

Light

The quality, intensity, and duration of light affect the growth of both host and pathogen. Plants grown in reduced light are often weak and etiolated. As a result, they are more susceptible to infection. Changes in day length, which profoundly affect plant growth, may decrease or increase susceptibility to various diseases. The photoperiod and light intensity also influence inoculum survival (some spores are killed by exposure to sunlight for 1 hr), abundance of sporulation, the ability of spores to penetrate the host, and the length of the incubation period.

Nutrition

Nutrition affects the rate of plant growth and the ability of plants to defend themselves against disease. For example, overfertilization with nitrogen results in rapid vegetative growth, the formation of tender succulent tissues, and delayed maturity, thus increasing susceptibility to pathogens that attack rapidly growing plants. On the other hand, lack of nitrogen results in slow growth and more rapid aging, which makes plants more susceptible to attack by those pathogens that attack weak, slow-growing plants.

The amounts of other major nutrients—phosphorus, potassium, and calcium—and of micronutrients available to plants also may affect their susceptibility to disease attack and resistance to disease.

Soil Type and pH

There are many types of soil, each composed of a solid, liquid, and gas phase in varying proportions. This complex environment—teeming with many different kinds of microorganisms, some beneficial and others harmful—has profound effects on both plant growth and disease development. Light sandy soils and low organic matter are especially favorable for the growth of many nematode pathogens, whereas heavier soils with high organic matter are less conducive to nematode diseases. Soil structure and texture affect water-holding capacity and soil temperature. As a result, the incidence of damping-off is increased in heavy, cold water-logged soils.

The acidity or alkalinity (as measured on the *pH* scale) also affects disease development, principally of root diseases. Some soil pathogens attack plants over a wide pH range, whereas some (club root of cabbage) do most damage in "sour" soils with a low pH and others (black root rot of clover) in "sweet" soils with a high pH.

Cultural Practices

The kind of crops (row vs. broadcast), cultivar used, number of plants per hectare, methods of cultivation (no-till vs. cultivated), and harvest method (hand or machine) also can affect disease incidence. Therefore, farmers should use the most efficient agronomic practices for a given crop that do not provide conditions favorable to disease outbreaks.

EPIDEMIOLOGY OF PLANT DISEASES

An *epidemic* develops when there is an increase in the amount of disease caused by a population of pathogens in a population of plants. *Plant disease epidemiology* is the study of the development of diseases in plant populations. This disease development occurs over time and over the area occupied by the host population and is influenced by the susceptibility of the host plant, the amount of pathogen inoculum present, the virulence of the pathogen, and whether the environment is favorable for epidemic development. The pathogen can be a fungus, bacterium, mycoplasma, virus, viroid, nematode, or parasitic plant. The plants can be part of any agricultural, horticultural, urban, forest, or natural ecosystem. The interaction between the host and pathogen can occur in any terrestrial, aquatic, or artificially created environment. The sum of all the host, pathogen, and environmental factors is the *plant pathosystem*.

Plant disease epidemiology, the science of studying plant disease epidemics, has both descriptive and quantitative phases. Initially we ask questions such as: where do epidemics occur, when can epidemics be expected to develop, and what factors play a role in disease development? As we gain more knowledge about a pathosystem, we begin to ask questions that are more quantitative in nature, such as: how many propagules of a pathogen are needed to initiate an epidemic, how much disease is present, how fast will disease develop, and how far can propagules of the pathogen move?

The disease progress curve (Fig. 4.3) is used to visualize the development of an epidemic in time. Several specific measurements or parameters characterize the disease progress curve: y_o represents the initial amount of disease and may indicate the amount of initial inoculum present; y_{max} is the maximum or final amount of disease that develops during the epidemic; the rate of change in the amount of disease over time is called r—the rate of epidemic increase or the rate of disease progress.

The amount of disease that develops in the population of plants depends on both y_o and r. If either y_o or r, or both, can be reduced, less disease develops. Most management practices for plant disease are aimed at reducing y_o, r, or both. An example of the effect of reducing r, the rate of disease increase, is shown in Fig. 4.4. The rate of disease increase is slower for epidemic B than for epidemic A, and less final disease is present at the end of epidemic B.

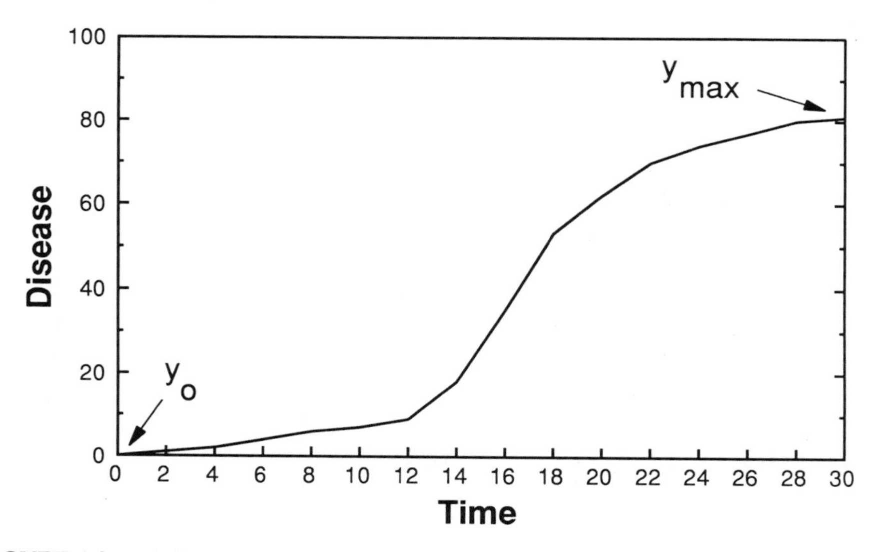

FIGURE 4.3. A disease progress curve over time, where y_o is the initial amount of disease, and y_{max} is the final or maximum amount of disease expressed as a percent.

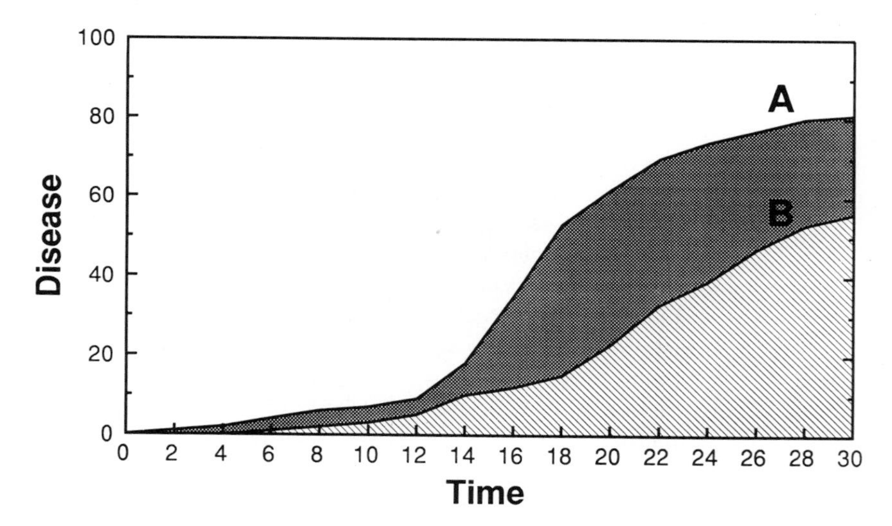

FIGURE 4.4. Two disease progress curves that show the effect of reducing the rate of disease progress. The rate of disease increase is lower for curve B than it is for curve A. More final disease is observed with a higher rate of disease increase.

The spread of pathogens over small to large distances and the spatial patterns of diseases also are important aspects of epidemic development. Sometimes the inoculum of pathogens spreads only with a field, but sometimes the inoculum is spread over regions or even between continents. For example, urediniospores of *Puccinia graminis* var. *tritici*, the fungus that causes stem rust of wheat, can be carried by wind from Texas to Minnesota within 1 to 3 days. Other fungi, such as *Phytophthora infestans*, the causal agent of late blight of potato, usually spread only within a field or between adjacent fields. If spores or other pathogen propagules are dispersed from a very small number of diseased plants over a field of previously healthy plants, a *disease gradient* develops from that original *disease focus.*

Sometimes the diseased plants are clustered together in several spots within a field or forest, and sometimes they occur at random. The arrangement of the diseased plants within a field or larger area is the *spatial pattern* of the disease. Diseases caused by soilborne pathogens are usually more aggregated or clustered in fields than diseases caused by airborne pathogens. The reason for this is that soilborne pathogens are not as easily or as quickly dispersed as are airborne pathogens. Thus, the spatial pattern of a disease depends on the specific type of pathogen.

5

Integrated Pest Management

Plant disease management is just one phase of farming. It is an integral part of the whole program of good farming, but it is not necessarily the most important part—unless diseases get out of hand every year and build up to the point where farmers can make no crop at all. And this has happened! Usually disease management is just one part of a complex system that must be performed so that adequate yields can be obtained to make some profit every year. Practices used to manage diseases must be integrated with the other practices necessary to produce a healthy crop. Disease management must fit in with efficient insect and weed management, rotations, irrigation, cultivation, harvesting, storage, marketing, and every other sensible practice that the farmer finds necessary to get the most profitable yields. Management of diseases is part of a system called integrated pest management (IPM).

Although farmers have used some IPM principles for years, recent emphasis has been on the economics of pest management practices. Farmers know instinctively that when they plant a crop they are gambling that they will reap a harvest (Fig. 5.1). They cannot always predict problems from diseases, insects, markets, and weather. They know that some things are beyond their control. But what farmer would not select new practices if there were a good chance of getting four dollars back for every dollar expended? Today we call this the cost/benefit ratio or cost/risk ratio. Most farmers will try new production practices, especially if there is a significant chance (high statistical probability) that they will be profitable.

EVOLUTION OF PEST MANAGEMENT PRACTICES

The history of the human species is filled with attempts to gain increasing control over the environment. About 8000 B.C., men and women learned how to cultivate grain

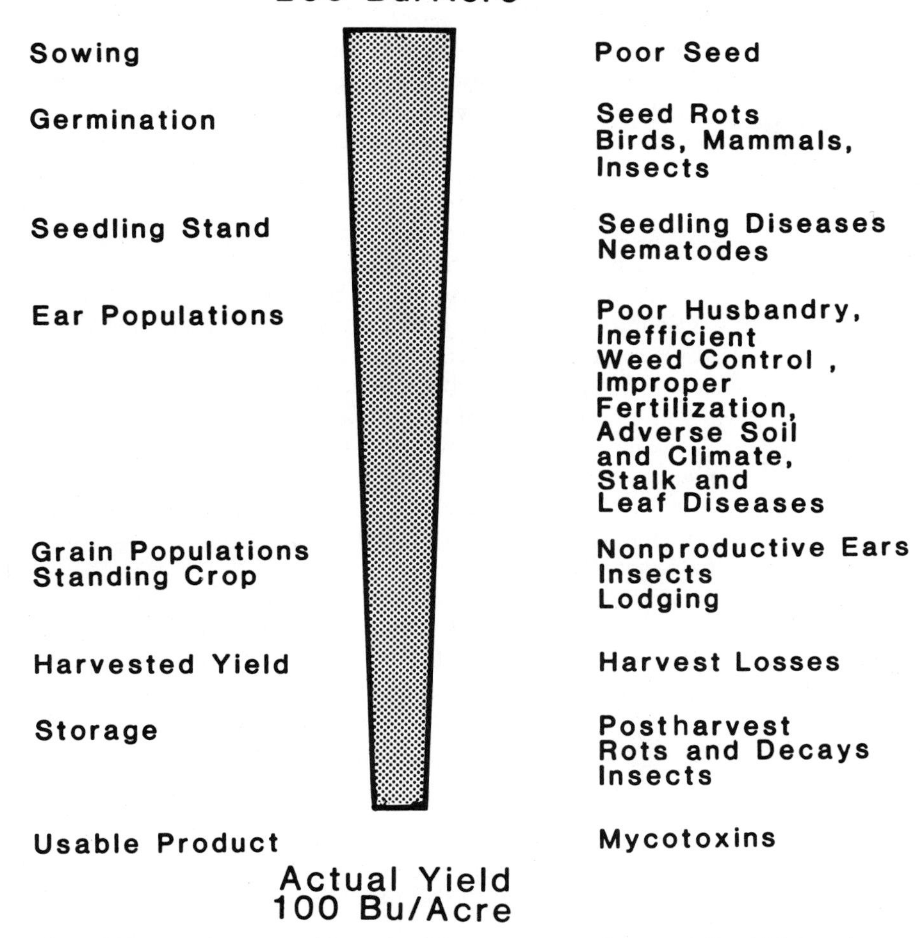

FIGURE 5.1. Why corn yields are low. No crop ever produces its full yield. Many things can happen to a crop between sowing the seed and using the harvest.

crops. It was a remarkable advance that freed family groups or tribes from a nomadic existence, but this subsistence agriculture was hampered by unsuitable weather, low soil fertility, and low-yielding varieties. Lack of machinery and ignorance of yield-increasing agronomic practices prevailed for hundreds of years.

Through the centuries, however, many cultural and physical practices evolved for protecting crops from plant pests through trial and error. Such measures as burning or plowing under crop refuse, rotating crops, planting in the most favorable season, using healthy seed, isolating crops, pruning, and managing water and fertilizer have been used for hundreds of years to improve the chances of making a profitable crop. As biological knowledge grew in the eighteenth and nineteenth centuries, and as pest problems became more severe as a result of more complex agriculture, the narrowing of the genetic base for major crops, and the introduction of pests into new areas, the search for more effective pest management methods intensified. The discovery in 1882 that spraying with Bordeaux mixture would reduce damage from downy mildew on grapes encouraged scientists to look for other chemicals that would kill pests. Since then many highly effective agricultural chemicals have been developed and widely used. These chemicals gave farmers a new ability to reduce pest populations at a relatively low cost. The widespread adoption of these chemicals for managing pests, along with the development of disease-resistant cultivars, led to a general decline in the use and study of many basic cultural and biological tactics for reducing pest populations to a level where profitable yields are possible.

By the late 1950s, however, it became apparent that unacceptable levels of crop loss from pests were still occurring despite tremendous improvements in crop production and protection methods. In addition, a variety of problems have caused concern about various agricultural technologies and future food production. In particular, the development of pests resistant to chemicals, crop losses and stress due to the phytotoxicity of pollutants, shifts in the prevalence of pest species, and the harmful effects of some pesticides in the environment and on nontarget organisms including humans have all raised questions about the efficacy and advisability of certain pest management practices.

Modern agriculture can overcome many of these problems, but its increased complexity and intensity demand unprecedented precision in the management of crop pests. Furthermore, the realization that plant diseases are natural components of agriculture that must be dealt with continuously fortifies the concept that diseases can

TABLE 5.1. Important pest management principles.

- Plant diseases and/or pests can never be eliminated, only managed at an economically acceptable level.
- Pest management requires knowledge and judgment: know the pests, know the environment, and know the crop.
- Provide the crop with every possible advantage. Use all available practices in integrated pest management.
- Continuous pest management is basic to efficient production.
- Epidemiology is basic to the development of a sound pest management strategy.
- Contain the disease at the least possible cost and minimize losses. Every dollar spent in pest management should return $3-$4. There are short-term gains and long-term gains.
- There are two main objectives: reduce the initial inoculum, and reduce the rate of increase and spread of the pest.
- Use proper practices. Develop and employ suitable pest management techniques. These include chemicals; resistant cultivars; seed treatment; crop rotation; proper planting date, seeding rate, and depth; proper harvesting and storage; plowing under of crop debris.
- Use effective logistics. Deploy resources at the proper time, in the correct amount, and the correct way. Trained personnel are essential.
- Beware of pollution. Use chemicals wisely. Preserve the environment.

never be "controlled" or eliminated. Effectively limiting losses from plant pests requires that they be "managed" in a process that incorporates the latest knowledge into the management program as rapidly as possible. The important principles of pest management are summarized in Table 5.1.

Many farmers are increasingly aware of the need for efficient farm management. Changes in farm structure, cultural practices, the price of food, farm inputs, interest rates, the limited choice of crops, special processing and marketing requirements, governmental restraints, and increased transportation costs—all affect farmers' decisions to use a certain practice. In addition, it is obvious that no strategy to increase food production on a sustainable basis can afford to disregard environmental and ecological principles. Efficient agricultural production in the short term cannot justify eventually destroying the ecological basis on which our entire food production system rests. The aim should be agricultural production without environmental destruction.

Continuous pest management undergrids efficient and profitable crop production. Every season of the year, and during certain seasons almost daily, the farmer needs to be actively fighting pests. All farmers must be taught to manage pests, for every farmer's practices affect the community of farmers. Some pests can be held at bay only when all farmers in the community participate in a continuous pest-management program.

LOW INPUT SUSTAINABLE AGRICULTURE (LISA)

Since the 1940s, agriculture in the United States has focused on increasing production. For years it seemed obvious that increased production resulted in increased profits. Booming export markets during the 1970s added emphasis to production because of global demand, and as a result many farmers borrowed money at high interest rates for expansion. Unfortunately, the export market did not meet expectations, land values did not continue to inflate, and the farm financial crises of the 1980s developed. Farmers could not pay their debts, and many lost their farms.

Since then, a growing concept that includes both a philosophy of farming and a set of agricultural practices has appeared. It is called alternative agriculture, organic farming, or low input sustainable agriculture (LISA). LISA is motivated by several economic and environmental issues including: (1) possible health risks from pollution of air, water, and foods by agriculture chemicals; (2) soil erosion and depletion of natural resources; and (3) the effect of conventional agricultural production systems on future generations' access to an "abundance of food, clean water, and a decent environment." Therefore, LISA includes two central ideas: (1) reduction in agricultural chemical input for the sake of food safety; and (2) environmental awareness, including resource conservation, land stewardship, and water quality. The basic concepts of LISA were incorporated into U.S. farm policy in 1990, and research has been funded on ways to reduce manufactured inputs in agriculture. For example, crop rotations and cultivation might be substituted for herbicides and pesticides, and manure, crop residues, and legumes might substitute in the crop rotation of inorganic fertilizer.

Several factors tend to make sustainable agriculture less profitable under 1992 conditions, although some farmers have achieved considerable success with low-input systems. First, federal farm programs do not give preference to low-input farming. Second, yields are likely to be lower because of both the problems of managing pests without chemicals and the use of land in the low-value rotation crops required for pest control and soil improvement. Third, low-input agricultural practices often require more management time and skill than conventional farming practices. Finally, reduced reliance on inorganic fertilizers and pesticides requires that farmers have more knowledge about the complex relationships among crops, weeds, insects, diseases, and soils.

IPM includes many aspects of LISA and has demonstrated that wise pest management often translates into increased grower profits. The adoption of IPM indicates that farmers are willing to reduce pesticide use if shown that it is economical to do so. We must protect our environment and increase the conservation of our resources for the future. However, agriculture must remain productive and profitable. Consequently it will take the combined efforts of scientists and farmers in many countries to improve world husbandry and keep the environment fit for human habitation.

THE IPM CONCEPT

The phrase *pest management* was first used in the late 1950s to describe the combined use of biological and chemical methods of insect control. It recognizes that pest species are a normal part of the environment and that (1) most crops can tolerate or compensate for certain levels of infection without incurring economic loss, (2) many pests play beneficial roles in the environment, and (3) eradication of pest species is seldom possible.

The term *integrated* was added later to indicate the need to use all relevant biological, chemical, and cultural tactics in the management of insects, plant pathogens, weeds, or any pest that lessens productive yields. Integrated pest management is a comprehensive and systematized approach that includes crop protection in the total cropping systems based on input from many disciplines.

Although IPM can be defined in numerous ways that describe the concept from different viewpoints, the following definition was adopted by the North Carolina State University Committee for Extension Agent Training in IPM: "Integrated Pest Management is the selection, combination and use of pest management actions on the basis of predicted consequences." It considers the economic, social, and political consequences of pest problems and seeks to utilize knowledge and techniques from all crop production and protection disciplines in the management of pests. Two important components of IPM are (1) an information-gathering system for *monitoring* insect, pathogen, and weed populations, and (2) the *planning* and *implementation* of a systematic crop management program. Integrated pest management seeks to contain pest populations at levels that permit farmers to stay in business. It uses procedures and methods that produce acceptable crop yields and cause the least harm to the environment including people. The ultimate goals are optimum crop productivity and profit, without harmful alterations in the environment.

In contrast to pest *control,* pest *management* involves the planned continuous application of selected, integrated practices to maintain pest populations at or below levels that permit profitable crop production. Successful IPM requires the use of various combinations of resources (and practices) that assist and strengthen each other as components of the overall strategy followed in the management of pests. The

TABLE 5.2. A successful, integrated disease management program.

- Select high-quality seed that is pesticide-treated.
- Use resistant cultivars.
- Assay soil for nematodes, and fumigate if necessary.
- Determine planting date based on weather forecast.
- Use proper row spacing and seeding rates.
- Use techniques of biological control, if available.
- Apply pesticides properly at the correct rate and correct time.
- Harvest crop efficiently; store it properly.
- Destroy crop residues.
- Rotate crops.
- Keep abreast of latest developments.

farmer must assemble the necessary equipment and resources so that they are available in adequate amounts at the correct time and place (Table 5.2). Advance planning is necessary for production and deployment of such items as equipment, fertilizer, pesticides, seed, and water. Safe, efficient, and effective management practices must be carried out. The farmer needs to know what each particular practice will do, how and under what circumstances it is to be used, and its relative effectiveness.

ECONOMIC THRESHOLDS

The intelligent integration of disease management practices with crop management decisions is a vital part of today's agriculture. Before making a decision the farmer should ask: "Is a given management measure necessary? Will it pay? What risks am I incurring if I do not use a given management practice? How do the short-term gains compare to the long-term gains?" A necessary component of decision making is the establishment of *economic thresholds* of disease, that is, the pest density at which management strategies and tactics should be used to minimize or prevent economic loss. Or, as the farmer asks: "Is this disease causing enough damage that I should spend time, money, and effort to manage it?" Pest losses must be based on measurements of both the quality and the quantity of crop production. As a result of the rapid increase in computer technology, more extensive use of cost-benefit analysis of alternative pest management strategies and tactics is probable in the future.

Economic thresholds depend, in part, on the market value of crops. In deciding whether to plant a given crop in a given year, farmers must be aware of economic developments not only in their own community but in other countries. Market demands, government policies, and current production levels in other parts of the country and of the world all affect the crop prices and, in turn, the level of disease infestation that warrants pest management efforts.

SOURCES OF INFORMATION

"Be not the first by whom the new is tried, nor yet the last to lay the old aside" might well be the motto for farmers in the complex world of modern agriculture. Successful farmers must undertake continuing education because every year sees a remarkable increase in the output of scientific data relevant to agriculture. The prompt acquisition and use of this information is one of the most vital components of farming, especially pest management, today. There is an unceasing supply of agricultural information available. Both public and private agencies make great efforts to assemble, digest, and dispense information to farmers and other interested citizens. Many countries now have regular programs on television and radio devoted to agriculture.

Newspapers publish agricultural sections, and numerous magazines are devoted to agriculture. Experiment stations, universities, and agricultural firms regularly publish articles, bulletins, circulars, handbooks, manuals, and periodicals. Television and radio programs on farm topics are broadcast daily in many communities. The extension services in some states have "teletips" that enable citizens to phone in and listen to a 3- to 5-min discussion on any of a large number of agricultural topics. Short courses, on-farm demonstrations, fairs, machinery exhibits, field days, and tours are designed to convey the latest information to anyone who cares to come, see, and listen.

It is only common sense that successful farmers should seek out the latest information on efficient agricultural practices.

Agricultural Consultants

For more than 100 years, ever since the land grant colleges were established in the United States, county extension agents have provided invaluable service and given good advice to farmers. County agents are trained in scientific agriculture and, as public employees, deliver and administer programs supported cooperatively by each state's agricultural extension service and the USDA. Traditionally, they have brought the latest information at no cost to farmers and advised them in the most efficient agricultural practices.

These extension agents seek out farmers to tell them of the newest discoveries. One reason why agriculture is strong in the United States is the service of thousands of extension agents who have devoted their professional lives to helping farmers. However, there are too many farmers and not enough extension agents available to supply the continuing advice and information necessary to meet each individual farmer's needs.

Modern farm managers cannot hope to fight all their battles alone. They need individual help from personnel trained in the various tactics of pest management. Fortunately there is another source of help available—private agricultural consultants who lease their expertise to farmers. These consultants perform such tasks as selecting proper cultivars, assaying the soil, monitoring pest populations, deciding where and if pesticides are needed, and correcting harvesting, storage, and marketing practices.

The agricultural consultant is a trained agronomist, economist, entomologist, pathologist, or member of a firm composed of skilled agriculturists, who helps the grower plan a production program, giving advice on such diverse decisions as cultivar selection, planting dates, soil assays, monitoring for pathogen, insect and weed incidence, pest management practices, irrigation, harvest dates, proper storage, and marketing. Usually the consulting firms charge on a per-hectare basis for their services and/or receive a bonus based on yield increases. Frequently, a farmer can pay for consultant services by omitting one unnecessary pesticide application based on information furnished by the consultant. As modern agriculture becomes more sophisticated, and farm enterprises increase in size and complexity, private consultants are assuming a vital role in helping agribusiness firms to avoid unnecessary risks and to be aware of the cost-benefit considerations involved with every decision made in growing a crop.

6

Management of Plant Diseases

Once the causal agent of a disease has been correctly identified, it is possible to develop plans to manage the disease. During the past 100 years, much research has been conducted on pathogens, diseases, and management methods. Today we can draw on this vast store of knowledge to help us in our efforts to manage plant pests.

Intelligent plant disease management is an economic necessity. Together with weed and insect management, it is a form of insurance that helps guarantee the farmer a profitable yield. It helps prevent wide fluctuations in yield. It prevents the destruction of lawns and ornamental plants. It helps prevent disastrous epidemics and famines.

There are at least five categories for the management of plant diseases:

1. Avoidance of pathogens by *exclusion* of the pathogen from a geographical area, either voluntarily or by legislation, and by *evasion* by the host plant, to prevent the pathogen from coming into contact with the host.
2. Eradication or removal of the pathogen from the host, soil, from other reservoirs of inoculum, or from a geographical area. The actual result may be a reduction of inoculum rather than the total eradication of the pathogen.
3. Protection of plants by environmental changes, biological methods, or physical methods that produce conditions less favorable for disease development or by the use of pesticides (see Chapter 7).
4. Use of cultivars with genetic resistance that resist the pathogen and disease development.
5. Chemotherapy with systemic chemicals to kill pathogens that are already in the plant (see Chapter 7).

These approaches to disease management are discussed in the following sections or in Chapter 7.

AVOIDANCE OF THE PATHOGEN BY EXCLUSION AND EVASION

Many plants grow in areas where certain pathogens do not occur. When this pattern is upset by moving plants, the plants may come into contact with a pathogen, and disease develops. Or the pathogen may be moved on seed, transplants, planting material, or equipment to areas where the pathogen did not exist and where there are susceptible hosts. When a plant pathogen is transported to a new country or environment that permits optimum disease development on an intensively cultivated crop, great devastation often results.

One of the most effective ways to manage disease is to keep the pathogen away from the host (*exclusion*) or to keep the host away from the pathogen (*evasion*). This type of management can take various forms. A governmental unit (county, state, or nation) can establish an embargo preventing the introduction of certain plants or plant products. Such *quarantines* are commonly used along with inspection. In 1912, the U.S. Congress enacted the Federal Plant Quarantine Act, which, among other things, prohibits the entry of plants into the United States. To enforce the act, inspection stations were established at all U.S. ports. Today it is virtually impossible to ship nursery stock from one country to another or from one state to another without the shipment being inspected and certified to be disease-free. Plants may be grown in certain areas or in certain ways to guarantee that they are disease-free.

Sometimes quarantines do not work. Airborne pathogens, for example, can be blown hundreds of kilometers or miles by the wind. Others are transmitted in or on seed. And sometimes people smuggle seed from one country to another. Some pathogens may be carried on clothing or in airplanes. In such cases it is difficult to stop the spread of plant pathogens.

If an area has a climate completely unsuited for survival of the pathogen or infection of the host, a quarantine may be unnecessary. Evasion—that is, growth of the host in areas free of the pathogen or under conditions unsuitable for the pathogen—may effectively contain plant diseases in some circumstances. For example, most bean seed in the United States is grown in the dry, irrigated regions of the West. Even though certain pathogens (especially bacteria) may be present, their spread and growth is restricted by the low humidities that prevail; as a result, the bean seed has less disease. Another example of evasion is time of planting. If cotton seeds are planted after the soil is warm, the seedlings can grow rapidly enough to evade the effects of several soilborne fungi that cause seedling diseases in cool wet soils.

ERADICATION OF PATHOGENS

Sometimes it is impractical to try to exclude a pathogen from a given area. However, other techniques can be used. When a pathogen has been introduced into a new area despite a quarantine, or before quarantines have been established, it is sometimes practical to eliminate or eradicate all infected host plants or those suspected of harboring plant pathogens. This helps prevent other plants from getting infected. Such eradication can occur across an entire state or region. Host eradication also is carried out routinely on individual plants in houses, greenhouses, home gardens, and farm fields to prevent the spread of numerous plant pathogens through elimination of the source of inoculum within the crop.

The practice of removing diseased, inferior, or abnormal plants from a field, greenhouse, or other planting is called *roguing*. This can be done by simply pulling out the plants and then burning, burying, or otherwise disposing of them so that inoculum cannot spread. Roguing is widely used to help manage virus diseases. It is most efficient when performed early in the growing season when only a small percentage of plants are infected.

With pathogens of annual crops—for example, cucumber mosaic virus (CMV), which overwinters mainly in perennial wild plants—eradication of the wild host sometimes eliminates the source of inoculum. Similarly, pathogens that require two hosts to complete their life cycle (e.g., cedar-apple rust) can be managed by eradication of the wild or less economically important host, which interrupts the life cycle.

Pathogens must survive from one growing season to the next to cause disease. Survival occurs in different ways or places called reservoirs. If these sources can be destroyed, the pathogen will not remain to attack the host the next year. There are many different kinds of inoculum reservoirs, each of which may require a different kind of sanitary measure to neutralize or destroy it.

Eradication is of little value against airborne pathogens because the wind will blow inoculum back into the clean area. Eradication from a small area is most successful against pathogens that spread slowly and have a low rate of multiplication; soilborne pathogens are a good example. Only one disease, bacterial canker of citrus, was ever reported to be eradicated. However, this disease reappeared in Florida in 1986. Careful inspections and destruction of plants are being used to eradicate this disease.

Eradication from Infected Host Plants

Many animal diseases can be cured by eradicating the pathogen from a sick animal, but not many plant diseases can be cured by eradicating the pathogen from a sick plant. The value of an individual plant determines whether it is worth the expense to try to cure it. An individual bean plant may not be worth curing, whereas a large camellia bush or stately elm tree has great sentimental, aesthetic, and economic value.

Pathogens may be removed from trees or bushes by pruning off infected branches. Large wounds, those more than 8 to 10 cm in diameter, should be avoided to prevent them from becoming ports of entry for new pathogens. Earlier recommendations were to paint with shellac or to use wound dressings, but research has shown that painting does not protect the wounds from rot. Current recommendations are to make smooth cuts near the base of branches but not so close as to remove the tissue that aids in rapid healing of the wounds.

Bulbs, roots, seeds, and tubers are usually treated with chemicals to prevent decay after planting from seedborne pathogens or those occurring in the soil. Chemicals are applied as dusts or slurries, or by soaking the plant material in water solutions. Many seed merchants routinely treat seed before sale. The use of treated seed is an important part of the certified seed programs of several crops. Chemicals are available that penetrate into the developing seedling or remain in a zone around the seed and selectively kill or inhibit plant pathogens. Growers who plan to treat seeds should ask the local extension agent or commercial sales representatives for the latest recommendations.

Plant materials, usually dormant buds, tubers, or woody tissues, may be heated to eradicate some pathogens. Certain viruses can be eliminated by heat treatment. Propagative material also can be freed of certain fungi and viruses by the use of meristem tissues or tip cultures. Carnations, chrysanthemums, and geraniums usually are propagated by cuttings. Oftentimes the top millimeter or so of the growing bud (*meristem*) is disease-free. Slicing off this part of the bud with a sharp scalpel and growing it on selective media results in a disease-free plant, which subsequently can be used to obtain disease-free cuttings.

Some newer fungicides have systemic properties and can move through plant tissues to the pathogen and eliminate it from a plant. Some of these chemicals now can be used to cure some plant diseases (e.g., metalaxyl for late blight of potato).

REDUCTION OF INOCULUM

Sanitation

Sanitation practices reduce the amount of inoculum available to start a disease outbreak. Another word for sanitation is cleanliness. Sanitation includes everything that can be done to reduce the amount of plant pathogens on a farm, in a field, greenhouse, or storage house, in a group of plants, or even in a single house plant. By reducing the amount of inoculum, the chances of spreading the causal agent to other plants are lessened. Sanitation practices that remove infected fruit from packing sheds reduce the chances of spores of a pathogen falling on healthy fruit.

Burning

Burning of plant residues is a form of sanitation. For example, it is important to get rid of debris from the previous crop as soon as possible. An old practice that has been used for centuries is to burn the dried remains of the preceding crop. Thus, fire is set to the straw and stubble of grain crops. Spores, bacteria, and viruses that happen to be on or in the crop refuse are destroyed by the heat. Insects and weed seeds also are killed. Burning, however, has three disadvantages: it leads to wind and water erosion; the organic matter in the stubble is lost; and smoke from burning increases air pollution. Also, sustainable agriculture programs encourage the use of organic residues to maintain soil productivity. More research is needed to identity positive and negative factors associated with residue management.

Controlled burning is the primary means of managing the brown spot needle blight of longleaf pine (*Pinus palustris*). Vigorous longleaf seedlings in the grass stage (from the second year until elongation begins, about the sixth year) are very resistant to fire. Carefully controlled burns of low intensity will scorch the needles, and thereby kill the fungus without seriously damaging the seedlings. Burning is used in grass seed fields in the northwestern United States to help control diseases and other pests. Recent air pollution problems are decreasing the use of this practice.

FIGURE 6.1. Destroying remains of the previous crop by plowing or disking reduces pathogen inoculum. After several weeks, the soil should once again be disked and sown to a cover crop to prevent water and wind erosion. *Courtesy F. A. Todd.*

Plowing or Disking

Another way to dispose of crop debris is to incorporate the debris into the soil, using a plow or disk, as soon as the crop is harvested (Fig. 6.1). Spores and insects are buried, the crop refuse begins to rot quickly, and some pathogens are destroyed. Organic matter is added to the soil as the remains of the crop are recycled for use by the next crop.

Some soilborne pathogens are found principally in the top 15 to 30 cm of soil or subsist on the plant litter in the top layers of soil. If this upper, infested layer is inverted with a lower, noninfested layer, disease may be averted, especially if the host plants are shallow-rooted. Special moldboard plows have been developed to invert the soil layers (Fig. 6.2).

Deep plowing is one of the principal methods recommended for control of the soilborne fungus *Sclerotium rolfsii,* which causes southern blight of peanuts, soybeans, and tomatoes (see Chapter 12). Deep plowing has some disadvantages. It is costly, it

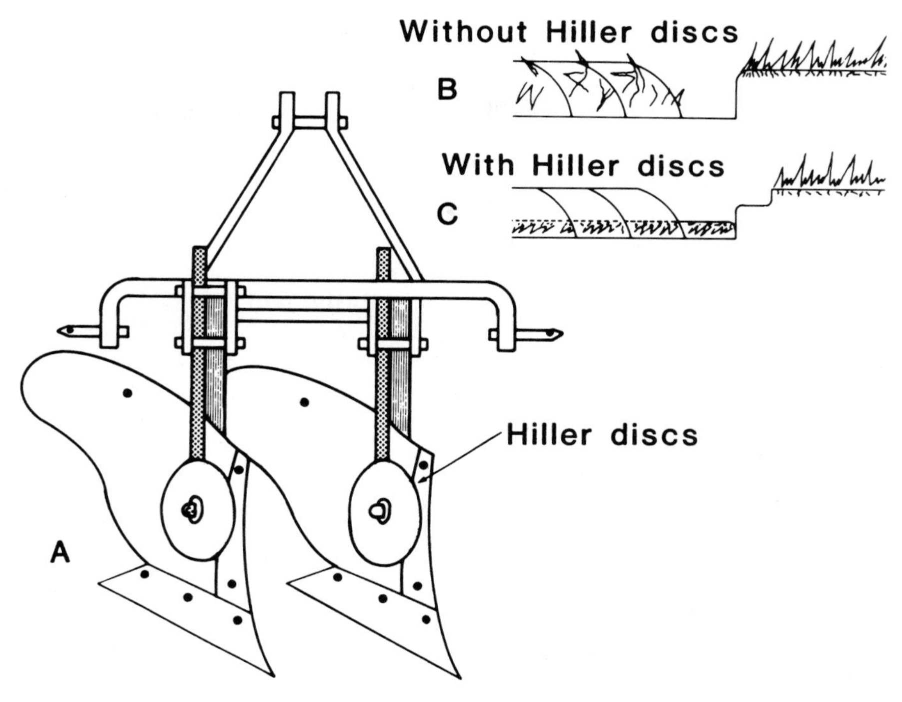

FIGURE 6.2. Moldboard plows equipped with heavy-duty concave disk coulters are used to bury the surface layer of soil, containing shredded crop litter, 8 to 14 cm below the surface of the soil. *Courtesy C. W. Averre and S. F. Jenkins, Jr.*

may disrupt the surface contour of fields, and in fields with shallow topsoils it brings infertile subsoil to the surface, where it is difficult to manage.

Soil and Root Media Pasteurization

One of the first rules of agriculture is to get the crop off to a good start. In addition to using healthy seed and/or disease-free transplants, growers should plant crops in pathogen-free soil. For many low-value, directly seeded field crops it is seldom feasible to eradicate pathogens from the soil. However, eradication of pathogens is practical and widely done in seedbeds used for growing transplants of high-value crops (e.g., flowers and bulbs), as well as in golf greens and greenhouse soils.

Heat, in the form of steam, is an efficient method of pasteurizing soil. Steam penetrates the soil quickly and imparts a large quantity of heat without raising the soil temperature excessively. The term *pasteurization* is preferred over *sterilization*. Sterilization refers to the destruction of all organisms in root media, whereas pasteurization indicates that only selected organisms are destroyed. Successful pasteurization eliminates the harmful soil pests but does not kill all beneficial organisms, which rapidly build up to normal levels after pasteurization. Pasteurization of soil and other root media to control pathogens, insects, and disease is a standard practice for greenhouse crops. It is generally done annually or between crops. This is necessary because of the threat of a rapid buildup of disease in greenhouses.

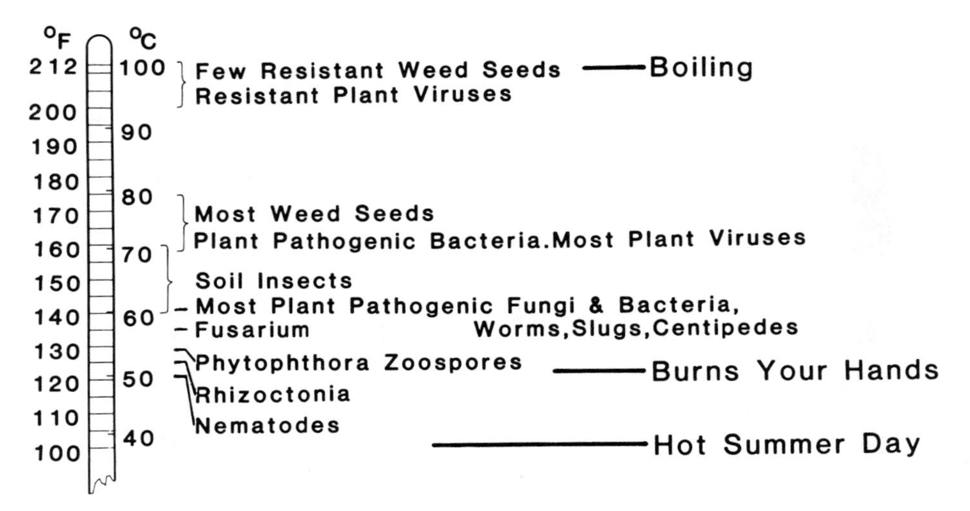

FIGURE 6.3. Temperatures necessary to kill pathogens and other plant pests. Most of the temperatures are for a 30-min exposure under moist conditions. *Adapted from K. F. Baker.*

PASTEURIZATION PROCEDURES

Each of the soil-inhabiting pests that injure plants has a temperature lethal to it (Fig. 6.3). Usually soil pasteurization is completed when the temperature throughout the soil has remained at 82°C (176°F) for 30 min. At this temperature most harmful organisms are destroyed, whereas many beneficial organisms are not killed.

The growth medium should be loosened before pasteurization and should not be dry but in good working condition. If it is too wet, more heat will be required to attain the proper temperature.

If many weed seeds are in the soil, wet the soil a week or two before pasteurization. The weed seeds will begin to germinate and will be killed at lower temperatures. Add fertilizers, limestone, peat moss, bark, or other amendments to the media as needed, and mix well before pasteurization.

Steam is distributed in ground beds through buried perforated pipes or tile from boilers or steam generators. Raised benches filled with root medium may be pasteurized with or without buried pipes, but a cover is necessary. Pipes are buried at half the soil depth if used in raised beds. Another method is to inject steam between the cover and the root medium through canvas hoses.

Potting media can be pasteurized in a wagon equipped with perforated steam pipes at the bottom (Fig. 6.4). A few pots of soil can be pasteurized by heating them in a stove or an oven. Pots and flats can be pasteurized in cabinets (Fig. 6.5) equipped with a steam outlet.

Covers are placed over the soil media during pasteurization to catch the steam and hold in the heat. Polyethylene, vinyl, or nylon covers are commonly used. Polyethylene is cheapest but lasts only a year. Nylon is the most expensive but can be reused many times.

Limited areas of field soil can be steam-pasteurized. The soil should be well prepared and not too wet.

FIGURE 6.4. Pasteurizing soil. Steam enters the aerator (foreground) while a blower introduces air. The two gases mix and then are conducted through a host to a covered soil wagon (background). With this device, soil can be pasteurized at 60°C (140°F) rather than at 100°C (212°F).

Steam is conducted from a boiler by a hose to a steam rake, which consists of a pipe header (10 cm in diameter and 4 m long) that is drawn by a cable across the field. Chisels, 45 cm long and spaced 25 cm apart, project into the soil. Behind each chisel is a 1-cm pipe that releases the steam into the soil. The rake crosses the field at a rate about 50 to 60 cm/min. A pasteurizing cover, sufficiently long to require 30 min to pass over any given point, is attached behind the header. The cover holds the steam in the soil and helps maintain the temperature at or above 70°C for at least 30 min. One hectare of soil can be pasteurized by a single rake in 80 to 140 hr.

AFTER-STEAMING PROBLEMS

Large quantities of manganese exist in many soils, and steam pasteurization converts additional manganese to a plant-available form. High levels of manganese are toxic to

FIGURE 6.5. Sterilizing pots and flats in a steam cabinet with steam from a greenhouse boiler. The cabinet doors are closed during the actual sterilizing process.

some plants and also interfere with uptake of iron by roots. Iron deficiency is commonly associated with high manganese levels. Fortunately, only a small but adequate amount of manganese is available for plant use in most soils; the majority of it is in a form unusable by plants. The longer soil is steamed, the greater the risk of manganese toxicity.

Root media that contain organic matter rich in nitrogen can release ammonium nitrogen during and after pasteurization. (Such media include compost, leaf mold, manure, and peat.) Normally, ammonium nitrogen is continuously converted to nitrate nitrogen by soil bacteria. During pasteurization, ammonifying and nitrifying bacteria are nearly eliminated, and it may take 3 to 6 weeks before populations build up to normal levels. During this time, toxic quantities of ammonia may develop. Root tips may be injured, and the plants remain stunted and wilted. Ammonium toxicity can be avoided by not overpasteurizing and by delaying use of the media until 2 weeks or more after pasteurization.

Storage House Disinfestation

Many crops must be stored for varying lengths of time before being sold. Often the storage house becomes contaminated with rot-producing organisms that survive between successive crops on the walls and floor of the storage house and on crates. These organisms may infect and rot the new crop. An important disease management measure is to thoroughly clean the storage house and surroundings of all crop remains, dirt, and other trash. The house should then be washed down with some disinfectant such as household bleach (sodium hypochlorite), according to directions on the label.

Refuse Site Destruction

Many crops must be sorted, graded, and/or trimmed before being sold. Imperfect, diseased, bruised, or improperly sized produce must be discarded. Sometimes large piles of rejected plant material accumulate. These cull piles serve as an inoculum reservoir for plant pathogens. For example, potato farmers often discard or throw away tubers that are discolored or rotten. The potatoes in such cull piles may be infected with the late-blight fungus. In the spring some of these infected tubers sprout, and the fungus produces spores on the new growth. The spores are blown to potato plants growing in nearby fields and start new infections or outbreaks of late blight. These culled potatoes can be destroyed by spreading them on the land and plowing them under, by burning, or by spraying with a herbicide to eliminate the primary inoculum for this disease. The damage from late blight is in part related to the time of infection—the earlier the infection, the more damage done. Thus, anything that delays the start of infection will help decrease losses.

Disinfestation of Machinery and Tools

Several plant pathogens can survive on or are spread by machinery and tools. For example, when Irish potato tubers infected with the bacterium that causes ring rot are cut to make seedpieces, the cutting knives can become contaminated and spread bacteria to healthy tubers. A few infected tubers may lead to contamination of all the seedpieces. Therefore, knives used for cutting seedpieces should be periodically disinfested with sodium hypochlorite, and contaminated containers disinfested by dipping in formaldehyde or sodium hypochlorite before use.

Tobacco mosaic virus (TMV) is in the juice of infected pepper, tobacco, and tomato plants. If such plants are bruised or cut by equipment used for cultivating, pruning, suckering, or harvesting, the virus can be spread to and infect healthy plants. This commonly occurs in greenhouse crops and causes great losses. This spread can be prevented by having workers dip their hands or pruning shears in milk after handling each plant.

Cropping Systems or Rotations

The practice of growing the same crop in the same field year after year is called *monoculture.* Perennial crops such as apples, peaches, grapes, forest trees, legumes, certain pasture grasses, and turf grasses occupy the same fields for many years. However, some farmers choose to grow annual crops such as corn, wheat, rice, or soybeans in the same fields year after year. Farmers who practice monoculture must be constantly on guard against disease outbreaks, for the continuous presence of a host may permit pathogens to build up to high levels. Epidemics then occur, and an orchard, vineyard, or forest stand may have to be replaced, or an annual crop may be lost.

The practice of growing different crops successively in the same field is called a *cropping system, cropping sequence,* or *rotation.* This practice aids in the biological control of plant pathogens by increasing the growth of antagonistic organisms in many cases. The cropping sequence may be completed in 1 year or extend over several years. For example, in frost-free areas or regions with long growing seasons, it is possible to plant annual crops year-round or to grow two or more crops in the same calendar year. Thus in California, Florida, and Texas, certain vegetable farmers alternately grow crops of tomatoes and sweet corn. In parts of the southeastern United States, it is a common practice to plant soybeans after small grains. In cold regions where crops cannot be grown year-round, completion of a cropping sequence may take 4 years. In the northeastern United States, a common 4-year rotation on dairy farms is to grow a crop of corn followed by oats the next year, then wheat, then hay in succeeding years. In the midwestern United States, alternate crops of corn and soybeans compose a 2-year rotation.

Crop rotation is an efficient and economical weapon for controlling plant diseases. It is a very old method of farming; Vergil, the Roman poet who lived in 30 B.C., praised the use of crop rotations.

Many farmers, however, cannot or will not use long cropping sequences of 4 years or more. They do not have enough land, or they prefer to risk the chance of disease rather than use their best fields only 1 year out of 4 for their principal cash crop. Some rotations are impracticable, as when certain fields cannot be irrigated, or are inaccessible, or are too steep or stony or unsuitable for a given crop in a sequence. Certain rotations may be uneconomical: allowing land to grow up in weeds for 2 years or more will help control many soilborne pathogens, but is less profitable than growing cultivated crops. Bare fallowing, the practice of periodically plowing or disking land during the growing season without seeding it, done for the purpose of destroying weeds and conserving moisture, also reduces the incidence of some diseases and can be used in a cropping system. But it is expensive to keep the land disked. Furthermore, the land is subject to erosion, and the farmer loses use of the field during the fallowing period.

As diseases continue to spread, many fields become infested with more than one pathogen. Therefore, the planning of suitable and effective rotations is more difficult on some farms than others. In those fields where more than one disease is present, the farmer must determine which disease is causing the most trouble and plan the rotation to combat this disease, even though such a rotation may produce conditions favorable for "minor" diseases.

Many factors must be considered in planning a satisfactory cropping system for a given farm or given region. Although it is difficult to outline any one cropping sequence to be used under a particular set of conditions, it is possible to list the factors that influence the choice of a cropping sequence. A good cropping system should have the following characteristics:

- Be profitable.
- Maintain the soil in good physical condition and reduce wind and water erosion of soil.

- Maintain proper soil productivity, keep a desirable microbiological balance, and utilize fertility efficiently.
- Help manage weeds, insects, and diseases.
- Be flexible, practical, and easy to follow.
- Include both row and sod crops.
- Be environmentally sound.

The effectiveness of disease management by crop rotation depends upon many factors but principally upon the life history and behavior of the pathogen, including (1) the persistence of the pathogen in the soil, (2) the rate of buildup of inoculum, (3) the means of spread of the pathogen, and (4) the degree of host specificity. In relation to the soil, plant pathogens may be classified as soil invaders or soil inhabitors.

Soil invaders include those pathogens that usually attack the aerial parts of plants but may reside temporarily in crop refuse or in surface layers of soil. The use of short rotations may be of great value in reducing the damage from these diseases. Included in this group are a number of leaf-spotting fungi such as *Alternaria, Cercospora,* and *Colletotrichum.*

Soil inhabitors are those pathogens that reside naturally in the soil. *Persistent* soil inhabitors are capable of indefinite existence in the absence of the host. Some soil inhabitors persist by means of long-lived, resistant spores or resting bodies. These organisms may be quite host-specific. Long rotations may reduce the inoculum level of these organisms but generally will not eliminate them. Examples of this type of soil inhabitor are fungi belonging to the genera *Fusarium* and *Phytophthora.* Members of a second group of soil inhabitors persist because of adaptation to the soil environment and can live as saprophytes in competition with other soil organisms. Many of these pathogens can persist indefinitely in the absence of the host plant, and may have a wide host range. Examples are pathogens that cause root and stem rots, seed decay, and seedling damping-off, such as *Pythium* and *Rhizoctonia.* Crop rotation is of little value in managing these pathogens.

Nonpersistent soil inhabitors can develop and reproduce only in the roots of suitable host plants, and are sometimes called *root inhabitors.* Various species of plant-parasitic nematodes are included in this group. Rotations involving the use of nonhost plants are useful in the management of these pathogens.

Some general principles about crop rotation are:

- Setting up a successful disease-management rotation involves (1) disease identification, (2) selection of locally adapted crops, and (3) effective arrangement of crop order.
- In general, longer rotations are more effective in managing diseases than short rotations; short rotations are better than no rotation.
- Specific crop rotations are not equally effective with all diseases.
- The sequence of crops in a rotation should not be rigid and inflexible; the crops in the rotation should be changed as often as practicable.
- A good rotation crop may lose its value, if grown too often, by permitting the buildup of populations or strains of pathogenic organisms that will attack it and other crops in the rotation.

- Rotations that affect the carryover of soilborne pathogens have other important effects on crop yields; thus, figures for crop yield and amount of disease are not always closely parallel. Those rotations that are favorable from the nutritional viewpoint may sometimes promote high yields despite extensive development of disease.
- The proper use of cover crops, a valuable part of crop rotation, prevents the loss of plant nutrients by leaching, washing, or blowing of the soil. In preventing the movement of soil by wind and water, cover crops reduce the spread of soilborne pathogens.

Cropping sequence alone is not the answer to disease management. However, when it is intelligently combined with soil fumigation, the use of resistant cultivars, and proper cultural practices (soil conservation, correct plowing, sanitation), disease losses will be minimized. With diseases that are prevalent and destructive, a grower should rotate crops in a field unless there is no alternative. In planning an overall pest management system designed to fit his or her own particular operation, each grower should consider carefully the benefits and/or limitations of rotation in his or her production program. In addition to the benefits to be derived from disease management, rotation of crops helps to prevent wind and water erosion, supplies organic matter to the soil, and helps maintain fertility and proper soil structure. These principles are an important part of sustainable agriculture, a system that is being used increasingly in agriculture today.

The applications of crop rotation in the management of specific diseases are discussed in the chapters dealing with each disease.

Greenhouse Management

Disease prevention and sanitation are especially important in greenhouses, which are designed to provide optimum conditions for plant growth. Their subtropical environment of warm temperature, high RH, and adequate water is also conducive to the growth of many disease organisms. In addition, continuous culture of the same or related crops year-round increases disease problems by providing an ever-present host for pathogens.

When high populations of plants (intensive cultivation) grow in a limited area, the chances of disease are increased. A disease is difficult to manage once it gets started in a greenhouse, because conditions are so favorable for development of epidemics. In such cases it may become necessary to discard and destroy an entire houseful of plants; in addition to the direct loss, the cleanup job is laborious and expensive.

Prevention plays an important role in disease management in greenhouses. Disease organisms may be carried in the soil and water, on pots and tools, on shoe soles and clothing. Windblown spores easily invade a greenhouse. Because they are virtually invisible, the presence of pathogens is known only when lesions or other symptoms of disease become visible.

Diseases are often the downfall of ignorant or careless greenhouse managers. Careful growers realize that disease is inevitable, and are continually on guard against them. An efficient disease management program is just as necessary as daily watering in greenhouse production.

A number of basic dos and don'ts can go a long way toward minimizing plant diseases in greenhouses:

- Use root media pasteurized with steam or methyl bromide.
- Use sterilized pots, containers, and equipment. All pots, flats, and plant supports should be sterilized before reuse.
- Sterilize the potting bench and keep it clean. Sweep it after use with a special broom kept only for this purpose. Swab it off once a week with disinfectant. Do not put contaminated potting media or pots on the potting bench.
- Store pasteurized potting media in special bins to avoid recontamination. Do not sweep soil from the floor into storage bins. Prevent dust from blowing from one bin to another.
- Avoid contaminated soil. Take all possible precautions against carrying foreign soil into greenhouses. Keep a flat pan containing a disinfectant solution at the door so that visitors can step into it to sterilize their shoes; or have visitors use disposable overshoes. To avoid contamination, do not place watering hoses on the floor; instead, fasten a clip to the side of the bench to hold hoses between watering.
- Do not water carelessly or hurriedly. Avoid splashing water from one pot to another. Do not overwater, as root-rotting organisms thrive in wet soils. Use automatic watering devices whenever possible. Construct walkways so that water puddles, which can harbor pathogens, do not form.
- Clean up plant debris. Do not let weeds grow under the benches. Do not throw poorly rooted cuttings under a bench. When disbudding or pruning, use a basket or bucket to hold plant parts until discarded. Workers handling tomato or pepper plants should dip their hands in milk before moving from one plant to the next to prevent spreading virus diseases. Impress upon employees that cleanliness is imperative. Devise sanitation practices that employees can follow. Think clean! Establish trash dumps or compost heaps far enough from greenhouses that spores, insects, or seeds cannot easily be blown back into the greenhouses.
- Do not keep leftover plants after the season. Dispose of the preceding crop as soon as possible so that it will not be a source of inoculum for the next crop.
- Use clean stock. Purchase bulbs, cuttings, or seedlings from reputable sources and use only certified seed. Discard any weak or diseased materials. Take cuttings for propagation only from healthy plants. Periodically sterilize the cutting knife by dipping it in alcohol and flaming. Work on a clean tabletop or clean newspapers.
- Maintain proper conditions. Many plant pathogens thrive when humidity is high, and most fungus spores require free water to germinate. Thus, steps should be taken to control humidity and condensation. Warm air holds more water than cold air; as air cools, the moisture-holding capacity drops until the dew point is reached, and water (condensation) forms on any solid surface. Condensation can be combated by opening ventilators for 15 to 20 min when the heat first comes on to allow warm water-filled air to escape. After the drier, incoming cold air is heated, the ventilators are closed. The relative humidity, which increases as air warms, can be reduced by watering early in the day.

- Develop a regular spray program for management of diseases, insects, and mites, and follow it carefully. Damping-off and root-rot diseases can be inhibited by drenching soil with fungicides. Leaf- and flower-spotting organisms—including mildews, rusts, and leaf spots—can also be prevented by regular application of fungicides. Precise recommendations should be obtained for one's own state.
- Greenhouse managers should subscribe to and read one or more trade publications to keep aware of new developments. Conferences and training seminars are helpful, as are gardening programs on radio and TV, especially for nonprofessionals with home greenhouses.

Weed Management

Weed management is a vital component of an efficient pest management program. Weeds can harbor plant pathogens and insects; and, more important, weeds reduce profits by using water and fertilizer that could be used by the crop, by interfering with harvesting machinery, and by encouraging pests.

Areas around greenhouses, cold frames, fields, and seedbeds should be kept weed-free. When the transplanting season is over, seedbeds should be disked promptly, to destroy all unused seedlings, and sowed to a cover crop such as soybeans or other legumes. Growers should become familiar with weed management practices recommended in their state.

Insect Management

Insect management is a necessary part of an integrated pest management program. Almost everyone is familiar with insect damage to plants such as that caused by the army worm, boll weevil, ear worm, Japanese beetle, scale insects, wire worm, and many more. Aside from the direct damage resulting from insect feeding or egg laying, some insects carry plant pathogens (see Table 4.1). The tiny holes made by insects provide entry points (*infection courts*) for microorganisms; oftentimes insects carry the pathogens that enter through the wounds.

The same principles that apply to disease management also apply to insect management (see Table 5.1). Therefore, it is necessary for growers to have an overall pest management strategy in order to keep losses to a minimum.

ADJUSTMENT OF THE ENVIRONMENT

Growers have learned to manipulate the environment by such practices as varying the time of planting to avoid disease, shading the soil by mulching or spacing of plants, using windbreaks to reduce damage from winds, and avoiding two much shade, which reduces the vigor of plants. Any practice aimed at improving the vigor of the plant often helps to increase its resistance to disease. Therefore, overall efforts to manage plant diseases should include sensible efforts to grow a healthy crop.

Soils should be prepared properly, and good soil drainage is essential. Proper irrigation should be used, which may include sprinkler irrigation to prevent frost damage. Erosion should be prevented by use of terraces, contoured rows, and strip cropping with sod and row crops. Soil pH should be maintained at the proper levels. Proper environmental storage conditions are essential; it is a waste to grow a good crop and then lose it with careless storage.

Sometimes it is impracticable or impossible to exclude or eradicate a pathogen from a given area. In such cases the crop must be protected from it by either physical or chemical means. Physical means, discussed in this section, include cooling, moisture management, drying, and radiation. The management of plant diseases by chemicals is discussed in Chapter 7.

Refrigeration

The most widely used method of preventing postharvest diseases of fleshy or succulent plant products is refrigeration. Temperatures of 0-5°C (32-40°F) inhibit the growth of many pathogenic microorganisms, thus preventing the spread of existing infections and the start of new ones. Most perishable farm products are cooled immediately after harvest, transported in refrigerated trucks, railway cars, or airplanes, and kept refrigerated until used by the consumer. One reason why the U.S. citizen can eat fresh fruit and vegetables year-round is the wide use of refrigeration by commercial companies and in the home to reduce plant diseases.

Moisture

High moisture levels in air and soil are conducive to many plant diseases. Some crops are regularly grown in low-humidity dry areas with surface irrigation to reduce or prevent certain leaf-spot diseases. Certain root-rotting fungi thrive in wet soils. The damage from diseases can be reduced by not growing crops in waterlogged soils or by using elevated beds. In special cases (golf greens and greenhouses), soil mixtures with good drainage are used.

Drying

The moisture content of crops such as corn, peanuts, sorghum, soybeans, and wheat must be carefully regulated if damage by storage fungi is to be avoided. These crops should be harvested at the proper stage of maturity and moisture content, and should be dried properly to the correct moisture content for storage. In many of the grain-producing areas of the world, grain is harvested in the dry season or during sunny periods. However, if wet weather prevails and harvest is delayed, or the grain cannot be dried properly after harvest, storage rots can markedly reduce the quality of the crop. Barley, corn, rice, and wheat should be stored below 13% moisture and soybeans below 12% moisture.

Fungi are a major cause of damage, spoilage, and heating of stored grains. Storage fungi cause loss of germination, bin-burning, bad odors, mustiness, and heating. Moldy grain makes feed less palatable and nutritious for livestock, and renders some foods unsuitable for human consumption. Some of the fungi that grow in grains and other seeds before harvest or during storage produce *mycotoxins* (mold-produced toxins). The common storage fungus, *Aspergillus flavus,* produces several toxins that are called *aflatoxins.* These toxic chemicals (*metabolites*) are produced by several fungi on corn, cottonseed, peanuts, and many other substances. Feeds contaminated with aflatoxins are toxic to cattle, ducklings, swine, and turkeys. Younger animals are more susceptible to injury than older animals. Only a few parts of aflatoxin per billion can cause pathological changes in susceptible animals. Therefore, it is risky to feed moldy grain to farm animals. It may make them sick, reduce weight gains, or even kill them.

Aflatoxins can be reduced or prevented from occurring in corn by growing adapted cultivars, harvesting the crop when it is mature and at proper moisture content, drying it promptly, if needed, and storing it in well-ventilated bins. Similar procedures should be used for harvesting other crops.

Sun drying of certain crops is still practiced in many parts of the world. Grapes are dried to make raisins, plums for prunes; coffee berries, dates, and figs are dried in the sun. Certain types of tobacco are sun-dried; apples or peaches are sliced and spread in the sun to dry. Needless to say, once sun-dried fruit absorbs excess moisture, spoilage can occur quickly at warm temperatures.

Radiation

Ultraviolet (UV) light, X rays, gamma rays, and particulate radiation such as alpha and beta particles destroy microorganisms and can be used to free plant material of pathogens. Unfortunately, the dosage necessary to kill pathogens also injures plant tissues; but we may expect a breakthrough in the 1990s so that fruits and vegetables can be stored at room temperature without spoiling. Another factor to be overcome is consumer fear. Many consumers believe that radiation of food will leave harmful particles on the food, and this myth will be a difficult one to dispel. With better radiation methods and consumer information, greater success can be expected in this area.

BIOLOGICAL CONTROL

Farmers have observed for centuries that there is less plant disease on certain soil types, called *suppressive soils,* whereas on other soil types (*conducive soils*), plant diseases cause significantly more damage. Until recently the farmers did not know why plants in certain fields had less disease, but they knew that less disease occurred. They also knew that plowing under green cover crops or adding manure or lime to soil, rotating crops, or flooding the land often would decrease disease and increase yields. These age-old observations coupled with attempts through the years to increase crop production by changing agronomic practices altered the microbial balance in soil and probably decreased plant disease through *biological control,* the control of one organism by another. All living things have enemies; the fact that a parasite feeds off a living host does not keep the parasite from being attacked in turn by other parasites and predators.

In 1928, Sir Alexander Fleming, an English scientist, noticed that a fungus, a *Penicillium* species, produced a toxin that killed bacteria. By 1939 this amazing antibiotic was purified and named penicillin. Shortly after Fleming's discovery in 1932, R. Weindling, a USDA scientist, published a report stating that a fungus, a

Trichoderma species that lives in the soil, parasitized other fungi and produced antibiotics that inhibited or suppressed the growth of several soil pathogens that caused plant diseases. Since these early discoveries, many scientists have demonstrated that one organism can suppress the growth of another with a subsequent decrease in the amount of disease.

The severity of a disease depends upon the interactions of the pathogens, antagonists, hosts, and the environment, as stated in Chapter 4. An epidemic depends on the coincidence of available inoculum of a virulent pathogen, susceptible host tissue, and a favorable environment. All of these factors can cause a complicated balance of factors to shift in favor of disease or health.

For the many plant diseases caused by soilborne pathogens, many interdependent variables operate in the complex soil environment and help to determine the amount of disease. The many antagonistic organisms that live in the soil may (1) produce antibiotics harmful to the pathogen, (2) compete for food in the soil, or (3) parasitize the pathogen. Basically the mechanisms that interfere with the growth of soilborne pathogens work in two ways: (1) by reduction of the amount of inoculum produced by the pathogen, and (2) by prevention or reduction of the rate of spread of the pathogen. Some specific mechanisms of biocontrol are illustrated in the following paragraphs.

Antagonists

It is not entirely clear how fungi such as *Trichoderma* or *Penicillium* or bacteria such as *Bacillus,* when added to soils, can suppress the amount of disease. In some cases, for example, small amounts of suppressive soil placed in soil near germinating seedlings can reduce the amount of seedling decay. We now know that some organisms in suppressive soils produce chemicals that are harmful to the soilborne pathogens that cause seed rots, damping off, or other diseases.

For example, penicillin interferes with bacterial cell wall formation, streptomycin interferes with bacterial protein synthesis, bacteriocins prevent the growth of species closely related to the producer, and siderophores inhibit absorption of certain necessary ions (iron) needed for growth by pathogens. Marigold (*Tagetes* sp.) roots exude terthienyls that are toxic to some nematodes and fungi. Seeds can be protected from soil pathogens by coating the seed with spores of antagonistic fungi that prevent pathogens from harming the seed or germinating seedlings. Undoubtedly many additional antagonistic compounds and organisms will be discovered as scientists continue to investigate biological control mechanisms.

J. Rishbeth, in 1963, developed the first practical use of an antagonist against a pathogen. He inoculated freshly cut pine stumps with a spore suspension of the fungus *Peniophora gigantea,* which prevents the infection of the stumps by *Heterobasidion annosum* and subsequent infection of nearby trees by spores of this destructive wood-rotting pathogen.

Scientists are busy identifying the organisms that produce suppressive chemicals and the mechanisms involved. We can expect great progress in the future as investigators combine biological control with sensible chemical management, IPM, and LISA to make farming more profitable, safe, and environmentally sound.

Fungistasis

Reproductive propagules of some fungi—spores, sclerotia, chlamydospores—can undergo an induced period of dormancy (called fungistasis) before germination.

The dormant propagules, even though enough food to grow is present, require an external shock or stimulus before germination will occur. This stimulus may be a root exudate, volatile gases such as methanol and ethylene, or unknown compounds from decomposing crop residues or other organic matter.

Pathogens Use Root Exudates as Food

Plant roots exude a surprising quantity and variety of organic compounds. The compounds that "leak" out of the roots into the rhizosphere may amount to 25 to 35% of the carbohydrate production of the leaves. Exuded chemicals are used rapidly by a thriving community of microbes in the root zone (rhizosphere), some of which are pathogens that use the host-plant exudates as a food base. By preventing pathogens from using these exudates as food, disease can be decreased. A number of factors can cause this complicated balance to shift in favor of beneficial or harmful members of the complicated microcosm of the soil. For example, the addition of large amounts of organic materials that contain large amounts of carbon and little nitrogen can favor beneficial organisms and inhibit certain pathogens. Alternatively, plants subjected to stresses such as prolonged drought, waterlogged soils, excessively high temperatures, compacted soils, or unbalanced fertilization may become more susceptible to disease because the exudates of nutrients from the plant roots can increase under these conditions, and pathogens may flourish.

Hyperparasites

A helpful aspect of biological control is the possibility of eliminating pathogen inoculum. Many different soil microorganisms parasitize conidia and oospores, other fungi attack sclerotia, and some nematodes are trapped and killed by fungi. Other hyperparasites prevent the production of inoculum. For example, wheat straw helps to reduce wind erosion of the soil, and the straw is colonized by saprophytes that prevent colonization by or kill pathogens that may grow in the straw.

Plant pathogenic fungi are themselves attacked by other parasites—fungi, bacteria, and viruses. For example, the fungus *Sporodesmium* forms thick mats of conidia and mycelia on sclerotia of *Sclerotinia sclerotiorum,* a fungus that causes white mold on snapbeans, sunflowers, and peanuts. The *Sporodesmium* uses the sclerotia as food and destroys them as a source of inoculum. Viruses have been shown to reduce virulence in *Cryphonectria parasitica,* the cause of chestnut blight, *R. solani,* a damping-off fungus, and other fungi.

Mycorrhizae

Competent farmers try to provide conditions for vigorous and healthy plant growth. There is increasing evidence that some useful fungi (mycorrhizae) that grow in or on plant roots increase yields. These symbiotic, beneficial fungi increase nutrient uptake, especially phosphorous by plant roots, and help protect the roots from pathogenic nematodes and fungi. The inoculation of seeds or roots of transplants with mycorrhizal fungi improves plant vigor and helps protect the plants from disease organisms and stress in general.

Cross Protection

Cross protection is another way to protect plants from disease. Oftentimes when plants are inoculated with a weakly virulent strain of a fungus or a virus, protection against infection by a more virulent strain of the pathogen develops. The virulent strains produce fewer, smaller lesions; multiplication is slowed down; and symptoms are delayed or prevented.

Plant pathogens are an example of survival of the fittest. Throughout evolutionary history, pathogens have developed ways to obtain food and energy from the host. Successful pathogens allow no excess accumulation of readily available nutrients (food is hoarded), discourage competitors, and convert nutrients as rapidly as possible into propagules that can serve as more inoculum.

Antagonists that compete with the hungry pathogen for nutrients and inhibit pathogen growth can be used in biocontrol. Most pathogens are composed of many strains. Thus, plants or soils can be inoculated or infested with weakly virulent strains that compete with more virulent strains for the food supply, and less disease results.

Abiotic Environment

The physicochemical makeup of the soil itself plays an important role in biocontrol. Soil composition is constantly changing. The oxygen, carbon dioxide, and water content, as well as temperature, pH, and nutrient availability, fluctuate and are changing from minute to minute and year to year. These changes make possible the diversity of microorganisms. Sandy soils are more favorable for root diseases caused by *Fusarium* than clay soils, perhaps because montmorillonite clay soils stimulate the development of bacterial antagonists. As we learn how to manage this dynamic ecosystem and its multitude of components, biocontrol will become more efficient and useful and less harmful to the environment.

Prevention of Frost Damage

Frost-sensitive crops (tomatoes, strawberries, flowers) can be protected from frost damage by spraying the leaves with strains of the bacterium *Erwinia herbicola*. These bacteria prevent the formation of nuclei that are needed for the formation of ice crystals, and frost cannot develop. This is a form of biocontrol using a microorganism to protect against an abiotic problem.

Weed Pathogens

Some weeds can be destroyed by inoculating them with weed-eating insects or weed-killing fungi. Some pathogens have been found that attack only specific weed species. If these pathogens can be established in an area, weed populations can be reduced or eliminated without chemical herbicides.

Transgenic Plants

Many fungi produce genetically controlled toxins that injure or kill host plant tissue. By introducing the recessive gene from avirulent strains of the fungus that do not produce the toxins and inserting the gene into a host plant, a disease-resistant cultivar can be developed. This is a form of biocontrol that involves genetic engineering (see Chapter 8).

Similarly, the gene controlling disease resistance can be introduced into a desirable cultivar transgenically to make it disease-resistant. This means that resistance genes from one species could be used in another species or even another genus. This topic is discussed further in Chapter 8.

Importance of Biological Control

Biological control is becoming more important in managing plant diseases because of concern about chemical pesticides. Biological control occurs naturally in many cases, but it has not been understood well enough until recently for investigators to know how to use it. Many microorganisms are antagonistic to plant pathogens, but it has been difficult to ensure survival of these organisms in sufficient numbers to be effective on the plant where the disease might occur. There has been progress in selecting strains of fungi and bacteria that have given good control of some diseases when applied to soil, seed, or plants. Biological control agents composed of bacteria have been formulated as insecticides and are widely used.

Effective biological control of plant diseases usually requires the application of multiple procedures, each operating in a different way or time. There is no single magical chemical or biological organism that, when added to the soil or applied to a leaf surface, will eliminate all diseases. Rather, we must discover and use as many as possible of the natural biological control measures that have evolved.

DISEASE-RESISTANT CULTIVARS

Disease-resistant cultivars are the least expensive, safest, and most practical way to combat diseases for many crops. The use of resistant cultivars is particularly appealing to those who must rely on expensive pesticide applications to protect large acreages of low-income crops such as wheat. As crop production costs increase, growers must diligently seek ways of reducing these costs. Also the use of resistant cultivars requires few changes in farmers' production practices; all they need do is substitute seed of a resistant cultivar for that of a nonresistant one. Often the cost of the seed is the same. However, it does take skilled scientists, time, and money to breed resistant cultivars.

The early phases in the development of a disease-resistant cultivar are usually tedious and expensive. In the United States, much of this work is performed by state experiment stations and land grant universities supported by tax dollars. Recent changes in patent laws have made it feasible for many private seed companies to employ plant breeders to develop resistant cultivars. These private breeders use breeding lines or *germplasm* sources developed in state experiment stations, USDA research centers, and their own facilities to build new cultivars for release to growers. The production of reliable resistant cultivars requires the cooperation of both public and private agencies. It is safe to say that without this cooperation the use of disease-resistant cultivars would have lagged far behind what it is today, and crop yields would be much lower.

Disease resistance in crops is the foundation of a disease management program. However, resistant cultivars are not available for all crops. Moreover, the use of disease-resistant cultivars may have some disadvantages. Cultivars adapted to one locality may not do well in another. Seemingly unimportant differences in environment may make the difference between a high-grade and a low-grade crop. Specific cultivars may respond differently to rainfall, the amount and kind of fertilizer, soil type, planting density, day length, length of growing season, and cultivation practices.

The use of disease-resistant cultivars also may lead to problems stemming from the genetic uniformity of such cultivars and the tendency for new strains of pathogens to develop or for a minor pathogen to become an important one.

Genetic Uniformity and Plant Diseases

When a new cultivar or several closely related ones resistant to only one strain (race) of a pathogen are introduced and planted over a wide area, there frequently is a rapid development and spread of other races of the pathogen that will attack the new cultivar. As a general rule for annual crops, a resistant cultivar lasts about 5 years. Because the resistance gene exerts selection pressure on the pathogen population, a new race or strain of the pathogen develops or is selected that can overcome the resistance gene. As a result of this process (i.e., the loss of effectiveness of the resistance gene), the host often is said to have had its resistance "broken down" or "worn out." Sooner or later an epidemic will develop and great losses result. This is what happened when southern leaf blight struck the corn crop in 1970. Genetic uniformity in crops, although desirable agronomically, is epidemiologically unsound and often catastrophic. Therefore, disease-resistant cultivars should be used in conjunction with other disease management measures.

Another example of the dangers of genetically uniform resistant cultivars—putting all your eggs in one basket, so to speak—is the Victoria-based cultivars of oats. Victoria was developed with resistance to crown rust and smut, two important oat diseases in the Midwest; it also gave good yields, had short stiff straw, and did not lodge easily, making it easy to harvest with a combine. Breeders in the United States used Victoria as a parent in developing some 30 oat cultivars. In the mid-1940s approximately 75 million ha of oats—two-thirds of the U.S. production—were planted to these cultivars. However, in 1944 a strain of the *Helminthosporium* fungus (now called *Bipolaris*) caused much damage to Victoria and related cultivars. In 1946 the epidemic of Victoria blight caused such severe losses to oat cultivars with Victoria in their pedigree that they had to be abandoned.

Sometimes, a cultivar resistant to a major disease turns out to be susceptible to a pathogen that usually causes only minor damage. After the resistant cultivar is planted over a wide area, the secondary pathogen may increase and become a major problem. This type of situation occurred after the blue mold (downy mildew) epidemic of tobacco caused a $25 million loss to the European tobacco crop in 1960. European breeders urgently requested and obtained seed of blue mold-resistant tobacco cultivars from other countries. Before these blue mold-resistant cultivars were introduced into Europe in the early 1960s, potato virus Y (PVY), the causal agent of vein banding, was a minor pathogen. Most of the European tobacco cultivars were tolerant to this virus, and vein banding caused only minor losses. However, the blue mold-resistant cultivars that were introduced and widely used since then are susceptible to certain strains of PVY. Vein banding became a serious tobacco disease in Europe, and tobacco breeders are urgently seeking ways to control this threat to the tobacco crop.

It is difficult enough to breed a usable cultivar resistance to one disease, and it is even harder to incorporate resistance to two or more diseases in one cultivar. One reason for this is that the disease resistance must be placed in a cultivar of a variety with acceptable qualities such as yield, quality, height, and maturity. Most often these agronomic or horticultural traits are given priority by plant breeders, unless the disease(s) significantly limits production every year.

Kinds of Disease Resistance

Some plants are *immune* to a particular pathogen; that is, the pathogen cannot cause disease even under favorable conditions. Other plants exhibit certain degrees of resistance ranging from near immunity to complete susceptibility. Tolerant cultivars can produce a good yield despite infection, whereas susceptible cultivars are greatly damaged or killed by the pathogen.

The ability of a plant to resist disease is an inherited characteristic. It is controlled genetically. Plants undergo sexual recombination and change genetically. The ability of a pathogen to cause disease also is inherited and controlled genetically. Some pathogens are composed of many strains, or races, each differing in ability to attack a given host cultivar. Disease expression in each host-pathogen combination is predetermined by the genetic material of the host and the pathogen. Cultivars resistant to only one pathogenic strain possess *race-specific* or *vertical resistance* (VR), whereas cultivars resistant to many pathogenic strains possess *nonspecific, horizontal,* or *general resistance* (HR). Usually vertical resistance is controlled by only one or a few genes. Horizontal resistance is controlled by many genes.

General resistance usually is only moderate (i.e., cultivars with HR may be damaged to some extent by a pathogen but not as much as similar cultivars lacking HR). Cultivars with VR generally exhibit a higher degree of resistance than do those with HR and are often easier to develop in a breeding program. However, it is more likely for new races of a pathogen to arise in response to a VR cultivar than to an HR cultivar. Thus, cultivars with VR tend to break down faster than those with HR (see above discussion under "Genetic Uniformity and Plant Diseases").

Any heritable characteristic that helps localize and isolate the pathogen at the point of entry (*infection court*), that reduces or neutralizes the harmful effects of toxic substances produced by the pathogen, or that inhibits pathogen growth contributes toward host resistance. Any characteristic that helps the plant grow and mature under conditions unfavorable for the pathogen also contributes to resistance. On the other hand, any characteristic that induces changes in the host that increase growth of the pathogen or the production of toxins harmful to the host, or that benefits the pathogen in any other way, contributes to susceptibility. It must be remembered that disease is the result of the interaction of a multitude of chemical processes initiated by both the host and the pathogen while in contact in a favorable environment. The genes of the host plant that control disease resistance do so by changing or adapting the physiological processes of the plant so that infection or disease development is neutralized or prevented from operating.

When the pathogen and host come in contact, a struggle for survival begins. If the pathogen wins, disease develops. If the host wins, there is little or no disease.

Developing Disease-Resistant Cultivars

There are four steps in breeding a disease-resistant cultivar: (1) obtaining a source of resistance, (2) transferring the resistance to a breeding line, (3) evaluating the lines for disease resistance, yield, and quality, and (4) releasing seeds or plant material with desirable characteristics to growers.

Sources of Resistance

All plants growing on the earth today are the result of natural selection and breeding of plants that evolved in different kinds of climates over millions of years. This evolution has produced countless genetically different forms, both wild and cultivated. The fact that these plants have survived indicates that some possess numerous genes for disease resistance.

An easy and quick way to obtain resistance is to select resistant plants from a cultivar already in use, save seed from it, and use this seed for the next crop. Sometimes this process is not possible. Then it is necessary to search wild or related species of the crop in question to find a source of resistance. This requires a diligent search of seed stocks and exhaustive testing to determine the quantity, method of inheritance, and type of resistance available. This approach requires that stocks of viable seed of all cultivars and wild species be maintained and available for future study. Seed collections of most U.S. crop plants are maintained by the USDA, state experiment stations, and universities and some seed companies. Scientists in other countries also maintain valuable seed stocks.

The genetic diversity needed to find sources of resistance is often found in the *center(s) of origin* for a plant. The center of origin is where the plant evolved. For example, potato comes from the Andes of Colombia and Peru in South America, and wheat comes from the Middle East. For most of our most important food crops, the centers of origin occur in areas of the world with the greatest population growth. In order to preserve the gene pool of the wild relatives of our crops, the centers of origin must be preserved. This will be increasingly difficult, because many areas rich in genetic diversity are being lost as land is needed for cities, homes, and food production.

Transferring Resistance

Once a source of resistance is found, the next step—a big one—is to transfer this resistance to a breeding line by hybridization. Crosses are made between resistant plants and susceptible plants and the offspring exposed to the pathogen so that resistant plants can be identified and saved. Sometimes successive crosses (*backcrossing*) with plants with desirable characteristics must be made. The success of the whole program depends upon maintaining and keeping only those individual plants that are disease-resistant and have desirable agronomic characteristics such as high yield and good quality.

One big disadvantage of the backcross method is that it takes 5 to 10 years or longer to incorporate or stabilize resistance, particularly to several diseases, in cultivars that also have acceptable yield and quality. Much time can be saved by the use of F_1 hybrids. In this method, a breeding line with dominant resistance is crossed with a closely related, agronomically desirable line; the seed from this cross is planted to produce a crop made up of a uniform population. Most of our corn cultivars are F_1 hybrids. The disadvantage of the F_1 method is that hybrid seed does not breed true; hybrid seed can only be obtained by special crossing procedures.

There are other genetic techniques that can be used to incorporate resistance into breeding lines, and new ones are being discovered. Advances in biotechnology will help investigators develop resistant cultivars more quickly in the future (see Chapter 8).

Evaluation

The third step in breeding disease-resistant cultivars is evaluating the lines in which disease resistance has been incorporated. This is time-consuming, laborious, and often expensive. Evaluation can take place in the greenhouse or in the field. During the backcrossing phase of the breeding program, when resistance is being stabilized into an acceptable cultivar, a greenhouse is almost indispensable. Since many plants will grow and flower year-round in a greenhouse, two or three generations (crops) may be grown annually. With suitable inoculation techniques and adequately controlled greenhouse conditions, successive backcross generations can be exposed to pathogens and the susceptible individuals eliminated in a reasonably short time.

Next comes field testing, which must be done to avoid the release of unacceptable or defective cultivars. After resistance has been stabilized, the various lines are grown in field plots. These field plots usually are of two types: disease-free or disease-infested. In the disease-free plots, the lines are tested for yield and quality in comparison with standard cultivars. In the disease-infested plots, the cultivars are tested for disease resistance under field conditions. Field plots may be located several hundred kilometers apart, in frost-free areas or tropical regions to speed up the testing process or increase the seed supply. In a critical year for disease development, extensive data can be obtained by having test plots at various locations. Then, too, by testing lines at numerous locations, unusual characteristics of any particular line show up more quickly than when only one location is used. The use of multiple test sites decreases the number of years that a line must be evaluated.

Release to Growers

The final step in breeding a disease-resistant cultivar is releasing it to growers. First, the supply of seed must be increased so enough is available for purchase by farmers. Great care must be taken to keep the seed stocks pure, and the cultivar must be continually tested to make sure disease resistance is maintained. There are several ways to furnish seed of a new cultivar to growers; the procedure varies somewhat by state and area. Usually the seed of the new cultivars is produced and sold by certified seed growers, as discussed in the remainder of this chapter.

CERTIFIED SEED AND SEED STORAGE

The use of poor seed is one of the most costly mistakes a grower can make. It is impossible to get more out of a crop than the potential of the seed used. Therefore, high-quality seed is a must. No farmer should try to save money by buying cut-rate or cheap seed of unknown quality, as seed costs average less than 3% of the total crop production expenses.

Down through the centuries most farmers have realized the necessity of saving seed for the next crop and improving the seed they planted for the present crop. Four thousand years ago the ancient Hebrews recognized the value of pure seed.

However, before 1900 many seedsmen and farmers had only a hazy concept of varietal purity and its importance. They produced and released supposedly new cultivars with little information about their adaptation, performance, or yield potential. Another serious problem faced by agronomy research workers at that time was the difficulty in obtaining supplies of pure seed of new cultivars produced in other states and countries. Often a new cultivar became so contaminated 3 to 4 years after release that its identity and value were lost to the seed-buying public.

About 1920, the International Crop Improvement Association, composed of members from the United States and Canada, was formed to get closer and more uniform cooperation between states, to coordinate standards, to minimize cultivar confusion, and to permit the sale of certified seed beyond state boundaries. From this beginning, crop improvement associations were established in every state. Usually these associations are designated by state law as the official agency for seed certification. In each state they are associated with the state university or department of agriculture. These associations establish and administer standards for certification of seed and inspect the production of certified seed under these standards (Fig. 6.6). Closely connected with the crop improvement association is a foundation seed producers corporation. Both organizations are nonprofit, self-supporting, open organizations

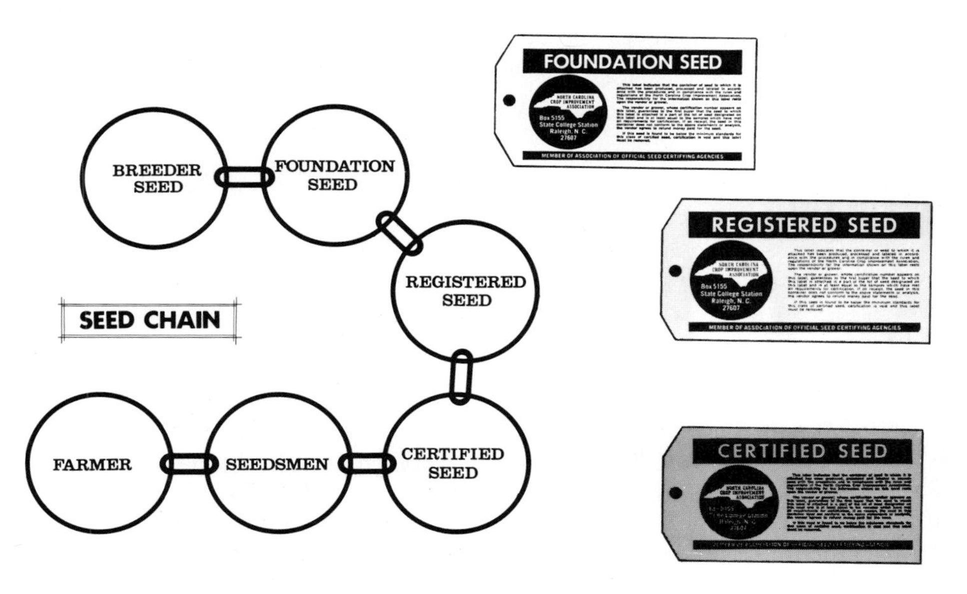

FIGURE 6.6. The certified seed chain. *Courtesy North Carolina Crop Improvement Association.*

composed of seedsmen and farmers or any persons or firms that desire to make seed growing a special branch of their farming operation. Membership in these organizations is voluntary.

What Is Certified Seed?

Certified seed is the final product of a four-generation scheme that has been devised to maintain the pedigree of superior crop varieties (Fig. 6.6). In the first step, *breeder seed,* which represents the true pedigree of the variety, is produced under the supervision of the plant breeder or owner of the variety. *Foundation seed,* the first-generation seed from breeder seed, usually is produced under contract by a foundation seed organization and is labeled with a white tag. *Registered seed* is produced from foundation seed for the purpose of increasing seed for one more generation before the production of certified seed. Registered seed is labeled with a purple tag and is not intended to be a commercial class of seed. In some states and for certain crops the step of registered seed may be eliminated, and certified seed is produced from foundation seed. *Certified seed* is produced from foundation or registered seed and is the final product of the certification program. Certified seed is labeled with a blue tag.

Certified seed is the best seed that is available to farmers. It is high-quality seed that, when planted under normal conditions, will give the desired stand of vigorous, relatively disease-free plants of the intended cultivar. Because of the four-step program in which it is produced, certified seed is of known origin, genetic identity, and purity. It has a known percent germination, and most of the time it is free of noxious weed seed. When feasible, it has been given chemical treatment for protection against pathogens. This reliable product is in high demand and is readily available in seed stores in agricultural communities through the United States. Certified seed is a big asset to farmers.

Storing Seed Properly

Because seed is produced for planting at a later date, it must be stored. If stored improperly, it may deteriorate, die, or lose vigor. Seed germination may be slow, and/or weak seedlings may be produced. Some seeds will actually rot if stored in moist environments.

Seed moisture is the most important factor affecting the storage life of seed. Each 1% decrease in seed moisture content nearly doubles the safe storage period. It is important to put dry seed into storage and keep it dry. Low temperatures also are desirable. Each 10°C decrease in temperature will roughly double the safe storage period of seed. Ideal storage conditions are low temperature and low RH, so the seed is kept cool and dry. However, if both moisture and temperature cannot be controlled, then it is more important to control moisture. A rule of thumb for safe seed storage is that the sum of the temperature and the RH should not exceed 100 units. For example, if seed is stored at 20°C, the RH should be kept below 80%.

Seed quality cannot be improved through storage, regardless of how good the storage facilities are. Most costly storage problems actually start in the field and result from improper seed harvesting and processing. Excessive delays in harvest, mechanical injury to seed, and delays in, or inadequate, drying followed by improper storage will result in seeds of low quality that germinate slowly and produce weak seedlings. The weak seedlings are often particularly susceptible to disease-causing organisms. Excessively high temperature should be avoided in drying seed to be used for planting.

The kind of seed to be stored, the time of storage, and the seed quality level to be maintained all must be considered in planning storage. Different kinds of seed have different lengths of storage life. Peanuts, soybeans, and some other legumes are relatively short-lived in storage and are not normally carried over from one planting season to the next. However, other legumes (clover and alfalfa), grasses (corn, sorghum, and related species), barley, rye, wheat, and certain vegetable and flower seed can be carried from one planting season to the next under proper storage conditions.

Seed that is carried over during the summer months in general storage needs protection from insects. A variety of insecticides can be used. A local county extension agent can provide a list of recommended chemicals.

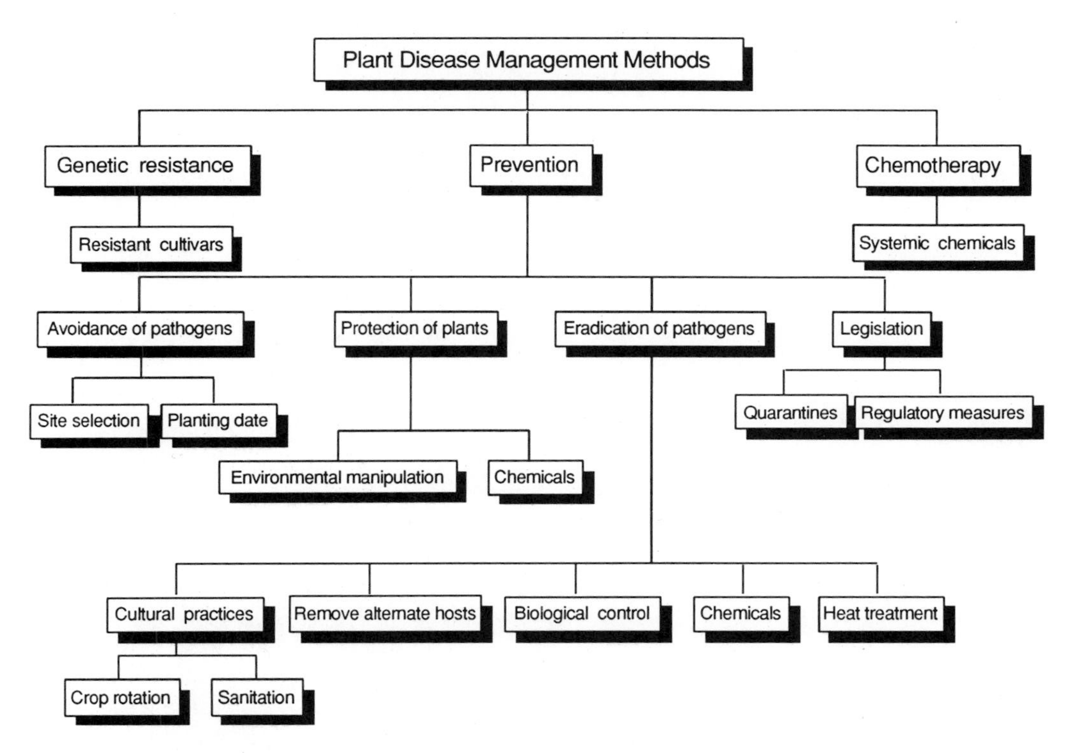

FIGURE 6.7. Methods of plant disease management.

Rules for Storing Seed

- Harvest seed when it is mature. Avoid injury and dry it promptly at the proper temperature.
- Store seed in a cool, dry place.
- Do not place seed on the floor or where moisture can be absorbed.
- Do not store seed in the same room with weed killers or volatile chemicals.
- Protect seed from insects and rodents.
- Run a test to determine the germination rate and percentage of any seed of doubtful quality.

SUMMARY

Many successful farmers, gardeners, and landscapers now use the five principal methods of disease management identified in this chapter (and in Chapter 7) to grow plants. Disease management practices are interrelated, and the use of several technologies in combination usually is more successful and efficient than any one method used alone.

The major types of plant disease management systems are outlined in Fig 6.7. Some commonly used specific practices are indicated for each type of method. As many crops and ornamental plants are not resistant to all diseases, disease prevention practices are frequently used. The success of these disease management methods depends on the accurate, rapid identification of the disease; the selection of practical and economical management methods; and the proper application and timing of the methods selected.

7

Chemical Management of Plant Diseases

Some people think that chemical management of plant diseases was invented just a few years ago, but it was 2000 years ago that the Greek poet Homer, author of *The Illiad* and *The Odyssey,* wrote about the "pest averting sulfur with its property of divine and purifying fumigation." Sulfur is still used today as a pesticide.

Chemical management of plant pests has been an increasingly important method during the last century. The use and misuse of chemicals was brought to the attention of the public in 1962 with the publication of Rachel Carson's *Silent Spring,* in which the hazards of chemical pesticides were emphasized. Agricultural scientists agree that pesticides can be hazardous, especially when used by ignorant or careless applicators; however, chemicals are essential in agriculture today. Without them our food supply would be threatened, and it would be impossible to grow certain crops profitably. Moreover, chemical application is the only economical means for managing some plant diseases.

Many plant scientists and chemists are working diligently to accumulate information to make chemical pesticides safer and even more effective. Agribusiness is a technical, skillfully organized industry devoted to providing food for an exploding world population, dedicated to the safe and effective use of pesticides. Each day researchers learn more about chemicals and their mode of action. New and safer chemicals are being developed, and it appears that pesticides will be used widely until effective substitutes are discovered.

The use of chemicals for plant disease management is merely one tool in managing plant pests, and it should be used in conjunction with the other components of IPM. When used improperly, chemicals are dangerous and expensive, and often will not reduce pest losses. The biggest problem with agricultural chemicals is the people who apply them. Many pesticide applicators (including homeowners) make significant application errors because of inaccurate calibration, incorrect mixing, worn equipment, and failure to read the product label. In 1990, these mistakes, resulting in both over- and underapplication, cost farmers $5 to $25/ha in added chemical expense, potential crop damage, and weed competition. In the United States alone, these mistakes could cost farmers a billion dollars annually. The increased possibilities of harm to the crop, applicators, farm workers, consumers, and the environment are self-evident.

The concept that chemical pesticides alone will control pests is no longer tenable, if it ever was. The application of pesticides to large acreages with little or no regard for their deleterious side effects or impact on nontarget species can no longer be tolerated. We must learn to manage all organisms (including humans) in the environment.

HOW PESTICIDES WORK

Certain chemicals are useful as pesticides because of their toxicity to pests. Chemicals that are used to manage plant diseases are more toxic to plant pathogens than to plants. Such chemicals either kill or inhibit germination, growth, and multiplication of the pathogen. Some pesticides are *nonselective,* that is, toxic to many pathogens; others are *selective,* affecting only a few or one kind of pathogen.

There is nothing magical about the use of pesticides. For effective results, a disease must be correctly identified and the correct chemical selected and applied (1) at the correct *time,* (2) in correct *amounts,* and (3) in the correct *way.*

Most of the older disease management chemicals can only protect plants from subsequent infection and cannot cure a disease after it has started. Such chemicals are called *protectants.* These chemicals protect only that part of the plant that was covered before the pathogen came in contact with the plant. Protectant chemicals are not absorbed into or translocated through the plant, but form a chemical barrier between the plant and the pathogen. Thus, when the plant makes additional growth and puts out new leaves, or rain washes off the chemicals, another application must be made.

Some of the newer chemicals in use today have *therapeutic, eradicant* or *curative* properties. These chemicals are absorbed and spread internally in the sap stream through the plant and are called systemic pesticides. The oxathiins, the benzimidazoles (benomyl, thiabendazole, thiophanate ethyl), the pyrimidines, and ethazol are examples of systemic fungicides.

Another new type of pesticide with some curative properties is the sterol inhibitors. These chemicals inhibit ergosterol biosynthesis in certain fungi. Many others are in the experimental stage and will be available in the near future.

An *antibiotic* is a substance produced by one microorganism that is toxic to another. Most antibiotics used today in managing plant diseases are absorbed and translocated systemically by the plant. Streptomycin, a well-known antibiotic, is used against several bacterial pathogens that cause blights, rots, spots, or wilts. Tetracyclines are also active against many bacteria and mycoplasmas. Cycloheximide is effective against several fungi. It is used for the control of turf diseases and is effective against powdery mildews. However, cycloheximide injures some plants, so its use is limited.

Growers should keep up with the latest developments in pesticides and use the most appropriate chemicals.

NAMES OF PEST MANAGEMENT CHEMICALS

Literally hundreds of chemicals are being used for crop protection. They are applied as aerosols, dusts, fogs, granules, mists, paints, pastes, smoke, soaks, and sprays. New ones are always being developed. Therefore, it is necessary to name pesticides properly and accurately. A committee composed of representatives from professional societies and industrial organizations was established in 1954 to prepare detailed specifications and procedures for naming pesticide chemicals. At present, every pesticide has three names:

1. *Generic, common,* or *coined name* for all compounds having the same active ingredient.
2. *Chemical name.*
3. *Trade name* (brand or manufacturer's name).

For example, the common name for the chemical tetrachloroisophthalonitrile is chlorothalonil. Two trade names for it are Bravo and Daconil 2787. Each pesticide differing from all others in the ingredient statement, analysis, manufacturer or distributor, name, number, or trademark is considered as a distinct and separate brand.

PESTICIDE SAFETY

Pesticides are beneficial when properly used but may be extremely dangerous when used improperly. Most pesticides are designed for specific organisms—animals, bacteria, fungi, insects, mites, nematodes, or weeds. Therefore, all pesticides should be handled carefully and all safety precautions observed (Fig. 7.1). Gasoline and table salt, when used in the wrong way, are also dangerous and poisonous.

Read the Label

The pesticide label contains the most important information you can learn about a particular pesticide. Therefore, before buying, mixing, applying, storing, or disposing of a pesticide, *read the label.* This means you must read to label at least five times. The information found on a pesticide label is outlined in Fig. 7.2.

Pesticide labels are required by law to carry certain signal words, such as *Danger* (skull and crossbones), *Poison, Warning, Caution,* and *Keep Out of the Reach of Children,* depending on the hazard of the particular product. Labels with a skull and crossbones indicate a product that is highly hazardous, whereas Caution on a pesticide label indicates a relatively nonhazardous chemical. Table 7.1 gives the meaning of the signal words.

Read the Material Safety Data Sheet

Material Safety Data Sheets (MSDS) provide essential information on hazardous chemicals, including pesticides. You should read the entire MSDS before using a

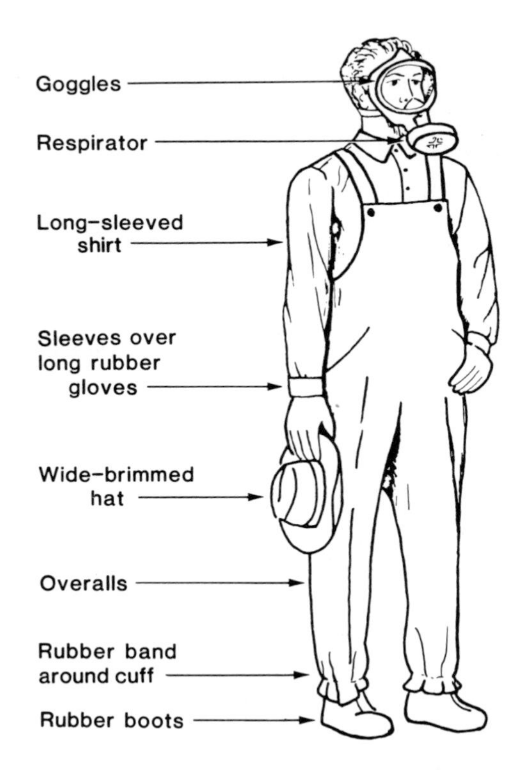

FIGURE 7.1. Protective clothing to be worn when applying pesticides. *Courtesy J. H. Wilson.*

pesticide. An MSDS should be available for each pesticide used and should be maintained in a readily accessible place for each stored pesticide. The availability of the MSDS can be of vital importance in the event of an accident with a pesticide.

The MSDS contains several categories of information. The order in which the information is presented may vary, but the overall information content will be similar among manufacturers. The MSDS lists the trade name, synonyms, chemical class and chemical name, and EPA signal word (e.g., Warning). The name, address, and emergency telephone number of the manufacturer are given. There are usually sections on physical characteristics of the chemical; first aid procedures; personal protection and precautions; spill, leak, and fire procedures; physical hazards information; reactivity data; health hazard information; and hazardous ingredients. Procedures for pesticide storage, pesticide disposal, and container disposal also are given.

Checklist of Safety Precautions

The following basic rules should be observed by all persons who use pesticides:

- Be careful. All chemicals are dangerous when handled carelessly.
- Read the label on the container.

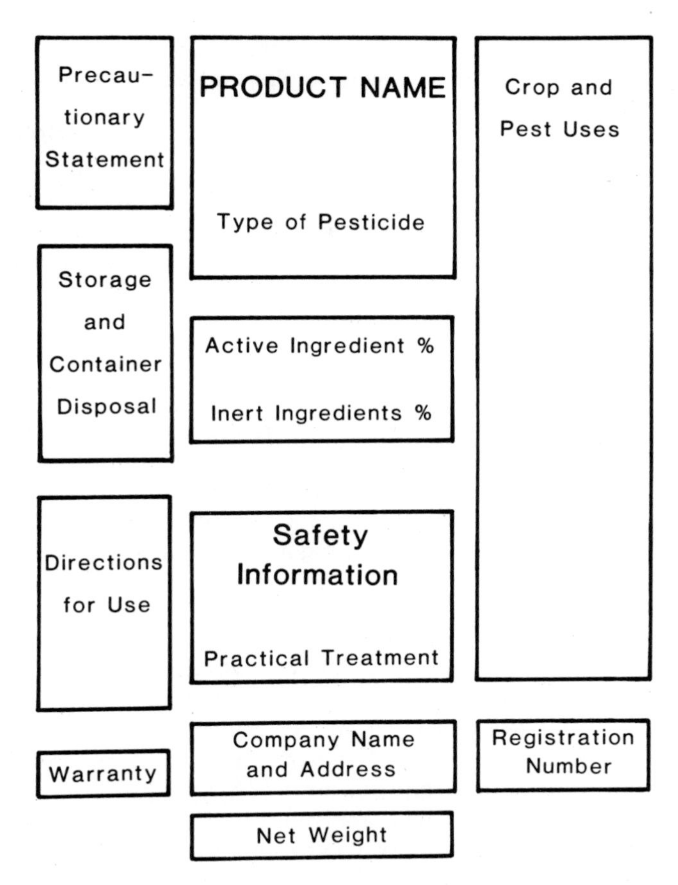

FIGURE 7.2. The label attached to a pesticide container describes how to use the product correctly and what special safety measures should be taken. *Courtesy USDA.*

- Review the information in the Material Safety Data Sheet (MSDS).
- Use pesticides only when needed.
- Do not eat, drink, or smoke when applying pesticides. Keep pesticides away from food and drinking water.
- Know what to do in case of an accident. Post a physician's phone number by the telephone.

TABLE 7.1. Signal words and meanings on pesticide labels.

Signal word[a]	Meaning	Lethal dose[b]
Danger Poison (in red)	Highly hazardous	Taste to a teaspoon
Warning	Moderately hazardous	Teaspoon to a tablespoon
Caution	Slightly hazardous	30 to more than 500 ml

[a] The statement "Keep Out of the Reach of Children" is required on all pesticides.
[b] For humans when taken by mouth (orally).

- If pesticides are spilled on the skin, wash them off immediately with water. Take off contaminated clothing and wash it before wearing it again, or dispose of it properly.
- Take time to explain the safe use of pesticides to employees.
- Check application equipment for leaks, clogged lines, nozzles, and strainers.
- Calibrate equipment frequently.
- Maintain detailed records of pesticides used and amounts applied.
- Cans, bottles, and drums that have contained pesticides should not be used for any other purpose. Handle and discard containers in a recommended manner.

Pesticide Residues and Tolerances

Modern analytical methods make it possible to detect very small amounts of pesticides in plant or animal products. Pesticides used on young plants, for example, may be found on the harvested portion of the crop. Quantities as small as one part per million (ppm) and even one part per billion (ppb) can be detected on crops. One part per million (ppm) is 1 mg in 1 kg, 1 lb in 1,000,000 lb, 1 oz in 31½ tons, or 1 oz in 63,000 lb.

What Is a Residue?

A *residue* is the amount of a pesticide that can be found on, or in, a harvested crop. Residues can come from pesticides applied directly to the crop, from a nearby crop by drift, from uptake from the soil and even from storage in an area previously treated with a pesticide. Most pesticides decompose in the soil, leaving residues that are neither toxic to plants nor taken up by plants in sufficient amounts to be harmful.

Pesticide applicators must do everything they can to avoid illegal residues. Many pesticides have warning statements on the label such as *Do Not Apply Later Than 7 Days or More Before Harvest.* These directions are given to prevent residue problems on some crops. If label directions are followed and residue problems develop, the manufacturer is responsible; if the label is not followed, the applicator or grower is responsible.

What Is a Tolerance?

A *tolerance* is the amount of pesticide residue that by law may remain on or in food. The tolerance for a pesticide is set by the U.S. Environmental Protection Agency and represents an amount of the pesticide that is many times less than the amount that would cause injury to consumers.

A tolerance may be set for each pesticide and also for each specific use of a pesticide. For example, the tolerance for Pesticide-X on beans might be 2 ppm and for alfalfa it might be 1 ppm. If a tolerance is exceeded, the product cannot be sold, and the grower may be liable.

PESTICIDE FORMULATIONS

Once the pest has been identified, and the proper pesticide has been selected, it must be applied to the target. Many factors affect our ability to place the pesticide on the target in the manner and amount needed for most effective results, with the least undesirable side effects, and at the lowest possible cost. The selection and use of equipment is of utmost importance. In modern agriculture elaborate and expensive application equipment sometimes is necessary. Moreover, without proper consideration of formulation, compatibility, and use of records, successful application is impossible.

Pesticides usually are not used in the pure chemical form. The *active ingredient* is usually mixed with inert substances. Such a mixture is called a *pesticide formulation*. A pesticide formulation must be prepared so the user can apply it in a safe, convenient, and effective manner with the proper equipment. Available pesticide formulations include dusts, emulsifiable concentrates, flowables, granules, liquids, soluble powders, solutions, wettable granules, and wettable powders. Combinations of pesticides are sometimes used when management of more than one pest is required. Adjuvants are added to some pesticide formulations to increase their effectiveness.

Dusts (D)

A dust formulation consists of a pesticide mixed with finely ground talc, clay, pyrophyllite, powdered nut hulls, or other such inert materials. Dusts must be uniform in particle size. Dusts are used dry and should not be mixed with water. Dust formulations require special equipment for application, and may be more hazardous to use than other pesticide formulations because dusts blow in the wind. Mechanical rubbing, and rain can remove dust from plant surfaces. A film of water on plant surfaces increases the sticking power of dusts. Thus, it is more efficient to dust at night when leaves are wet with dew or when there is no wind. Dusts may have to be reapplied at frequent intervals to provide adequate protection. For these reasons dust formulations are seldom used.

Emulsifiable Concentrates (EC)

An emulsifiable concentrate (EC) contains a water-insoluble pesticide that forms an *emulsion* when mixed with water. An emulsion is one liquid dispersed throughout another type of liquid. Many active ingredients in pesticides are not soluble or have low solubility in water but are soluble in other liquids; in an EC the active ingredient may be dissolved in an oil with an emulsifying agent added. The EC can be mixed with water to form a "milky" emulsion. Emulsifiable concentrates are not abrasive and are relatively harmless to equipment. Some plants are sensitive to some of the solvents or emulsifiers used in these formulations. An EC must be continually agitated during application to prevent settling out of the active ingredient.

Flowables (F)

A flowable formulation is a finely ground wettable powder that is sold as a thick suspension in a liquid.

Granules (G)

A granular pesticide formulation usually is prepared by *impregnating* (soaking up) a liquid pesticide to granules made from clay, corn cobs, walnut shells, or other porous materials. Granules are prepared to a standard size that are described by mesh size. For example, all of the particles in a 20/40 mesh granule will pass through a screen with 20 openings per inch, but none will pass through a screen with 40 openings per inch.

Granular pesticides are often applied to the soil. Because granules do not cling to plant foliage, they can be applied over plants and fall to the soil. Application of a pesticide formulated as granules is usually safer than liquid or dust formulations of the same pesticide because the applicator is exposed to less pesticide in the case of granules.

Soluble Powders (SP)

Soluble powders are powdered materials that will dissolve in water.

Solutions (S) or Liquids (L)

Pesticides formulated as solutions usually contain the pesticide and a solvent. The solvent may be a petroleum product or water; the type of solvent used to prepare the solution will depend upon the nature of the pesticide and the use of the formulation. Solutions prepared for special uses should not leave unsightly residues. Many of these preparations with petroleum products will damage living plants if not used properly.

Wettable Powders (WP)

Powdered pesticide formulations that contain wetting agents are called wettable powders. Their active ingredients usually are more concentrated than those in dusts. With agitation, wettable powders form a suspension in water that should not settle out too quickly. Wettable powders are less likely than emulsifiable concentrates to damage sensitive plants, but are more abrasive to pumps and nozzles.

Wettable Granules (WG) or Dispersable Granules (DG)

Wettable granules are similar to wettable powders and are replacing wettable powder formulations. Wettable granules are safer for applicators because they cause less exposure to dust than do powders.

Combinations of Pesticides

Plant diseases and insects often attack a plant at the same time, and sometimes growers may wish to protect crops against several diseases. Therefore, for reasons of economy the use of multiple combinations of pesticides is a necessary practice used by many growers in commercial agriculture. The utilization of multiple pesticide combinations, often called *tank mix, piggyback,* or *one-shot* applications, is a laborsaving shortcut. Two fungicides may be combined, or a fungicide and an insecticide, or a fungicide and a growth regulator plus minor elements. However, only compatible pesticides can be used in combinations or mixtures, according to label directions.

Compatible chemicals can be mixed without affecting the effectiveness, physical properties, or phytotoxicity of either chemical. Incompatible chemicals lose effectiveness or injure plants when mixed and thus should not be mixed. *Physical incompatibility* often results when "hard" water is used for spray mixtures. In such cases the pesticide tends to fall out of suspension. Pesticides also may be *chemically incompatible,* and react when combined so as to lose the effectiveness of one or all components. For example, most organic pesticides should not be mixed with alkaline compounds with a pH higher than 7.0. Chemical incompatibility frequently is the cause of poor performance of some multiple pesticide combinations. When two or more chemicals are mixed and applied at the same time, each must end up in the correct place (*placement compatibility*) to be efficient. Some fungicides must be uniformly distributed over the plant to get efficient disease management, whereas certain insecticides need to be directed only at the top of the plant; mixtures of such chemicals would lack placement compatibility. Before combining pesticides, read the directions on the label and avoid the risk of plant injury. Pesticides should be mixed only if a particular combination is specified on the label. Never combine pesticides of unknown compatibility.

Adjuvants (Additives)

Many plants have smooth leaves with waxy coatings; others have leaves covered with hairs. Smooth cabbage or onion leaves, young apples, or fuzzy peaches are difficult to cover properly and uniformly with sprays. The spray may "ball up" and run off, leaving unprotected areas on the leaves or fruit. Therefore, some chemicals require an *adjuvant,* which is a material added to a formulation to improve its physical or chemical characteristics. Adjuvants are commonly used to improve the spreading ability or the "weathering" of the pesticide after application.

Adjuvants include detergents, dispersing agents, emulsifiers, foam suppressants, penetrants, safeners, soaps, spreaders, stickers, and wetting agents. Most pesticides already have the necessary adjuvants in the formulation. Therefore, additional adjuvants should be used only when called for by the pesticide label.

SOIL TREATMENT

A big step forward in the management of soilborne pathogens and root diseases occurred about 1950 with the development of *fumigation* techniques to inject chemicals into the soil to control nematodes, fungi, and bacteria. Some crops cannot be grown profitably in specific areas unless the soil is first treated with chemicals. Many fields planted to crops that return a high net income per hectare (for example, cotton, ornamentals, strawberries and vegetables) are routinely treated before planting. Flower beds, greenhouse soils, golf greens, and lawns often are treated before the seed is sown or the crop planted. Fumigation pays off for tree crops, too. Preplant fumigation is recommended as a standard practice before planting apple, banana, citrus, or peach trees. As with all pesticide applications, care must be taken with soil treatment not to contaminate the environment, especially groundwater.

Fumigants

Two types of pesticides, *fumigant* and *nonfumigant,* are used to treat soil. A fumigant produces a gas, vapor, or fume in the soil. Some soil fumigants are applied directly to the soil; others are applied under gasproof covers made of polyethylene or similar materials. Fumigants may be used to control a single pest or a number of pests. The latter, called *multipurpose fumigants,* are most useful in fields where the soil is infested with more than one pathogen.

All fumigant pesticides are applied as gases or have a gas phase after application. The gases diffuse through the air in the soil or dissolve in soil water. These toxic chemicals are absorbed through the cell walls of fungi and bacteria or through the skin of nematodes.

Factors Affecting Fumigant Performance

Some fumigants are phytotoxic and may injure or kill crops planted in the soil. Therefore, many fumigants must be applied as *preplant* treatments, after which a waiting period of 10 to 21 days is required between fumigant application and planting date. This permits the fumes to evaporate and diffuse out of the soil. Several factors influence fumigant performance and the required waiting period between application and planting.

Temperature. Fumigants must become gases to be effective. If the temperature is high, the conversion to a gas is rapid; if it is low, the conversion is slow. However, if the temperature is too high or too low, the conversion may be too rapid or too slow to control pathogens. The label will indicate the proper temperature range for applying a particular fumigant (10-25°C for most fumigants).

Moisture. Fumigants that are applied in the soil are affected by its moisture content. Fumigants applied in soil with a moisture content that is too high or too low may not give efficient pest management. In dry soils, fumigants may escape too rapidly; in wet soils, water occupies most of the air space so the fumigant moves very slowly. Soil fumigants perform best under average soil moisture conditions.

Soil Condition and Type. Fumigants move faster in loose and sandy soils than in clay or compact soils, other factors (temperature and moisture) being comparable. Fumigants applied to a cloddy or poorly prepared soil are likely to be ineffective. Highly organic soils and soils with plant residues (stalks, vines, leaves) may "tie up" the fumigant. Proper planning and soil preparation can reduce this problem.

Timing and Aeration. Proper timing is important in applying most fumigants. For example, with seedbeds the temperature is usually higher and fumigants perform more efficiently in the fall than in the spring. The exposure time, or time required to kill pathogens, varies with the fumigant applied, temperature, soil type, crop, and other factors.

Aeration time, or the time required to allow the fumigant to leave the soil before planting, is important to avoid injury to the crop. Proper aeration is also necessary to assure the adequate dispersal of the fumigant before workers enter the area.

Other Factors. Some crops are very sensitive to certain fumigants. Even small quantities of a fumigant or fumigant residues can harm them. *Read the label* to find out if the fumigant can be used before a crop is planted or on living plants, and the limitations, if any, on the crop in question.

Broad-spectrum fumigants applied under covers can temporarily harm nitrifying bacteria essential for converting ammonium to nitrate nitrogen. Nitrate fertilization may be required to correct this condition.

Applying Fumigants

The skill and the accuracy with which pesticides are applied are just as important as selecting the correct pesticide. For effective use of soil fumigants, careful attention to application details is required.

Before fumigation, the soil should be prepared in the same manner as for planting a crop; that is, the soil should be moderately loose and free of undecayed roots or other partially decomposed crop residues. Fumigants perform best in fairly moist soil; results may be unsatisfactory when fumigants are applied to dry soil. Moderate rain following application is beneficial. Heavy rains soon after application chill the soil and delay escape of the fumes; hence the waiting period before planting must be extended. The soil temperature should be above 5°C (41°F), preferably between 15° (59°F) and 27°C (81°F). Fumigants should not be applied on windy days because the evaporation rate is too high.

In today's modern agriculture there are many types of equipment for applying farm chemicals, ranging from hand applicators or one-row equipment for the homeowner or small farmer to sophisticated multirow tractor-drawn applicators capable of treating many hectares in an 8-hr day. All farmers should learn the proper care maintenance, calibration, and use of equipment used to apply farm chemicals.

LOW-VOLATILE LIQUID FUMIGATORS

Low-volatile fumigants can be handled in nonpressurized containers. The fumigators for these fumigants may be gravity- or forced-pressure-fed and are designed for row or *broadcast* applications. Most row fumigators are the constant-head gravity type, whereas broadcast fumigators are force-fed types. The forced-fed type requires no agitation, and the liquid is delivered into the soil through an injector. Positive-displacement piston-type applicators also are used for broadcast application, but their accuracy is greatly reduced at high speeds.

Chisel-Injection Method. In the chisel-injection method, used for broadcast (overall) treatments, the liquid fumigant is applied in bands not more than 30 cm apart and 20 to 25 cm deep (Fig. 7.3). These large-scale applications are made with power-driven applicators of the continuous-flow type that deliver from 5 to 10 streams into the soil simultaneously. Applicators consist of a tank to hold the fumigant and a

FIGURE 7.3. A broadcast applicator for liquid fumigant nematicides. Nematicide from the tank flows through a regulating apparatus to each of the six curved shanks and is injected into soil at a depth of about 20 cm. Sometimes a drag is used behind the rig to firm and seal the soil. *Courtesy A. L. Taylor and J. N. Sasser.*

system of tubes to convey it into the soil, the tubes passing down the back edge of narrow cultivator teeth or "chisels." One or more pumps are included in the mechanism, and the liquid is delivered under pressure. The chisels are mounted on a horizontal bar or framework, which can be raised or lowered. The applicator may be mounted directly on a tractor or carried on a trailer. A drag is pulled behind the applicator to firm the top layer of soil, thus sealing in the gas. Irrigation water is often needed when low-volatile fumigants are applied to growing plants such as turf or orchards.

Row Method. The row method of application, in which only the row area is fumigated, requires about half the amount of fumigant needed for the chisel method. Usually one or two streams of fumigant or a band of nonfumigant about 25 to 30 cm wide is applied for each row. The area between rows is left untreated. The crop is planted in the treated area, and by the time the roots have grown into the nontreated area, the plants are large enough to escape serious damage. Many vegetable and field crops are treated by the row method.

Row treatment if less expensive than other types of treatment, but has several disadvantages:

- It can be used most efficiently only as a spring treatment.
- It must be carefully timed and fumigant applied at least 2 weeks before the recommended planting date for the locality. Bad weather may upset these calculations.

FIGURE 7.4. Fumigation of soil with methyl bromide. After the soil is prepared, this tractor-drawn chisel applicator applies the methyl bromide and lays and seals the plastic cover, all in one operation. *Courtesy Reddick Equipment Co.*

- Because only the row area of the soil is fumigated, the pest population may build up faster than with broadcast treatments.

HIGH-VOLATILE (GAS-TYPE) FUMIGATORS

High-volatile fumigants, such as methyl bromide, must be handled in closed pressurized containers. The fumigators are designed for force-feeding these fumigants from a pressurized tank. The tank is precharged with enough pressure to empty its contents, or pressurized gas is fed into the tank during application to displace the fumigant.

The precharged tank cylinders are approximately two-thirds filled in order to have room for the precharging gas. The fumigant is injected into the soil and covered with plastic to prevent its escaping. Although these applicators are referred to as "gas type," the fumigant must remain as a liquid in the tank, pressure lines, manifold, and metering devices for accurate application.

High-volatile fumigation often is used for treating seedbeds. Some applications can be made by growers with the proper pesticide licenses. Many farmers, however, who must fumigate several hectares of seedbeds prefer to hire commercial applicators to treat the soil, using large rigs that inject the fumigant and lay the cover simultaneously (Fig. 7.4). However, the same rules apply; the soil must be prepared properly and the correct fumigant applied at the correct time, in correct amounts, in the correct way.

Nonfumigants

Another type of soil treatment chemical was introduced about 1960—the *nonfumigant* nematicides, some of which also kill insects. Nonfumigant nematicides dissolve in the soil water and are distributed through the soil. Some of these nematicides are systemic in plants; they are absorbed by the plants through the roots after application to the soil or can be foliarly absorbed if applied to the leaves. Presumably the chemicals are taken up by the nematode during feeding, or they may also enter through the nematode's skin (cuticle) in contact with plant tissue. Nonsystemic nematicides remain in the soil water and enter through the cuticle of the nematodes. A big advantage of nonfumigant nematicides is that there is no waiting period; the crop can be planted immediately after the nematicide is applied.

Granular Application

Nematicides in granular form are applied to gardens, pastures, row crops, and turf. Depending on the chemical, granules are usually safer to handle (less hazardous to applicators), more convenient to apply, less phytotoxic, and/or more selective than other formulations. In some cases, granules are preferred because of their more effective placement of the chemical. As with other pesticide formulations, however, granules must be handled carefully because most are toxic to humans.

Granular application equipment includes small hand-operated applicators and large truck- or tractor-mounted or tractor-towed units. Granules may be applied in a band or broadcast uniformly (Fig. 7.5). They are distributed from a hopper by

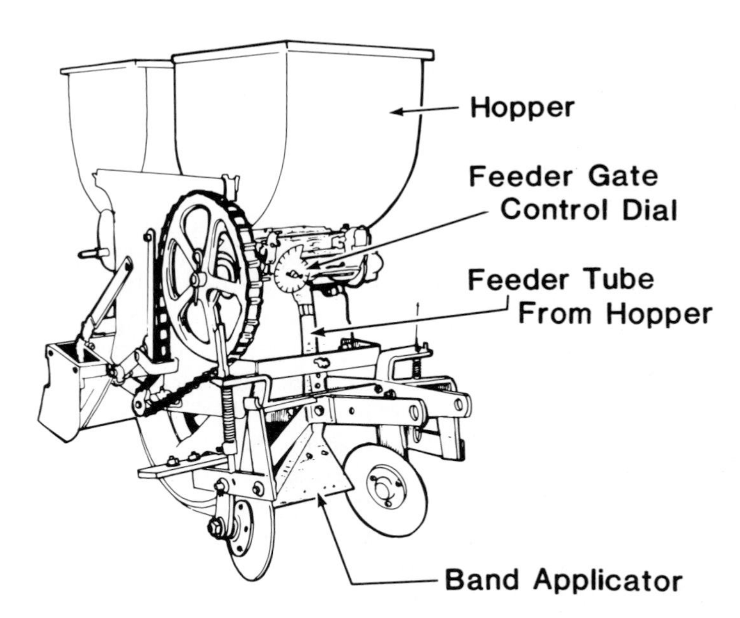

FIGURE 7.5. Diagram of row-type applicator used to apply granules of chemical pesticides or biological control agents at planting time. *Courtesy American Association for Vocational Instructional Materials.*

deflecting baffles for band or broadcast application, or by a centrifugal thrower for broadcast spreading. The granules are either incorporated into the top 10 to 20 cm of soil by disking or spread on top of the soil (or turf) and watered into it.

Granules seldom are used for foliar application because they will not stick to the leaves. One advantage of using granules is that no mixing is required. However, they generally cost more than other formulations.

A granular applicator should be easy to clean and fill. It should have mechanical agitation over the outlet holes to prevent bridging and to keep the flow rate constant. Application should stop when drive stops even if the outlets are still open. All directions on the pesticide label should be followed carefully when granular pesticides are used.

Liquid Application

Liquid nonfumigant nematicides may be diluted and sprayed uniformly over the soil surface and then disked into the soil. For treating gardens or small areas, hand sprayers can be used; for large fields, low-pressure (boom-type) tractor sprayers often are used (see Fig. 7.8).

SEED TREATMENT

Many pathogens are carried on or in the seed, and seeds often are planted in soil infested with pests. To protect the seed and the developing seedling from attack, various protectants can be applied as dusts or sprays; or the seed can be soaked in a pesticide solution and then allowed to dry. Tubers, bulbs, and roots also may be chemically treated to protect from diseases or insect pests. Usually a bright-colored red, blue, or yellow dye is included in the chemical mix to stain the seeds, indicating that they have been treated. Seed treatment is a specialized business and is best performed by skilled operators. Therefore, this operation usually is conducted by seed-distributing agencies rather than by individual growers. Because some chemicals are effective against some pathogens and injure certain kinds of seed, specialized equipment is required to properly treat the seed. The cost of treated seed is usually so small (only a few cents per hectare) in relation to the value of the crop that farmers cannot afford *not* to use this form of insurance.

Caution: Chemically treated seed should *never* be used for livestock feed or food for human consumption. Sacks, bags, or other containers should be handled or discarded according to label directions. Do not handle treated seed with your bare hands.

POSTHARVEST TREATMENT

Some agricultural products must be stored for several days to several months, during which time storage diseases may develop. A number of chemicals can be used to reduce postharvest or storage diseases. Some may be applied just prior to harvest; others, in the wash water during processing, or by using wrapping paper impregnated with fungicidal materials or by periodically fumigating the storage house with volatile chemicals.

Chemical residues can be a problem in the use of postharvest chemicals. Before any material can be used, the Food and Drug Administration (FDA) sets a limit on the amount of pesticide residue that can be present on materials intended for consumption. Exhaustive tests must be conducted before a chemical can be labeled as postharvest treatment. Nonetheless, many postharvest diseases are successfully controlled by chemicals, and the search for new, safer chemicals continues.

TREATMENT OF GROWING PLANTS

Chemicals used for plant disease prevention are applied in the greatest volume as sprays or dusts to leaves, flowers, or fruits. At present, more fungicides are used than bactericides. Most chemicals are protectants and must be present on the plant surface before a fungus spore or bacterium arrives there. Therefore, it is important that the entire plant be covered with the chemical. Growing leaves, branches, and fruit must receive repeated applications to protect the new growth.

Reducing Treatment Costs

The application of pesticides to plants can be an expensive and time-consuming task that must be done carefully if efficient and safe pest management is to be achieved. Pesticide applicators are always on the lookout for ways to reduce costs.

For certain diseases, spray "schedules" have been developed that tell the farmer at what intervals pesticides should be applied, depending on the disease, frequency and duration of rains, RH, temperature, and time of year. For other diseases (apple scab, late blight of potato, peanut leaf spot), recording devices have been developed that automatically monitor such critical factors as temperature, RH, rain, and wind. These "little gray boxes" are placed in the crop area and collect and process data, using a program on a microcomputer; based on such data, a forecaster can tell farmers whether it is necessary to spray. This practice sometimes results in great savings in the cost of chemicals and the expense and time of application.

Another way to reduce treatment costs is to make more pesticide stick to the plant surfaces. Equipment is being developed to charge each particle electrically. With such *electrostatic* dusting or spraying equipment, the negatively charged particles are attracted to the positively charged plant surface, so that more uniform coverage is obtained and less material is required. Electrostatic application fits in well with *ultra-low-volume* (ULV) application and may be one of the improvements available in the near future.

Another type of sprayer that is being developed recycles the chemical that does not stick on the plant. Spray nozzles are directed so that the extra spray is caught and returned to the spray tank rather than being wasted on the soil. This type of sprayer will permit more efficient use of chemicals and reduce the hazard of excess chemicals in the environment.

In the early 1950s, airblast sprayers came into general use by growers with large acreages. These sprayers had powerful engines that could envelop a tall tree with a fog of pesticide in a few seconds or cover up to 20 rows of cotton in one sweep. At that time fuel and pesticides were cheaper than they now are. At about the same time, lightweight, durable compact pumps driven by the power takeoff (PTO) systems of tractors and more durable nozzles came into general use for conventional-type sprayers. Currently, a concern with this type of sprayer is the amount of pesticide solution that reaches nontarget areas.

Application costs also have been cut by reducing the amount of inert carrier or water used to dilute pesticides or by eliminating the water altogether. Not only is it expensive to haul water back and forth across a field either in a sprayer or in an airplane; it is often difficult to have clean water convenient for conventional spraying techniques. The application of small quantities of pesticide, as little as 10 liters/ha, is referred to as *ultra-low-volume spraying*. Eliminating the diluent lessens the expense of applying pesticides enough that application of pesticides by airplane, helicopter, or multirow dusters or sprayers to low-value crop in large fields (e.g., wheat, corn, and other grains) becomes feasible.

The first concentrate or ULV sprayers were introduced in the early 1960s. These sprayers made it possible to reduce the volume of water used per hectare by breaking up the spray into smaller droplets through the shearing effect of high air velocity. Later, centrifugal energy nozzles led to the development of *controlled droplet application* (CDA), in which sprays are applied in a narrow range of droplet sizes (usually less than 100 μm) selected according to the target at which the pesticide is aimed. Droplet size is determined by the rotational speed of a spinning disk inside a hollow cone in the sprayer head.

The droplets produced by conventional pressure sprayers range in size from 20 μm to 400 μm. Many of the very small droplets drift away or bounce off leaves and are lost. The large droplets are ineffective in covering leaf surfaces; so certain areas of the leaves are unprotected. In contrast, CDA produces spray droplets of uniform size that are highly efficient in reaching and covering plant surfaces, even under windy conditions. The uniform droplets give maximum penetration of the plant canopy and cover both upper and lower leaf surfaces. In addition to the savings in pesticide costs, CDA results in smaller amounts of chemical residues in the crop.

At the present time, ULV tractor-mounted or aerial equipment is widely used in some areas. Small hand-carried, battery-operated sprayers for the small farmer also are available. Further refinement of ULV application will undoubtedly lead to more efficient chemical management.

In the remainder of this chapter, various types of spraying and dusting equipment, general application guidelines, calibration of application equipment, and environmental considerations related to the use of pesticides are described in detail. Much of this information applies specifically to the treatment of growing plants with sprays or dusts. However, some of the information applies to other types of treatment or to the use of pesticides in general.

PESTICIDE SPRAYERS AND DUSTERS

Sprayers are the most commonly used pesticide application equipment. There are many types and sizes, ranging from small handheld sprayers for home and garden to large tank models for spraying crops, golf courses, greenhouses, nurseries, orchards, roadways, tall trees, and vineyards. Proper spraying requires skill and accuracy. Moreover, it is expensive. Therefore, careful attention should be paid to selecting the proper equipment to do an efficient job. In addition, proper care and maintenance of valuable spray equipment is a necessary component of efficient application.

A large sprayer for commercial use is a long-term capital investment. It should be selected carefully, based on the following considerations:

- Does the size, design, and operation of the sprayer fit its intended use?
- Is the sprayer designed so that it is easy to fill, clean, handle, operate, and adjust?
- What is the quality of material and construction of the tank, pump, agitator, and strainers?
- Are spare parts and repair service quickly available?

For proper and safe operation, a sprayer must be in good condition in the hands of a well-trained licensed operator. The operator's manual for all spray equipment contains instructions on how to use and care for it. To keep a sprayer at peak performance, it must be serviced and adjusted before each use (see below, Calibration of Equipment section). To prevent problems during future use, a sprayer should be cleaned and stored properly after each use (see below, Cleaning and Storing Equipment section), and checked and serviced regularly.

Hand Sprayers

Hand sprayers are used for spot treatment and home gardens. Most are displacement-type sprayers; that is, air is pumped into a pressurized tank, thereby displacing the spray liquid and forcing it out through the nozzle. Most hand sprayers are economical, simply constructed, and easy to use, clean, and store. The only way to agitate the spray liquid is by shaking. There are five types of hand-operated sprayers:

1. *Aerosol* sprayers have sealed containers of compressed gas and pesticide. Capacities usually are less than 1 liter and the container cannot be reused when empty.
2. *Trigger-pump* sprayers are small, refillable units with a trigger-action pump for pressure.
3. *Hand-pump* sprayers are lightweight, push-pull, refillable units with a capacity of about 1 liter.
4. *Compressed air* sprayers vary from 4 to 20 liters in capacity. Air is compressed by a piston-type hand pump. Both hand-carried and backpack models are available. Air pressure is not constant, and the repeated pumping required to maintain pressure is tiresome and time-consuming. More expensive constant-pressure sprayers are equipped to feed carbon dioxide or compressed air into the spray tank through a regulator valve.
5. *Garden hose* sprayers are operated by pressure from a water system. When the water is turned on, concentrated spray liquid is siphoned out of a container of about 1-liter capacity attached to the water hose. The pesticide is diluted by and mixed with the stream of water as it discharges. These sprayers must be equipped with an antisiphon device to prevent possible contamination of water supplies.

Small General-Use Sprayers

The *power wheelbarrow sprayer,* the *estate* or small wheel-mounted sprayer, and the small *skid-mounted sprayer* are classified as small general-use sprayers. These sprayers are designed to handle spraying jobs too large for hand equipment, such as large garden, greenhouse, nursery, and small tree spraying.

Estate sprayers ar small two-wheel mounted units equipped with handles for moving. Trailer hitches are available as special equipment. Smaller models have 60- to 120-liter tanks and pumps that deliver 6 to 12 liters/min at pressures up to 250 psi. Larger models have 800-liter tanks and pumps that deliver 12 to 16 liters/min at pressures up to 400 psi. Power is supplied by a 1.5- to 3-horsepower air-cooled engine. Spray material in the tank is mechanically agitated. Standard equipment includes a hand gun and a hose. Spray booms are available for some models.

General-Use or Multipurpose Sprayers

General-use sprayers are designed to handle most of the spraying needs on general farms from low-pressure weed or row-crop spraying (30-40 psi) up to medium-pressure field-crop spraying (60-80 psi).

These sprayers may be tractor-mounted, pull-type or self-propelled (Fig. 7.6). Roller-type pumps that limit pressure to about 80 psi for continuous spraying often are used. Centrifugal pumps are available for low-pressure spraying when larger volumes

FIGURE 7.6. Boom-type, tractor-mounted, general-purpose sprayers can be adapted to apply many different kinds of pesticides. Note protective equipment worn by the operator. *Courtesy W. A. Skroch.*

are needed. Tank sizes range from 200 to 800 liters (50-200 gallons). The spray material usually is agitated hydraulically.

This type of sprayer is used to apply a wide range of pesticides and fertilizers. Good-quality materials for the tank, pump, and other components are essential to reduce corrosion problems. A general-use sprayer normally is equipped with a four or six-row boom, but special application booms or hand guns also may be used.

High-Pressure Sprayers

High-pressure sprayers are used primarily for treating vegetable and fruit crops. The high pressures required break up the pesticide liquid enough to assure complete coverage of the fruit and foliage.

High-pressure sprayers are similar in design to multipurpose sprayers but have piston pumps. The pumps can deliver up to 100 liters/min at 400 to 500 psi. All hoses, pressure relief valves, nozzles, and valves must be high-pressure types. Field booms are designed for spraying of specific crops, such as tomatoes, grapes, and blueberries. Hand guns can be used for spraying large fruit trees.

Airblast Sprayers

Sprayers that use air as a vehicle to transport spray droplets from the point of release to the point of application are called *airblast* sprayers (Fig. 7.7). High-volume fans

FIGURE 7.7. An airblast sprayer. A high-speed, fan-driven air stream breaks the spray material into fine droplets as it comes from the nozzles. Sprays may be directed to one or both sides and higher or lower, as desired.

supply the air; adjustable air vanes permit spraying on one or both sides of the sprayer as it moves between rows of trees or across a field. Nozzles operating at low, moderate, or high pressure deliver the spray droplets into the high-velocity air stream. The high-speed air breaks up larger droplets and transports the smaller droplets for thorough coverage.

Low-Volume Air Sprayers (Mist Blowers)

A special class of air sprayers, called low-volume sprayers or *mist blowers*, is characterized by the following features:

- High air velocities (120-200+ mph) and somewhat lower air volumes than in conventional airblast sprayers.
- Metering devices, which may or may not be conventional nozzles, that operate at low pressures and depend on the high-speed air for liquid breakup.
- Use of much lower volumes of water than what is required by conventional airblast sprayers.

A considerable saving in time and labor is possible with low-volume sprayers because less water is handled. However, equipment calibration is more critical, favorable weather for spraying is more essential, and pesticide coverage on some crops may be less satisfactory with low-volume sprayers than with conventional airblast sprayers.

Ultra-Low-Volume (ULV) Sprayers

ULV spraying applies a chemical concentrate directly without the use of water or any other liquid carrier. Many ULV ground sprayers use a fan that delivers high-speed air to help break up and transport the spray droplets. The main advantage of ULV sprayers is the labor and time saved because of the elimination of water. The main disadvantages are increased risk from spraying concentrate and the limited number of pesticides that are cleared for ULV application.

SPRAY EQUIPMENT PARTS

The overall quality of a sprayer and its suitability for various uses depend upon the quality and type of its component parts. The critical parts of a sprayer are discussed in this section. Figure 7.8 illustrates the parts of a typical boom-type sprayer.

Tanks

Tanks should be made of rust-resistant steel, fiberglass, or polyethylene to avoid rust, sediment, plugging, and restriction problems. Tanks from 220 to 600 liters (50-150 gallons) are available on mounted sprayers; 800-liter (200-gallon) or larger tanks come on pull-type sprayers. Tanks need a large covered opening in the top with a removable strainer to make filling, inspection, and cleaning easy. A drain plug in the bottom is necessary for complete drainage and collection of the rinsate during cleaning of the tank.

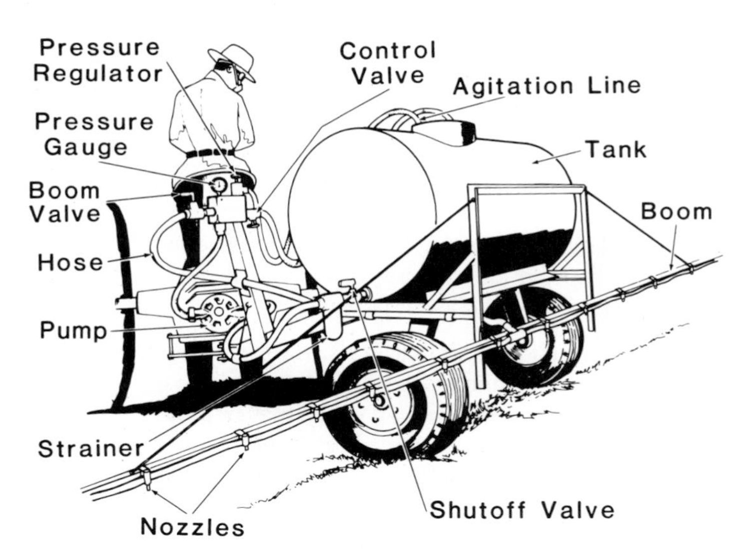

FIGURE 7.8. Diagram of boom-type sprayer showing major parts and accessories. *Courtesy American Association for Vocational Instructional Materials.*

Tank size should be large enough to keep the refill frequency to a minimum. A calibrated sight gauge on the tank makes it easy for the operator to check the level of fluid at any time. Plastic and fiberglass tanks may have calibrated markers on the outside of the tank. The fluid level is visible through the tank wall.

Pumps

The pump is the heart of the sprayer. It must have a pumping capacity of sufficient volume to supply the nozzles and hydraulic agitator and to maintain the desired pressure. Pump parts must be able to withstand the corrosive effects of the spray materials and have a reasonable life.

The following six types of power-operated pumps are commonly used in pesticide sprayers:

1. *Piston* (reciprocating). On upstroke, the piston pulls liquid into the pump chamber. On the opposite stroke, the fluid is forced out of the chamber through a surge tank to allow for a continuous and steady flow.
2. *Diaphragm.* This is similar to a piston pump except that the piston is replaced by a flexible diaphragm.
3. *Roller-impeller.* Rollers are fitted into slots formed on the periphery of the impeller (a solid wheel fitted onto a shaft). The slots allow the rollers to follow the off-center contour of the pump housing. Fluid fed into the inlet side is forced out by the squeezing action of the rollers.
4. *Flexible-impeller.* This is similar to a roller-impeller pump except that it has a

TABLE 7.2. Typical sprayer pump characteristics.

Pump type	Available construction material	Available port sizes (in.)	Maximum practical pressure (psi)	Capacity of ¾-in. port[a] (gal/min, 40 psi)	Self-priming	Effect of abrasive material	Reparable
External gear	Bronze Cast iron	½-1	75	6	Partly	Severe	No
Internal gear	Cast iron	¾	75	6	Partly	Moderate	No
6 roller-impeller	Cast iron Ni-resist	¾-1½	75	6-8	No	Moderate	Yes
7 or 8 roller-impeller	Ni-resist Cast iron Bronze	¾	75	10	No	Moderate	Yes
Centrifugal	Cast iron	1¼-2	50	40[b]	No	Little	Maybe
Diaphragm	Cast iron	¾	75	5	Yes	None	Yes
Piston	Cast iron	¾-1¼	400	6-10	Yes	Little	Yes

[a] Pump at 540 revolutions per minute (RPM).

[b] 1¼-in suction × 1-in. discharge centrifugal pump. Pump RPM = 3250 at 50 PTO.

series of rubber paddles attached to an off-center shaft. As the paddles turn, the liquid is forced out of the pump.

5. *Gear.* Fluid enters the pump, and gear teeth force liquid into the discharge outlet.
6. *Centrifugal.* Liquid enters through the center of the rotating impeller and is flung outward by centrifugal force and forced out the outlet.

When wettable powders are used, the resistance of the pump to wear is very important. A pump that is subjected to moderate wear, such as a roller pump, should be sized to provide a reserve capacity about 50% greater than needed for the nozzles and the agitator. This reserve capacity will extend the life of the pump.

A pump will be damaged if it is operated dry, or with a restricted inlet. Dry operation occurs when the pump is operated with the tank empty, or if it is operated unprimed. Pump damage from not being primed is reduced when the suction line enters the tank at the bottom. A pump may also be damaged by improper mounting. If the pump is mounted directly on the tractor PTO shaft, do not attempt to bolt it down. A torque bar that extends about 15 cm from the pump is chained to a stationary part of the tractor to prevent the pump body from rotating. With piston pumps two torque bars should be used. One is chained stationary, and the other is attached to a spring on the opposite side of the pump from the chain to keep the chain tight.

Table 7.2 lists capacities, pressure ranges, wear characteristics, and other information that can be used to select the proper sprayer pump.

Agitators

Agitation is essential for a good uniform spray supply rate and can be accomplished by *mechanical* or *hydraulic* means. For large orchard and vegetable sprayers a mechanical paddle mounted lengthwise in the bottom of the tank and driven by power is the most efficient agitator.

Most field sprayers have hydraulic agitation of the *bypass* or *jet* type. The bypass type uses the return hose from the pressure relief valve to agitate the liquid. The bypass agitator hose must extend to the bottom of the sprayer tank. Bypass agitation is not sufficient on field sprayers with tanks larger than 220 liters (50 gallons) unless a centrifugal pump is used.

Gear, piston, and roller pumps require a jet agitator on 440-liter (100-gallon) or larger tanks. The jet must be located in the bottom of the tank and held in place on a pipe (*never on a hose*). The jet agitator must be connected to the pressure side of the sprayer and must operate at the operating pressure of the sprayer. Do not attach jets to the bypass line. A 220-liter tank requires one or more jet outlets; 400- to 600-liter tanks should have three or more outlets on a manifold. The manifold should be placed horizontally about 5 to 10 cm from the bottom of the tank. The tank capacity and operating pressure will determine the minimum jet size.

The orifices in a jet agitator must be sized so that the pump can maintain the operating pressure when all the nozzles are operating. In addition, some reserve capacity is needed to compensate for pump and nozzle wear.

Hoses

Hoses should be made of neoprene or other oil-resistant materials and strong enough to tolerate peak pressures. The hose test pressure should be twice the operating pressure. The suction hose should be made of two-ply fabric. It should be larger than the pressure hoses (boom, jet agitator, and bypass) because it must provide the total pump flow through the suction hose, suction strainer, fittings, and valves. The pressure drop in pressure hoses, fittings, valves, and strainers should not exceed 10% of the total operating pressure. Hose clamps of the worm-screw type are the most reliable and are required when the operating pressure exceeds 100 psi. When the pressure exceeds 300 psi, use two worm-screw clamps or other special clamps.

Keep hoses from kinking or being rubbed. Rinsing often, inside and outside, will prolong the life of hoses. Remove and store hoses during the off-season, or at least store the sprayer out of the sun. Always replace hoses at the first sign of surface deterioration.

Strainers

Strainers (filters) in sprayers prevent foreign materials from entering and putting wear on the precision parts of the sprayer; they also prevent nozzle tips from clogging. There should be a 12- to 25-mesh screen strainer in the filler opening, a 15- to 40-mesh screen strainer in the suction or supply line to the pump, and a 50- to 100-mesh screen strainer between the pressure relief valve and the boom. A 25- to 100-mesh screen strainer is needed with some nozzle assemblies. Strainers should be 50 mesh or larger for wettable powder applications.

The strainer screen should be large enough to prevent excessive pressure drop and plugging. *The suction strainer should have at least 6 cm^2 of screen area per liter per minute of flow.* Clean the strainers after each use and replace them if deterioration is evident. Strainers are the best defense against nozzle and pump wear and nozzle clogging. Use nozzle screens as large as the nozzle sizes permit. The screen opening should be smaller than the nozzle opening.

Pressure Gauges

Pressure gauges are important because pressure affects the amount of spray applied and droplet sizes. Use the lowest pressure possible for each job because high pressure uses extra power, wears equipment faster, and increases drift. Pressure gauges should be handled with care and equipped with a gauge protector when one is spraying corrosive materials or using a piston-type pump. Select a gauge that can be read accurately at the pressure to be used. Check gauges frequently for accuracy, and do not use a gauge under too much pressure.

Pressure Regulators

Pressure is regulated with a pressure relief valve that opens when the operating pressure exceeds its setting. For low-pressure spraying (10-40 psi), a diaphragm pressure relief valve is recommended.

When the desired operating pressure on a sprayer cannot be reached, check the pressure relief valve by observing the flow from the return hose. If there is flow in the return hose before the operating pressure is reached, the relief valve is not seating and must be checked for wear, trash, and broken springs. Repair or replace defective relief valves IMMEDIATELY.

On units with a centrifugal pump, the pressure may be regulated with an adjustable orifice valve (*globe* or *gate valve*). This device will make the sprayer a constant-rate sprayer (that is, the spray rate is independent of ground speed) during reasonable changes in ground speed as long as the tractor gear is not changed.

Cutoff Valves

There should be a quick-acting cutoff valve (or valves) on every sprayer between the pressure regulator and the nozzle. One valve may be used to cut off the entire operation, or a combination of two or three valves may be used to control the flow to different sections of the boom. Flow cutoff valves should be easily reached from the operator's seat.

Nozzles

Selection of nozzles that meet the requirements of each particular spray application is important in the effective use of agricultural chemicals. Nozzles help control the rate, the uniformity, the thoroughness, and the safety of pesticide application.

Select nozzles that will provide the proper particle size and application rate within the recommended range of pressures. The spray pattern and nozzle location should provide excellent coverage of the target area but allow little spray to escape to other areas. *Be sure that all nozzles on a boom are the same and the recommended type for the job.*

To meet various spray requirements, nozzles are classified by spray delivery patterns, spray angle, discharge rate, and composition. Eight types of nozzles, based on their spray patterns, are common:

1. *Flat fan nozzles* (Fig. 7.9A) are used for broadcast or boom spraying. Because the spray rate tapers at the edges, these nozzle patterns must overlap 30 to 50% for even distribution.
2. *Even flat fan nozzles* (Fig. 7.9C) are used for band spraying with a uniform spray pattern throughout. They should not be used for broadcast spraying.
3. *Whirlchamber (nonclog) nozzles* are available as wide-angle (120°) hollow cone nozzles. Clogging is minimized, and less drift occurs because of the relatively low boom height and large droplet size. Coverage remains fairly constant with changes in boom height.
4. *Flooding nozzles* (Fig. 7.9F) may be used to apply herbicides and fertilizer solutions. These nozzles operate at low pressures and have wide spray patterns (up to 160°). The pattern width varies with pressure and height.
5. *Boomless (or cluster) nozzles* (Fig. 7.9E) are used for wide-swath spraying. Only one or a cluster of nozzles may be used. The boomless nozzle does not

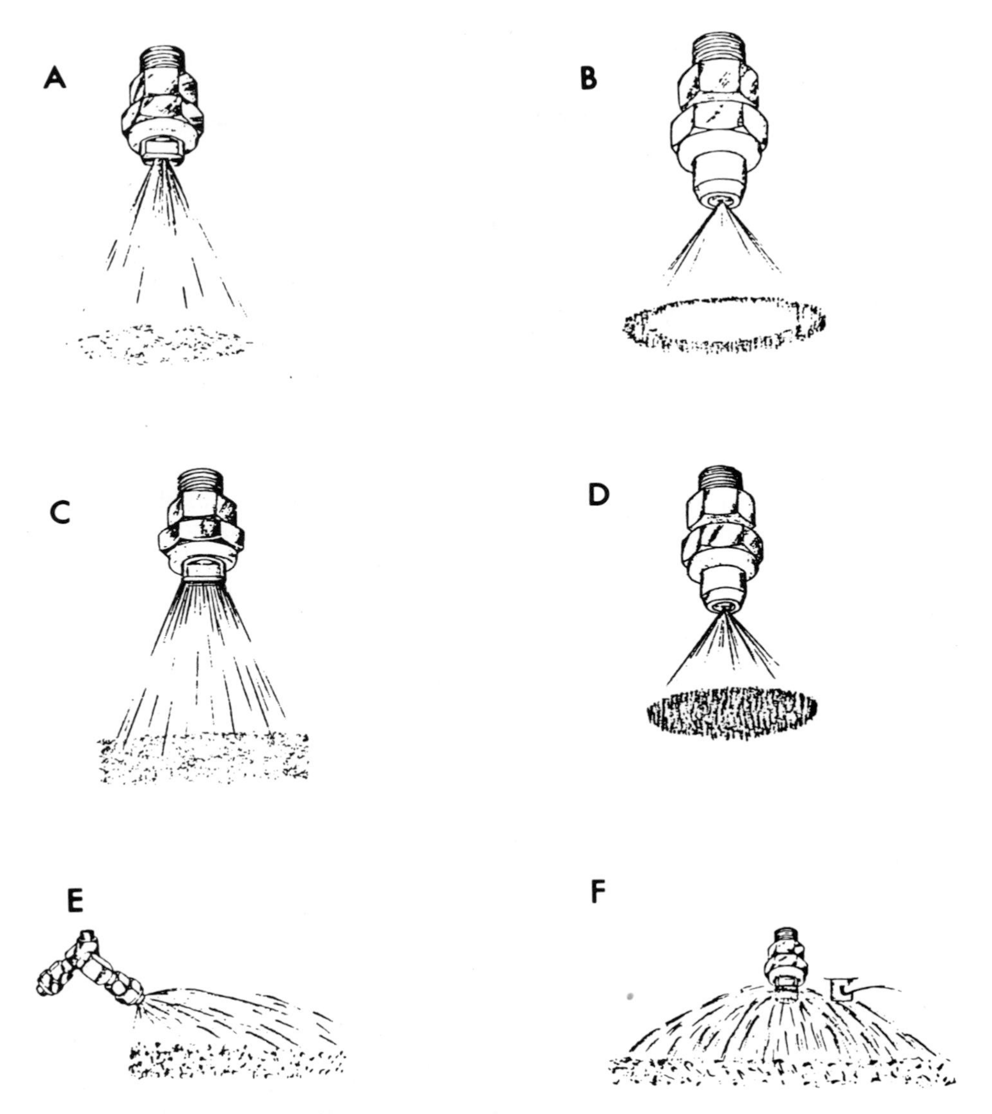

FIGURE 7.9. Spray patterns of various nozzle types. (A) Flat fan. (B) Hollow cone. (C) Even flat fan. (D) Solid cone. (E) Boomless. (F) Flooding. *Courtesy USDA.*

give as uniform coverage as other types, and the spray pattern is more affected by wind than that of a boom-type nozzle.

6. *Hollow cone nozzles* (Fig. 7.9B) are designed for moderate to high pressures and are used to obtain thorough coverage of crop foliage and very uniform distribution.
7. *Disk-core cone nozzles* are used to produce hollow or solid cone (Fig. 7.9D) spray patterns. The combination of disk and core used determines the spray

TABLE 7.3. Nozzle types and operating pressures for different uses.

Weeds	Insects	Diseases
Flat fan, 20-40 psi	Flat fan, 20-60 psi	Flat fan, 40-60 psi
Wide angle flooding, 10-40 psi	Hollow cone, 60-120 psi	Hollow cone, 60-120 psi
Whirlchamber, 7-20 psi	Disk cone, 40-100 psi	Disk cone, 80-400 psi

pattern and angle. These nozzles are used for spraying abrasive materials at high pressures.

8. *Solid cone nozzles* (Fig. 7.9D) are used for hand spraying, spot spraying, and foliar application of pesticides at moderate pressures.

The correct operating pressure should be selected for each spray job. Because drift often is a problem in spraying, the pressure should be as low as possible for the nozzle to operate. Insecticides normally are sprayed at moderate pressures; fungicides are sprayed at high pressures to achieve adequate coverage. Table 7.3 lists the type of nozzle and operating pressure recommended for various types of spraying.

Spray nozzles are made of the following materials for the reasons indicated:

- *Brass*—commonly used, inexpensive.
- *Stainless steel*—noncorrosive, expensive.
- *Aluminum*—resistant to moderately corrosive chemicals.
- *Tungsten carbide tips*—used for very abrasive materials such as wettable powders.
- *Plastic*—used for nonabrasive materials, corrosion-resistant, inexpensive.

During spraying operations, nozzles should be watched for clogging and changes in nozzle patterns. If nozzles clog or other trouble occurs in the field, be careful not to contaminate yourself while correcting the problem. Shut off the sprayer, and move it to the edge of the field before dismounting it. Wear protective clothing while making repairs.

Dusters

Although dusters once were very popular for applying pesticides, the use of dusters has declined in favor of sprayers. It is difficult to get uniform coverage of foliage with dusts, and drift is a greater problem than it is with sprays.

Hand dusters are for professional use in structures and can be used for spot treatments, gardens, and other small jobs. A duster may consist of a squeeze bulb, bellows, tube, or shaker, a sliding tube, or a fan powered by a hand crank.

Power dusters use a powered fan or blower to propel the dust to the target. Sizes range from knapsack or backpack types to those mounted on or pulled by tractors.

A power duster should be easy to clean and give a uniform application rate. With both hand and power dusters, the dust cloud should fall well away from the user and be directed to the target.

PESTICIDE APPLICATION GUIDELINES

Preapplication Checklist

All applicators should do the following *before* making any pesticide application:

- Identify the disease.
- When in doubt, always ask the advice of an authority on diseases and pesticides.
- Use only recommended pesticides for the problem.
- Make sure the pesticide applicator's license is current and of the appropriate type.
- Use the least hazardous pesticides at the lowest rate that will do the job.
- Read the label.
- Know the signs and symptoms of pesticide poisoning.
- Know what to do in the event of an accident.
- Have a physician's phone number posted by the telephone. In an emergency, time is extremely important.
- Take time to explain the safe use of all pesticides to employees. Make sure they understand.
- Users of organophosphorous and carbamate pesticides should contact a physician before using these materials, and request a blood test. Inform the physician about the types of pesticides that will be used and make sure she or he has the current phone number for the nearest Poison Control Center.
- Check application equipment for proper operation.
- Calibrate equipment frequently for proper output (see below, Calibration of Equipment section).
- Check respirators for cleanliness and proper fit.
- Check gloves and other protective clothing before each use.
- Make sure plenty of clean water, soap, towels, and clean clothing are readily available in case of an accident.
- Have a responsible person receive pesticides and properly store them.
- Warn people in the area that pesticides are to be used, and see that livestock and pets are removed from the area to be treated.
- Do not apply pesticides to food, feed, and water containers.
- Check to make sure that the time intervals between date of application and harvest will comply with those given on the label.
- Keep accurate records of pesticide purchase and application.

Ground Application

Ground applicators should do the following *before* and *during* any application:

- Make sure equipment has been properly calibrated (see below, Calibration of Equipment section).
- Mix only the amount of chemical necessary for the job.
- Use proper precautions for the chemical being used.
- Have two people working if highly toxic materials are being used.
- Stay out of drift of the chemical or dripping foliage.

- Wear appropriate protective clothing.
- Inform workers to stay out of the treated area until the reentry time is up.

Fumigation

Fumigators should do the following *before, during,* and *after* any application:

- Make sure equipment has been calibrated (see below, Calibration of Equipment section).
- Post property before a fumigation.
- Have available and *use* a fumigation safety kit, an effective gas mask for the fumigant to be used, and a chart of instructions for artificial resuscitation.
- Take all safety precautions necessary to protect the public.
- Keep records of fumigation jobs.

Greenhouse Applications

Applying pesticides in greenhouses presents special problems. In a normal greenhouse operation, employees must work inside, and personal contact with plants and other treated surfaces is almost a certainty. In addition, unauthorized persons may attempt to enter the premises. Ventilation in greenhouses frequently is kept to a minimum to maintain desired temperatures; as a result, fumes, vapors, and dusts may remain in the air for considerable periods of time after pesticide applications. Pesticides applied to plants and other surfaces in greenhouses do not generally break down as rapidly as those outside because exposure to environmental factors such as rain and UV light is less inside.

The following precautions should be observed whenever pesticides are used in a greenhouse:

- Select the least hazardous chemicals that are effective for managing the particular pest problems encountered.
- Use gas masks and waterproof protective clothing as specified on the label when using hazardous pesticides.
- Post warning signs on the outside of all entrances when fumigants or other highly toxic pesticides are being applied.
- Read and follow label instructions carefully.
- Do not enter a treated building without a gas mask until it has been aired for the length of time recommended on the pesticide label.
- Avoid skin contact with treated plants and other surfaces for the time specified on the label. Where this is impossible, workers should wear protective clothing and wash frequently.

Postapplication Checklist

The following should be done *after* any pesticide application:

- Do not enter premises until the proper time interval has elapsed.
- Clean application equipment thoroughly (see below, Cleaning and Storing Equipment section).

- Clean and store protective equipment after each use.
- Change clothing and wash it after using a pesticide.
- Store unused chemicals properly in original container with the label intact.
- Dispose of unwanted pesticides and containers properly.
- Keep accurate records on pesticide use.

Pesticide Use Records

Pesticide records can help users protect themselves and their investments. Records can help in the following ways:

- To determine what crop can be planted after pesticides have been used.
- To improve pest control practices by comparing results obtained with various chemicals or equipment used.
- To plan for future pesticide needs and reduce carryovers.
- To establish proper use in case of residue questions in marketing a crop.
- To establish proof of use of recommended procedures if indemnity payments are involved.
- To determine if errors were made.

The following information should be kept when pesticides are used: crop and cultivar treated; pest or pests treated; location and acreage treated; time of day, date, and year of application; type of equipment used; pesticide used, amount used per hectare, stage of crop development; pest population (low, moderate, high); weather (temperature, wind, rainfall); harvest date; and results of application. A copy of the Material Safety Data Sheet (MSDS) also should be kept for each pesticide used.

Keep notes in a pocket notebook. Write down information as it happens. Transfer it later to a permanent record sheet.

CLEANING AND STORING EQUIPMENT

Sprayers

A sprayer should be thoroughly flushed and washed at the end of each day. *Never leave unused pesticides in the sprayer.* The rinsate must be collected and applied on a crop for which it is labeled or collected and treated as a hazardous waste.

Screens in the sediment chamber and nozzles should be removed and cleaned frequently with a small brush. In some cases this may need to be done daily. Wire or nails should *never* be used to unplug nozzles because the precision surfaces may be distorted and alter the spray pattern. Use a wood splinter, if necessary, to remove an obstruction.

The pump, tank, and all fittings should be thoroughly washed several times with water after each spray job. Flush and drain the booms, hoses, and manifolds. Disassemble and clean nozzles.

For long periods of storage the sprayer should be cleaned thoroughly and stored under a shed or in a building.

Granular and Dust Equipment

Dusters and granular equipment should be emptied and cleaned at the end of each day. Carefully estimate the material needed to avoid excess material in the equipment at the end of the job. After removing all material, clean the equipment according to the operator's manual; then oil or grease the bearings to prevent corrosion. Store the equipment in a dry place.

Fumigators

After using a fumigator, place the remaining fumigant in the original container and seal it. Clean the equipment according to the manufacturer's directions and store it in a dry place.

CALIBRATION OF EQUIPMENT

Calibration—the process of adjusting equipment to apply a specified rate of material—is necessary to assure that each pesticide is applied according to label directions. Too much pesticide is dangerous; too little will not do a good job. Only by calibrating correctly can efficient results be obtained. Careful attention to calibration procedures will prevent serious errors in application rates.

There are many ways to calibrate pesticide application equipment. The preferred methods differ according to the kind of equipment being used. Extension service personnel and firms that sell applicators can provide instructions about how to calibrate equipment. Additional methods for calibrating sprayers can be found in other publications.

Factors Affecting Application Rates

The rate of application of a pesticide depends on the application width, the discharge or flow rate of material, and the ground speed. The application width, in turn, depends on nozzle spacing and row spacing. The flow rate is affected by nozzle size, as well as by the pressure on the nozzle and the density and viscosity of the pesticide being applied, as follows:

1. The *pressure* on a nozzle or orifice influences its output rate but only in proportion to the square root of the pressure. This means that the pressure must be increased four times to double the output.
2. The *density* (weight per liter) of the liquid affects the discharge rate through a nozzle or orifice at a given pressure. The heavier a liquid is, the slower its discharge at the same pressure. Most nozzles are rated for water, which has a density of 1 kg/liter. In applying lighter (oils) or heavier (fertilizers and fumigants) liquids, adjustments must be made for the density of the material.
3. The *viscosity* or resistance to flow of a material affects the flow rate. Thick liquids flow much more slowly than water. Most liquid pesticides have a viscosity very close to that of water, and viscosity usually is not a factor.

Ground speed is an important variable affecting the application rate of most equipment. For equipment that has a constant discharge rate (constant-pressure sprayers), the application rate is inversely proportional to the speed; that is, when the applicator is driven twice as fast, the application rate is reduced by one-half.

Application equipment can be calibrated to deliver a specified rate of material by adjusting one or more of the variables that affect application rate. Before calibrating any piece of equipment, set it up to do the desired job according to the manufacturer's instructions or other information sources.

Sprayer Calibration Methods

Before calibrating or using a sprayer, fill the tank with water and operate the sprayer to make sure that all parts are working properly. Remove all nozzles, boomcaps, or plugs and thoroughly flush out the system. Then select nozzles of the proper type and size for the job. Check the discharge rate of each nozzle using a watch and measuring container. If the discharge rate of a particular nozzle varies more than 10 to 15% from the average, replace the tip.

There are two basic ways of calibrating a sprayer—spray methods and collection (pint jar) methods—although numerous variations are possible.

Spray a Hectare (Acre) Method

Probably the most accurate calibration method is to spray a measured hectare or acre and measure the quantity of material applied. This method can be used for all large acreage spray jobs. Calibration should be done in the field that will be sprayed, not on a road or other hard surface, because a sprayer will travel faster on a road than in a field at the same throttle setting. The following steps explain how to calibrate a sprayer by this method:

1. Completely fill the tank with water.
2. To determine how far to drive in the field, use one of the following formulas:

$$\text{Distance (m) to drive for 1 ha} = \frac{10{,}000}{\text{sprayed width (m)}}$$

$$\text{Distance (ft) to drive for 1 acre} = \frac{43{,}560}{\text{sprayed width (ft)}}$$

   Sprayed width is the distance between the nozzles on each end of the boom plus one between-nozzle spacing (one-half of a nozzle spacing on each end of the boom).
3. Spray the measured hectare or acre at the desired speed and pressure.
4. Refill the tank, carefully measuring the quantity required to fill it. This amount is the application rate per hectare or acre.
5. If the measured application rate is different from the desired rate, correct by changing speed and/or nozzle size and determine the amount applied again. Repeat this step until the rate is correct.

Collection (Pine Jar) Methods

The pint jar methods are used to calibrate sprayers for row and broadcast ground application. In these methods of calibration 1 pt (473 ml or 16 oz) of liquid is collected from one nozzle. In Method 1, pesticide is collected from the sprayer while it is parked; in Method 2, material is collected while the sprayer is moving.

PINT JAR METHOD 1

1. Select a nozzle with an average discharge rate.
2. Determine the average nozzle spacing (effective boom width divided by number of nozzles).
3. Measure the time it takes to collect 1 pt (473 ml or 16 oz) of water from the average nozzle while parked.
4. Beginning about 6 m (20 ft) behind a marker, drive in the field at the speed to be used when spraying. Begin timing the sprayer (tractor) when it passes the marker, and drive for the same length of time that was required to collect 1 pt of water when the applicator was parked.
5. Measure the distance between the beginning and end markers. Compare this distance to those in Table 7.4 to determine the application rate.
6. If the rate is not acceptable, change nozzles and/or speed and recalibrate.

PINT JAR METHOD 2

1. Select a nozzle with an average discharge rate, and securely attach a pint jar under the nozzle to collect all the material from that nozzle. A jar lid with a hole for the nozzle will help retain all the material in the jar.
2. Starting at a marker, operate the sprayer at the desired speed and pressure until 1 pt (473 ml or 16 oz) is collected from the one nozzle.
3. Measure the distance required to fill the pint jar.
4. Determine the average nozzle spacing (effective boom width divided by the number of nozzles).
5. Compare this distance with those in Table 7.4 to determine the rate.
6. If the rate is not acceptable, change the speed and/or nozzles and recalibrate.

Fumigator Calibration

Fumigants are metered from an applicator with orifices to injectors. These orifices serve the same purpose as spray nozzles. The orifice size is selected in the same manner as a nozzle. First, determine the required flow rate in liters or gallons per hour, or liters or gallons per minute, using a spray nozzle formula. Then, using an orifice (spray nozzle) catalog or a flow formula, determine the size of orifice needed.

Low-Pressure Fumigators

A low-volatile fumigator may be calibrated by carefully collecting the material applied over a given area. There are two methods of collecting the fumigant: while the applicator is moving and while it is stationary.

The fumigant must remain as a liquid until it passes through the metering orifices. With low-volatile fumigators this is a problem only when the metering pump is operating at high speeds. Even though low-volatile fumigants are slow to vaporize, they will form vapor in the metering pump at high speeds due to the sudden vacuum created on the intake stroke. If this happens, the application rate is drastically reduced.

TABLE 7.4. Calibration table for sprayers (pint jar method).[a]

Nozzle spacing[b] (in.)	Distance required to catch 1 pt per nozzle at various application rates[c]							
	5 gal/A	7½ gal/A	10 gal/A	12½ gal/A	15 gal/A	20 gal/A	25 gal/A	35 gal/A
6	2178 ft	1452 ft	1089 ft	871 ft	726 ft	545 ft	436 ft	311 ft
8	1634	1089	817	653	545	408	327	233
10	1307	871	653	523	436	327	261	187
12	1089	726	545	436	363	272	218	156
14	933	622	467	373	311	234	187	133
16	817	545	408	327	272	204	163	117
18	726	484	363	290	242	182	145	104
20*	653	436	327	261	218*	163	131	93
21	622	415	311	249	207	156	124	89
22	594	396	297	238	198	149	119	85
24	545	363	272	218	182	136	109	78
30	436	290	278	174	145	109	87	62
36	363	242	182	145	121	91	73	52
42	311	207	156	124	104	79	62	44
48	272	182	136	109	91	68	54	39

[a] To use this table, find the value of the average nozzle spacing of the sprayer being calibrated in the left-hand column, then read across the row to the value closest to the distance traveled in the test procedure; finally read up the column to find the corresponding application rate in gallons per acre. For example, if a sprayer has a nozzle spacing of 20 in. and travels 218 ft during collection of 1 pt of material (or water) from one nozzle, the application rate is 15 gal/acre. The speed is accounted for in the distance.

[b] When nozzle spacing is not uniform or when more than one nozzle is used per row, use the average spacing. If three nozzles are used per row and the row spacing is 42 in., the nozzle spacing would be 42 ÷ 3, or 14 in.

[c] To convert rate to liters per hectare, multiply the table value by 9.3.

METHOD 1

1. Measure 30.5 m (100 ft) in the field.
2. With the applicator equipment running at the desired speed and engaging the soil, collect the fumigant applied in the 30.5 m (100 ft) section from one or more soil tubes (the entire applicator output may be collected).
3. Determine the application width being collected. For example, if the material is collected from 6 tubes spaced 20.3 cm (8 in.) apart, the application width is 121.9 cm (48 in.). If the material for one row is collected, the application width is the row spacing.
4. Compare the amount of material collected with those quantities in Table 7.5 to determine the application rate.
5. If the rate is not acceptable, adjust orifices and/or pressure and recalibrate.

METHOD 2

1. Measure 30.5 m (100 ft) in the field.
2. While operating the equipment with the fumigant shut off, record the time required to travel 30.5 m (100 ft).
3. Determine the application width for the orifice or orifices to be used. For row application, use the row width and collect all material for one row.
4. While the application equipment is not moving, collect fumigant for the length of time required to travel 30.5 m (100 ft) in the field. Compare the quantity collected with the amounts in Table 7.5 to determine the application rate.
5. If the rate is not acceptable, adjust orifices and/or pressure and recalibrate.

High-Pressure Fumigators

To calibrate high-volatile applicators dispensing methyl bromide, *the container must be weighed before and after applying fumigant to a measured plot.* The material cannot be collected because it would immediately volatilize. The following steps are involved in calibrating a high-pressure fumigator:

1. Weigh the container of fumigant.
2. Apply the fumigant to 30.5 m (100 ft) of row for row treatments, or 93 m^2 (1000 ft^2) of area for broadcast treatments.
3. Reweigh the container to determine the amount of fumigant used.
4. To calculate the application rate for row treatments, multiply the amount used (in pounds) per 30.5 m by the following values, depending on row width:

Row width	*Multiply by*
30 in. (76.2 cm)	175
36 in. (91.4 cm)	145
42 in. (106.4 cm)	121
48 in. (121.9 cm)	105

TABLE 7.5. Calibration table for low-pressure fumigators.[a]

Application rate[b] (gal/A)	Quantity collected per 30.5 m (100 ft) at various application widths									
	61 cm (24 in)		76.2 cm (30 in.)		91.4 cm (36 in.)		106.7 cm (42 in.)		121.9 cm (48 in.)	
1	0.6 oz	17.4 ml	0.7 oz	21.7 ml	0.9 oz	26.1 ml	1.0 oz	30.4 ml	1.2 oz	34.8 ml
3	1.7	52.1	2.2	65.0	2.6	78.2	3.1	91.2	3.5	104.3
5	2.9	86.9	3.7	108.4	4.4	130.3	5.1	152.1	5.9	173.8
7	4.1	121.6	5.1	151.7	6.2	182.5	7.2	212.9	8.2	235.1
9	5.3	156.4	6.1	195.1	7.9	234.6	9.3	273.7	10.6	312.8
12	7.1	208.5	8.8	260.1	10.6	312.8	12.3	365.0	14.1	417.1
15	8.8	260.7	11.0	325.2	13.2	391.0	15.4	456.2	17.6	521.3

[a] To use this table, find the column headed by the value corresponding to the application width of the fumigator being calibrated, then read down the column to the value closest to the amount of fumigant (in ounces or milliliters) collected during the test procedure; finally read *left* across the row to find the corresponding application rate.
[b] To convert rate to liters per hectare, multiply the table value by 9.3.

TABLE 7.6. Minimum metering pressure for methyl bromide fumigators.

Fumigant temperature		100% Methyl bromide (psi)	70% Methyl bromide 30% chloropicrin (psi)
(°C)	(°F)		
10	50	5	0
20	68	15	5
30	86	25	15
40	104	40	25

5. To calculate the application rate for broadcast treatments in pounds per acre, multiply the amount used (in pounds) per 93 m^2 by 43.5; multiply by 107.5 to obtain the rate in pounds per hectare.
6. If the rate is not acceptable, adjust orifices and/or pressure and recalibrate.

Remember that speed, pressure, and applicator injector spacing will affect the rate of application.

The fumigant must remain as a liquid until it passes through the metering orifices. A highly volatile fumigant must be pressurized enough to prevent the formation of vapor until after the fumigant passes through the metering orifices. The minimum pressure required to maintain a fumigant as a liquid depends on its temperature.

Table 7.6 lists the minimum metering pressures required to maintain methyl bromide fumigants as liquids at different temperatures. This table is based on a pressure loss of about 2 psi in the system due to tube friction. If an excessive pressure drop is present because of undersized fittings, tubes, and so forth, the pressures indicated in Table 7.6 must be increased.

In metering highly volatile fumigants at low pressure, it is extremely important to keep the fumigant container cool (out of the sun) to prevent vapor formation in the metering system.

Granular Applicator Calibration

Granular applicators may be broadcast or row types. Calibration is done by collecting and weighing the material applied for a measured distance. Follow this procedure:

1. Collect the sample as follows:
   - *Row applicator:* Collect the quantity applied per row for 30.5 m (100 ft).
   - *Broadcast tube applicator with spreaders:* Collect the material applied per tube for 30.5 m (100 ft).
   - *Manual centrifugal (slinger) broadcast applicator:* Place the spreader in a large paper or plastic bag and collect the material applied during operation for 30.5 m (100 ft) at normal speed.

TABLE 7.7. Calibration table for granular applicators.[a]

Application rate[b] (lb/A)	Quantity collected per 30.5 m (100 ft) at various application widths				
	30.5 cm (12 in.)	61 cm (24 in.)	91.4 cm (36 in.)	106.7 cm (42 in.)	121.9 cm (48 in.)
5	5 g	10 g	16 g	18 g	21 g
10	10	21	31	36	42
15	15	31	47	55	62
20	21	42	62	73	83
25	26	52	78	91	104
30	31	63	93	109	125
40	42	83	125	146	167
50	52	104	156	182	208
60	63	125	187	218	250
75	78	156	234	273	312

[a] To use this table, find the column headed by the value corresponding to the application width of the applicator being calibrated, then read down the column to the value closest to the weight of the granules collected during the test procedure; finally read *left* across the row to find the corresponding application rate.
[b] To convert rate to kilograms per hectare, multiply the table value by 1.1.

- *Power centrifugal (slinger) broadcast applicator*: With the hopper empty, determine how long it takes to move 30.5 m (100 ft). Then fill the hopper with granules; with the applicator stationary, collect the material dispensed during the length of time required to move 30.5 m. If the applicator cannot be calibrated in a stationary position, collect the material in a suitable container while operating it over a specified distance.

2. Determine the application width (row spacing or tube spacing as appropriate).
3. Weigh the sample with a gram scale or a very accurate ounce scale. (Do not include the weight of the container.)
4. Compare the weight of the sample collected with those quantities in Table 7.7 to determine the application rate.
5. If the rate is different from the desired rate, adjust the equipment and recalibrate it.

FORECASTING PLANT DISEASES AS AN AID TO PESTICIDE SCHEDULING

Since time immemorial, farmers have worried about the weather and used all means at their disposal to protect their crops by predicting future weather and disease outbreaks. Farming is a risky business, and accurate weather and disease forecasts often lead to increased profits.

Disease forecasts are useful to farmers in the practical management of disease in their crops. The ultimate purpose of forecasting is to minimize disease losses and optimize pesticide use. Farmers must balance three factors in disease management:

FIGURE 7.10. This solar-powered microcomputer and environmental monitor identifies weather conditions favorable for the development of several diseases and recommends when pesticides should be applied. *Courtesy Neogen Co.*

risk, cost, and benefit. For example, if forecasts indicate that the risk of disease is low, there is no need to invest in the expense of spraying with pesticides; if the forecast predicts a high disease risk, then it may be wise to start a spray program. Forecasting provides the means for determining when and where a given management practice should be applied. If a fungicide is not needed, why go to the expense of applying it? Also, use of pesticides only when needed can help protect the environment.

One of the spin-offs of the space program is the use of satellites to monitor weather patterns around the world, permitting forecasters to make more accurate weather predictions, both local and regional, for the next few hours, the next week, or longer. Also, rapid progress in computer science and equipment now permits huge volumes of information to be analyzed quickly and efficiently. The small "gray boxes" placed in farmers' fields or orchards to record temperature, rainfall, light, wind, and RH data are really small computers that help in predicting disease outbreaks (Fig. 7.10).

Undoubtedly, with future advances in computer science, sophisticated equipment, mathematical modeling, and more accurate collection of weather data, disease forecasting services will be increasingly valuable to farmers and will help in safeguarding the production of crops and the environment.

Forecasts Based on the Weather

The incidence of most plant diseases is highly dependent upon the weather, and some early forecasting methods were based solely on weather patterns. For example, late blight of potato occurs during rainy weather, with high RH and cool temperatures (below 22°C). Conversely, there is little chance of late blight breaking out during periods of hot, dry weather. Therefore, successful late-blight forecasting can be accomplished by keeping accurate temperature and RH records at critical times during the growing season. Forecasts based on weather can indicate when diseases are and are not likely to occur.

Forecasting based on weather assumes that the pathogen is always present in a locality, and disease will break out if suitable weather develops. Sometimes the pathogen is *not* present, in which case the disease will not break out even under favorable conditions; this is one of the drawbacks of weather-based forecasting.

Forecasts Based on Inoculum Potential

Some forecasts are based on the amount of the pathogen available to infect the host. Spore traps are used to determine the presence and numbers of airborne spores; for soilborne pathogens, assays of soil or plant tissues determine how many propagules are present in the soil or plant litter and debris. These data then are used to calculate the chances of disease outbreaks. Forecasts based on the amount of pathogen present assume that weather favorable for disease development will occur. This assumption does *not* always hold true.

Forecasts Based on Type of Pathogen and Rate of Increase

As a result of Vanderplank's contributions to the study of epidemiology (see Chapter 4), another type of disease forecasting is available. If the type of pathogen present and its rate of increase under different weather conditions are known, the chances of disease occurring can be predicted with some accuracy, based upon measurements of the amount of inoculum present in a given area.

PROTECTING THE ENVIRONMENT

Anyone who applies pesticides has an obligation for the safety of others and the environment. Our human environment is part of the whole *ecosystem*. This term describes the entire interacting system of plants and animals—of all kinds from bacteria to humans—living in the physical world. Every plant or animal is affected by other plants and animals in the environment and by the physical characteristics of the environment.

Under a given set of conditions, the different species of plants and animals tend to come to an equilibrium among themselves, the so-called *balance of nature*. Pesticides are used to force the balance to be favorable to humans to provide them food and freedom from disease. Ecological theory says that the greater the number of species, and the greater the differences among these species, the greater the stability of the overall system and the less it will react to changes in the population of any one species.

Human activities often reduce the diversity of natural species. For example, in commercial agriculture a single crop is planted over a wide area, replacing the original mixture of many plant species. Such a system is not stable, and we must expect that it will never be easy to maintain a balance favorable to humans. We have learned that some pesticides not only reduce the numbers of *target* pest species, as intended, but also change the numbers of such desirable *nontarget* forms as birds, changing the balance of nature in unintended, and sometimes unfavorable, ways. With our need for food, we can never return to a truly natural condition. Instead we must reach the most reasonable balance we can achieve, considering all human needs.

We cannot do much about controlling such aspects of the environment as light, rain, temperature, and wind, but we can control some other things, including the use of pesticides. Many people consider pesticides a tool for preserving or improving the environment. Others feel that pesticides cause pollution. As a weed is a "plant out of place," a pesticide sometimes can be a "tool out of place." Correct use prevents pollution by pesticides; incorrect use can damage the environment and lead to lawsuits, fines, and the payment of damages by the applicator.

The majority of contamination incidents are a result of the mishandling of agricultural chemicals in storage and at mixing sites. The following safety tips will help you to avoid contaminating water:

1. Store agricultural chemicals away from water sources (wells, ponds, streams).
2. Read product label instructions, and choose the lowest effective rate.
3. Extend with a hose the distance between the well head and the sprayer fill area.
4. Prevent spills and backsiphoning of agricultural chemicals by:
   - Never leaving a filling operation unattended.
   - Filling tanks with water and then moving away from the water source before adding chemicals.
   - Keeping the water supply hose above the spray tank and out of the liquid.
5. Pressure-rinse or triple-rinse containers immediately upon emptying them. Add the rinse water directly to the spray tank mix.
6. Whenever possible, use rinse pads to capture and recycle equipment rinse water.
7. Calibrate spray equipment, and check it frequently.

Controversies and problems related to the use of pesticides are likely to be greater when agricultural land lies close to urban areas than in all-agricultural areas. Pesticide damage to nontarget organisms can cause problems in any community, but the denser the human population in the area adjoining treated areas, the greater the risk of injury to humans and domestic pets. Furthermore, contamination of a reservoir used for a community's water supply and public recreation, by careless pesticide application, can cause major disruptions in a community and, quite possibly, severe sanctions against the offending applicator.

Reducing Damage to the Environment from Pesticides

Using pesticides in a way other than as directed on the label can (1) injure plants and animals, (2) leave illegal residues, and (3) damage the environment (including water supplies). Any pesticide can cause harm if not chosen and used with care.

Not all pesticides act the same after application. Most are classified as either nonpersistent or persistent. *Nonpersistent* pesticides break down quickly and remain on the target or in the environment only a short time before being changed into harmless products. Some are highly toxic; others are fairly nontoxic. *Persistent* pesticides break down slowly and may stay in the environment without change for a long time. Often this characteristic is good for pest management because it provides long-term management with a single application. Most persistent pesticides are not broken down easily by microorganisms and are only slightly soluble in water.

Some persistent pesticides that can injure sensitive crops planted on the same soil the next year seem to be of little hazard to the environment beyond the treated soil. Other persistent pesticides can accumulate in the bodies of animals, including man. Such pesticides may build up to a level that is harmful to the animal itself or to a meat eater that feeds on it. These are called *accumulative pesticides.*

The chances that pesticides will cause environmental damage increase with movement off the target. This can occur by various means:

- Drifting out of the target area as a mist or dust.
- Evaporating and moving with air currents.
- Moving on soil through runoff or erosion.
- Leaching through the soil.
- Being carried out as residues in crops and livestock.

When the plants on land adjoining a treated area are sensitive to the pesticide being applied, it is very important that drift be controlled (see below, Controlling Pesticide Drift section).

Direct Kill of Nontarget Organisms

Pesticides sometimes are applied over a large area to such target organisms as fungi, insects, and weeds. Many nontarget plants and animals within the treated area may be susceptible to a pesticide and can be harmed by it. For example, herbicides may kill nearby crops and landscape plants; some insecticides are likely to kill any bees or other pollinators that are in a field during application. Runoff from a sprayed field can kill fish in a nearby stream or pond. Life in streams can be wiped out by careless tank filling or draining and improper container disposal.

To reduce the danger to nontarget organisms, the most selective and least toxic pesticide that is effective for a given pest problem should always be chosen.

Contamination of Soil and Water

A number of factors affect the risk of soil and water contamination by pesticides. On crops where repeated pesticide application is necessary, pesticides (especially those that are persistent) may build up in the soil. On row crops where tilling is common and water drains into aquatic areas, runoff of contaminated soil particles and pesticides is likely. Pesticides should not be applied just before predicted heavy rains or just before some types of irrigation.

Although some fungicides and herbicides are more effective when applied before a rain, heavy rains cause runoff and tend to wash the pesticide away from the target area. The runoff can carry the pesticide into sensitive areas where crop injury may result. Runoff also may reach farm ponds, streams, and waterways, causing contamination, fish kills, or injury to domestic animals such as dairy cows. The use of proper soil conservation practices, and the planting of grasses especially, will help prevent the runoff of pesticides and fertilizers.

The use of persistent pesticides may limit future planting options. *Carryover* of certain pesticides in the soil from one year to the next means that only crops that the pesticide will not kill or contaminate can be planted. Even pesticides directed at plants or animals can move to the soil. Pesticides may be washed or brushed off. They may be worked into the soil with dead plant parts.

Many combinations of these and other factors may lead to problems associated with soil and water contamination by pesticides. As the risks of contamination increase, applicators may need to consider changing the time of application, pesticides, or the normal application method.

Controlling Pesticide Drift

Pesticide drift is the movement of spray or dust particles or chemical vapors a considerable distance from the place of release or discharge. Most frequently it is associated with agriculture and is considered undesirable because it can injure or leave undesirable residues on nearby crops and livestock, pollute the environment, and injure humans. It also wastes chemicals and causes uneven distribution.

Spray or Dust Drift

A number of factors influence the movement of particles discharged from application equipment. The most important are *weather* and *particle size.* As a general rule, the application of pesticides should be avoided when the wind is blowing faster than 12.8 km/hr (8 mph). When spraying near the property of others or in public areas, be conscious of the possibility of drift and take necessary precautions. Do not spray near these areas on windy days.

A weather condition called *inversion,* where warm air overhead traps cool air beneath, can cause air movement and drift. Inversions usually occur in early morning or late afternoon.

Droplet size influences the length of drifts. For example, with a 4.8 km/hr (3 mph) wind, visible spray particles (50 μm or more) may drift as much as 100 m (109 yd), whereas invisible vapor (10 μm or less) may drift several kilometers (a mile or more). Because dusts are lightweight with a small particle size, they often present a greater drift hazard than sprays.

Other variables that influence droplet size are *pressure* and *nozzle orifice.* Too often farmers assume that high pressures (100 psi, for example) will given more effective pest management by driving the spray material down into the leaves. However, high pressure reduces the size of spray particles, causing more drifting; the higher velocity also causes more turbulence with less effective coverage. Many fungicides and insecticides must be applied at high pressure (100-400 psi) to reach and manage the intended pests. However, even where high pressures are needed, excessive pressure should be avoided to minimize drift. Increasing the nozzle orifice (opening) size can help reduce drift because large openings require less pressure, resulting in larger droplet sizes and hence less drift.

A wide range of chemical formulations is available for spray applications, and the spray particle size can be altered to fit application needs. With the increasing popularity of low-volume and ULV spraying, drift problems may increase. However, the recently developed *foam* technique that applies pesticides in a foam is helpful in solving drift problems. Reported benefits of foam include extended application periods (wind not a factor), less material used (more complete coverage, no overlapping), cleaner equipment, and fewer reloading trips (less water used).

Vapor Drift

Vapor drift, unlike spray or dust drift, is related directly to the chemical properties of the pesticide. Most problems are the result of accidental or unintentional misuse. If applicators understand the chemical properties, they can control and avoid damage from drift.

Vapor drift can be due to vapor leakage, which can be stopped by properly sealing fumigant or other volatile materials after application and by applying these materials with vapor-tight equipment. Soil sterilization in a greenhouse that ends up damaging plants in an adjacent greenhouse is one example of vapor drift. Ester formulations of phenoxy herbicides may volatilize and drift under high-temperature conditions. Amine or acid formulations should be used where vapor drift might cause problems.

Facts to Keep in Mind

When other conditions are the same, the following generalizations apply:

- Smaller droplets or dust particles drift farther than large ones.
- There is more chance of drift with airblast sprayers than with boom sprayers.
- There is more chance of drift with high-pressure sprayers than with low-pressure sprayers.
- Low-volume concentrates are more likely to drift than high-volume dilutions.

SUMMARY

Agricultural chemicals are an important part of an agricultural production system, but these products should be used according to label directions to achieve desirable results. The use of sound production systems and careful pest identification should precede the selection and use of agricultural chemicals. Forecasting systems for diseases should be used, when possible, to schedule applications of pesticides. Consideration of these factors and careful application techniques may help reduce the amount of pesticides used and will help to protect the environment.

8

Biotechnology

> Biotechnology, a collection of techniques, tools, and procedures, is bringing about profound changes in plant pathology. The potential impact of these changes on our understanding of biological systems and in the successful production of food and fiber is immense. In the arena of genetic engineering alone the stable introduction of foreign genes into plants represents one of the most significant developments . . . in agricultural technology.
>
> Genetic engineering of plants and microorganisms, protoplast and tissue culture, the sequencing of RNA and DNA, the use of monoclonal and polyclonal antibodies and molecular probes, are some of the new frontiers. These frontiers represent new avenues of research that could not previously be pursued and, as a result, our horizons expand and our view of what may be possible in managing plant diseases changes.*

Throughout recorded history the domestication, breeding and improvement of animals and plants has depended on the fundamental observation that living organisms both resemble and differ from their parents. Within the past century scientists have gained an increasing understanding of heredity and change among living things—the science of genetics, which has developed and expanded rapidly. It is now established that each inherited trait or function of an organism depends upon a unit of inheritance, a specific *gene,* which directs its development. In microorganisms (bacteria) that commonly reproduce asexually and whose single cells lack a defined nucleus, a single gene controls each specific trait. In higher plants and animals, two genes exist with respect to each trait: one inherited from each parent.

Some plants have 10,000 or more genes. Each gene is a tiny segment of a

---

*C. Lee Campbell, editorial in *Phytopathology News,* 1991, Vol. 25, p. 30.

threadlike component of the cell called a *chromosome*. The chromosomes, located in the nucleus, are composed of deoxyribonucleic acid (*DNA*) and protein. DNA is a wondrous chemical that can reproduce exact copies of itself. Many genes make up one chromosome. Each gene is a tiny piece of DNA that regulates the reproduction of a specific protein, which in turn governs the composition and development of a specific trait. Thus, the fidelity and the diversity of inherited traits are locked within the DNA of chromosomes. The cytoplasm also contains other organelles, such as chloroplasts and mitochondria, floating as tiny blobs in the cytoplasmic "soup." Some genetic material is located in the cytoplasm. The inheritance of the following important traits has been linked with cytoplasmic genes:

- Male sterility
- Flower morphology
- Resistance to fungal toxins
- Organelle morphology
- Albinism and variegation
- Fraction 1 protein production
- Nitrate growth response

By using the accumulated knowledge of the laws of genetics and the techniques of controlled crossbreeding and selection, plant breeders have greatly improved the usefulness of many farm crops. Breeders have increased crop yields and quality, incorporated more effective resistance to disease and insects into cultivars, expanded the geographical limits of crop production by increasing crop tolerance to stress conditions (cold, day length, drought, heat, salinity), and overcome natural breeding barriers such as sterility and abortion. Increases in crop yields in the United States between 1930 and 1989 are shown in Table 8.1. The degree to which applied genetics has contributed to these yield increases is impossible to determine because there have been great improvements in farm management, fertilizers, herbicides, irrigation, and pest management. In the case of corn, however, it is estimated that the use of hybrid corn accounts for as much as 50% of the harvest increases. Development of disease-resistant cultivars also accounts for a portion of the yield increases. By cross-pollinating a low-quality, scab-resistant apple cultivar with a high-quality, scab-susceptible cultivar and selecting through several generations, pomologists have developed a scab-resistant apple of high quality.

Breeding a disease-resistant cultivar usually is a long and laborious process. After a source of resistance is found, 10 years or more may elapse before crosses are made and suitable cultivars selected. The incorporation of resistance to several diseases into a desirable agronomic cultivar becomes increasingly difficult and tedious. (See Chapter 6, Disease-Resistant Cultivars section.)

Each major agricultural crop has numerous relatives, both wild and cultivated, that contain many desirable characteristics. For example, corn is not susceptible to attack by the fungus that causes stem rust of wheat; all presently cultivated corn cultivars are immune to wheat rust. Obviously this immunity is inherited and transmitted from one generation to the next. If some way could be found to transfer rust immunity from corn to wheat and incorporate the resistance into usable cultivars, then wheat crops would be protected from attack by this pathogen, which annually reduces worldwide yields.

TABLE 8.1. Average yields of major crops in the United States—1930 and 1989.

	Yield per acre[a]			Increase (%)
	1930	1989	Unit	1930-1989
Wheat	14.2	32.8	Bushels	131
Rye	12.4	28.1	Bushels	127
Rice	46.5	122.0	Bushels	162
Corn	20.5	116.2	Bushels	467
Oats	32.0	54.4	Bushels	70
Barley	23.8	48.6	Bushels	104
Grain sorghum	10.7	55.4	Bushels	418
Cotton	157.1	614.0	Pounds	291
Sugar beets	11.9	19.1	Tons	61
Sugarcane	15.5	35.4	Tons	128
Tobacco	775.9	2016.0	Pounds	160
Peanuts	649.9	2426.0	Pounds	273
Soybeans	13.4	32.4	Bushels	142
Snap beans	27.9	69.8	Hundredweight	150
Potatoes	61.0	289.0	Hundredweight	374
Onions	159.0	365.0	Hundredweight	130
Tomatoes				
Fresh market	61.0	253.0	Hundredweight	315
Processing	4.3	27.0	Tons	527

Source: USDA (1989 Agricultural Statistics).
[a] To determine yields per hectare, multiply table values by 2.47.

However, until the last few decades, the transfer of desirable traits from one species of plant to another was possible only if they could be mated by cross-pollination. It still is impossible to crossbreed or hybridize unrelated plants such as soybeans and tomatoes by cross-pollinating them. In addition, many plant species are difficult to manipulate genetically by conventional methods. Therefore, more rapid and convenient ways to improve plants are being sought. Classical or conventional genetics will continue to play the major role in animal and plant breeding, but new genetic techniques will assume ever-increasing importance.

Since 1960, scientists have made great progress in finding ways to genetically change and improve plants, including microorganisms (bacteria and fungi). This new technology, a form of *biotechnology*, deals with ways to change and/or grow plants in test tubes in the laboratory, rather than by use of conventional plant breeding techniques that involve crossbreeding. As a result, the development of superior strains of food crops and trees can proceed at a faster pace. Scientists hope to accomplish within a few years in the laboratory what usually takes 10 to 40 years to do by traditional methods. Biotechnology includes *genetic engineering*, by which the hereditary apparatus (DNA) of an organism is altered so that the cell can produce more or different chemicals or perform completely new functions. These altered cells can then be used in industrial production. When genes from other organisms, for example, the gene for producing the coat protein of a virus, are inserted into plants, the plants are called *transgenic plants* (Table 8.2).

TABLE 8.2. Examples of traits transferred to or selected in plants through tissue culture or gene transformation techniques.

Resistance to pathogens	Host plant	Technique
TMV-resistant	Tomato	Gene transformation
TMV-resistant	Tobacco	Protoplast fusion In site selection Gene transformation
Black root rot (*Thelaviopsis basicola*)	Tobacco	Protoplast fusion
Potato virus Y tolerance	Potato	Gene transformation
Wildfire (*Pseudomonas syringae*)	Tobacco	Gene formation
Cucumber mosaic virus	Cantaloupe	Gene transformation
Tolerance to herbicides	**Host plant**	**Technique**
Glyphosate tolerance	Tomato	Gene transformation?
	Tobacco	Gene transformation
	Cotton	Gene transformation
Glufosinate tolerance	Alfalfa	Gene transformation
Bromoxynil tolerance	Cotton	Gene transformation
Resistance to insects	**Host plant**	**Technique**
(Lepidopteran) Hornworm insects	Tomato	Gene transformation
Budworm	Cotton	Gene transformation

VEGETATIVE PROPAGATION

For centuries many agricultural plants, including banana, Irish potato, strawberry, sugarcane, and others that usually do not produce true seed, have been routinely propagated vegetatively by the rooting of cuttings, that is, by *cloning*. Also, some economically important plants (bulbs, chrysanthemum, fruit trees, hops, Irish potato, sugarcane) that are *heterozygous* and do not breed true must be propagated by vegetative means so that the desirable characteristics of each cultivar will be perpetuated. Cloning can be accomplished with (1) organs (flowers, leaves, roots, stems), (2) tissues (endosperm, epidermis, mesophyll, nucellus), and (3) single cells.

Organ Culture

When pieces of living organs (leaves, roots and stems) of some plants in the form of adventitious buds, cuttings, rhizomes, shoots, stolons, and underground stems are cut or broken off and placed in water or another suitable rooting material under favorable growing conditions, they often produce roots and eventually a new plant. This ancient process, sometimes called *organ culture,* has been widely used for hundreds of years. However, it is laborious and time-consuming to grow plants from cuttings or rhizomes, even though conventional methods have improved since the 1940s by the use of rooting hormones, controlled light, and mist.

Adventitious Shoots and Callus Tissue

Adventitious shoots are often formed on roots, lower parts of stems, or trailing leaves of some plant species (Fig. 8.1). If small bits of these plant tissues, called *explants,* are first dipped in alcohol or sodium hypochlorite to kill any contaminating bacteria or fungi and then grown in flasks, they may produce many adventitious shoots that will grow into new plants. In other cases, the explant may form a cluster of cells, called a *callus,* that consists of an unorganized mass of dividing and enlarging cells. If culture conditions are readjusted when the callus appears, the cells can become specialized into tissues and organs and eventually into a complete normal plant. Each callus can

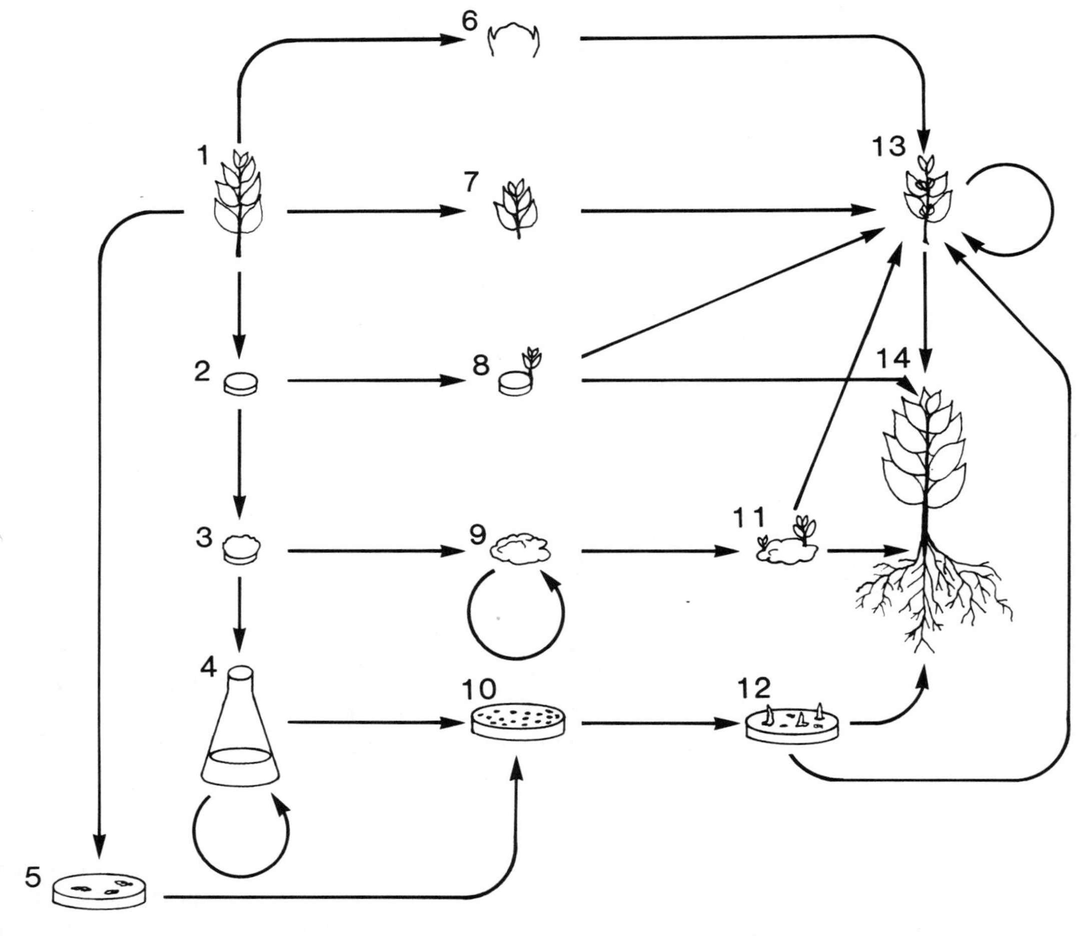

FIGURE 8.1. Diagram showing relations between the various methods of *in vitro* propagation. 1—parent plant; 2—organ explant; 3—callus induction; 4—cell suspension culture; 5—liberation of protoplasts; 6—meristem culture; 7—shoot-tip culture; 8—adventitious shoot; 9—callus multiplication; 10—plated cells or protoplasts; 11, 12—shoots from callus or cells; 13—axillary shoot proliferation; 14—rooted plantlets for use in greenhouse or field.

be further subcultured. Routinely, as many as a thousand identical plants can be produced from each gram of original explant.

Immature embryos are well suited as donor tissues in the regeneration of cereal plants (corn, millet, oats, sorghum, and wheat). Embryos of these crops can be easily dissected out of the parent plant, are easily preserved in storage, and produce vigorous cultures that yield large numbers of regenerated plants. Thousands of "daughter" plants can be easily produced from a single embryo. Another effective and widely used donor tissue for grasses and cereals is the immature flower.

Tissue and Single-Cell Culture

Many plants can be propagated from meristems, tissues, or single cells. However, both tissue culture and single-cell culture require more precise techniques than does organ culture. Since 1960, increased scientific understanding and more refined laboratory techniques have led many commercial enterprises to concentrate on cloning crop and ornamental plants under controlled conditions. More than 300 species of plants now are propagated this way.

Meristem Culture

The shoot apex (growing point or *meristem*) of plants is the region that produces all the aboveground parts of the plant. If the meristem is cut off aseptically and placed in a nutrient broth or gel under proper conditions of light, temperature, and moisture, it will begin to grow. In some species, each meristem in culture grows into one plantlet. In others, the meristem forms multiple plantlets.

In small flasks or jars of about 250-ml (½-pt) capacity, meristems can be grown in large numbers on shelves in environmental chambers where light and temperature can be controlled. After the plantlets have grown large enough, they can be potted in ordinary potting soil and allowed to grow until they are large enough for use. The complete cycle from plant to meristem to plant permits plant production on a far more massive scale and in a far shorter time than is possible by conventional breeding techniques.

Cloning is particularly suited to securing daughter plants from valuable trees from which it is difficult to secure seed. Through meristem cultures and other cloning techniques, thousands of plants can be secured from a single "mother" tree.

Oftentimes valuable plant stocks used for propagation become infected with viruses that make them stunted, weak, or otherwise unsuited for commercial use. In former years these plants could not be used as parent material because new plants derived from such "mother" plants also would be infected with virus. It is now known that cells of the growing tips (meristems) of many virus-infected plants are free of virus. Culture of meristems from some virus-infected plants will result in new plantlets that are free of virus. In turn, meristems from these virus-free plantlets can be grown.

By using this method, growers have produced virus-free asparagus, banana, Irish potato, strawberry, sugarcane, and sweet potato. Several other crops, including chrysanthemum, clover, dahlia, lily, tobacco, and tulip, also have been freed of virus by employing meristem culture. These virus-free meristems can be stored until needed in test-tube cultures or at freezing temperatures to protect them from reinfection.

Another advantage of this method of propagation is that cultures of virus-free tissue can be sent to other countries and used immediately without having to be held in quarantine for long periods to establish that they are disease-free. Tissue culture provides an ideal method of storing and rapidly multiplying virus-free material.

Tissue Culture and Germplasm Preservation

The continuing search for ways to improve current cultivars and commercial crops makes preservation and maintenance of large collections of germplasm necessary. In this way, all characteristics of possible value can be saved for future investigation and study. Seed storage is a widely used and efficient storage method, but is not always feasible for the following reasons:

- Some plants do not produce seeds.
- Some seeds remain alive for a few months only.
- Some seeds are genetically mixed and not suitable for maintaining true-to-type breeding stocks.
- Seeds of certain species infected with seedborne pathogens die after a few months in storage.

Meristem and tissue cultures provide the scientist with additional ways to preserve germplasm. Cells, tissues, or small plantlets can be maintained at low temperatures or on starvation media. However, these techniques are not suitable for long-term storage of several years or more. As a result, storage at freezing temperatures has been investigated. Buds, embryos, meristems, plant cells, pollens, protoplasts, shoot tips, and whole seeds have been shown to remain alive when held at −20°C in an ordinary home freezer, although progressive changes in quality occur. This deterioration can be prevented if a suitable protectant, such as dimethyl sulfoxide or glycerol, is added, and storage is at a very low temperature, preferably that of liquid nitrogen (−190°C), which stops living processes. This storage method is receiving increasing attention as a useful tool in germplasm preservation.

REGENERATION OF PLANTS FROM CELLS

Using Protoplasts from One Species

Normal plants of some species can be regenerated from single cultured cells. Animal cells have no rigid wall and can be grown directly without prior modification. Plant cells, on the other hand, contain rigid cellulose walls cemented together by *pectin* and hemicellulose material. Before individual plant cells can be made to grow and divide, the cell wall must be removed, usually by enzyme action. These naked cells, called *protoplasts*, may be isolated from flowers, leaves, roots, or stems from elite individual plants. Leaves are often used because they are the easiest tissues to convert into mass protoplast populations. By growth of protoplasts in aseptic baths of hormones, minerals, sugars, and vitamins, individual cells can be stimulated into growing into whole plants in the laboratory under controlled conditions.

Protoplasts have several characteristics that make them suitable for study:

- Large, relatively homogeneous populations of protoplasts, of known genetic composition, can be easily obtained from leaf mesophyll cells or other tissues from single plants.

- Protoplasts can be treated experimentally and cultured in the laboratory with high efficiency.
- The absence of a cell wall permits protoplasts to absorb genetic materials from the solution in which they are suspended.
- Protoplasts in solution can fuse with each other.

Researchers have noted that some tissue cultures derived from single protoplasts taken from virus-infected leaves are free of virus. This finding may indicate that, in addition to meristem cells, some viruses are unable to invade every cell in an infected leaf. Thus, protoplast culture may be used as a method to obtain virus-free tissue from desirable plants.

Conversely, scientists can inoculate cultures of protoplasts with minute amounts of a plant pathogenic virus to study virus replication and physiology. In addition, antiviral compounds, bactericides, and fungicides can be evaluated by testing their effect on protoplast cultures infected with specific pathogens.

Protoplasts from several cereals and grasses have been induced to produce mature plants. Theoretically, all the protoplasts from the leaf of a plant should be genetically identical, but sometimes changes occur, or can be induced to occur. Tissues or plants derived from different protoplasts vary in their ability to resist or grow in the presence of pathogens, toxins, or herbicides or to absorb certain nutrients. These mutant plants with desirable or unusual traits can be identified and used.

Selected cells resistant to the toxin produced by race T of *Bipolaris maydis*, the fungus that causes southern corn leaf blight (SCLB), have been regenerated into mature plants. These plants are resistant to SCLB and can be used to produce seed resistant to the disease. In addition, plants of Irish potato grown from individual protoplasts proved to be resistant to late blight caused by *Phytophthora infestans;* others matured earlier, and others produced more uniform potatoes. These results indicate that protoplast culture enables scientists to direct plant development selectively to desired ends, and saves space and time.

Similarly, protoplasts from other plant tissues can be used for regeneration. Cells from fertilized eggs, embryos, endosperm, epithelial tissues, or root cells can be manipulated and subjected to various stress conditions. As a result, desirable plants that require less water to grow or can withstand temperature extremes or high salinity can be selected.

Regeneration of Haploid Cells

The techniques described above all pertain to *diploid* cells—those with two sets of chromosomes, one set derived from the male parent and the other from the female parent. *Haploid* sex cells, those with only one set of chromosomes, can also be used for regenerating new plants. For example, cells from immature anthers or pollen cells containing one set of chromosomes can be isolated and regenerated. Also, cells from unfertilized eggs (ovules) can be grown in culture to produce haploid callus tissue and, eventually, haploid plants. This task is often a difficult one, but the addition of growth hormones (auxins, cytokinins, and gibberellins) at the correct time in correct amounts has made the production of haploid plants possible. These plants then can be treated at the appropriate time with *colchicine* (a chemical that induces the cells to double their number of chromosomes) to produce diploid plants that are homozygous. In this genetic state, the plants can be used as new cultivars or as parental material to crossbreed with others to produce new cultivars with desirable characteristics. A new cultivar of barley was produced in 5 years by this method of doubling the chromosome number of haploid plants produced in tissue culture, and diploid tobacco plants derived from haploid plants obtained from anther tissue are used routinely as breeding material.

Fusing Protoplasts from Different Species

Often it is difficult to crossbreed distantly related plant species because natural barriers and sterility problems prevent sexual fertilization, embryo development, and/or seed formation. The fusion of asexual or vegetative cells to form hybrids and the subsequent regeneration of these hybrids into plants is an alternative method for the production of interspecific crosses in plants. Much information is available on cell fusion and/or somatic hybridization of plant cells.

Essentially, five steps are involved in this process:

1. Isolation of living protoplasts capable of division from the parent plant tissues.
2. Fusion of the two parental protoplasts.
3. Fusion and integration of the two parental nuclei with subsequent mitosis and formation of hybrid cells.
4. Detection and isolation of the hybrid cells.
5. Regeneration of tissues and then plants from hybrid cells.

Thus, it is possible to take a solution of protoplasts of, let us say, soybeans and mix them with protoplasts of sunflower. The mixture of protoplasts then is incubated in a nutrient broth under suitable conditions of light, temperature, and air, and in a few days' time some (a very low percentage) of the protoplasts will stick together or fuse. Even the nuclei fuse. The chromosomes from each "parent" pair, and subsequently normal mitotic divisions occur. The cell number increases with the chromosomes from both the soybean and the sunflower contributing to the characteristics of the regenerated tissue which begins to form. After being allowed to undergo a few divisions, the cell mixture is diluted. Microdroplets are placed in small glass dishes with numerous small wells. Those wells containing single hybrid clusters can be scanned with a microscope, identified, and allowed to grow. By fusing protoplasts of tomato and potato, scientists have created a hybrid called a "pomato."

Unlike sexual crosses, the cytoplasms of the parental types become integrated during protoplast fusion. Because of this unique feature of somatic hybridization, certain heritable factors associated with the cytoplasm can be transferred. The transfer of male sterility, a desirable feature of economic importance in plant breeding, from one species of tobacco to another by protoplast fusion has been accomplished.

Successful protoplast fusion has been performed by using barley and soybean, corn and soybean, potato and tomato, tobacco and soybean, and sorghum and corn protoplast mixtures, and the list of successes keeps growing. It is hoped that scientists will soon find ways to regenerate plants from these protoplast "hybrids." Somatic hybridization by fusion of plant protoplasts promises to be of great use in transferring genes governing resistance to viral diseases from one parent to otherwise susceptible species. Indeed this has already been demonstrated: somatic hybrid plants produced from the fusion of protoplasts of susceptible *Nicotiana tabacum* (tobacco) and resistant *N. rustica* (a relative of tobacco) were resistant to tobacco mosaic virus.

Transplanting Cell Organelles into Protoplasts

Although inducing entire protoplasts from different plant species to fuse is possible, it is still a rare event and difficult to accomplish. Isolation of a part of the protoplast of one cell and its transfer to another is much easier, and the physiological shock to the protoplast receiving the material thus is greatly reduced. Furthermore, even when protoplast fusion occurs, hybrid production may be precluded because of chromosome imbalance, loss of unpaired chromosomes, or lack of nuclear fusion.

To avoid some of the difficulties involved with cell fusion, scientists are conducting experiments with nuclei isolated from one kind of plant and injected or transplanted into the protoplasts of another kind. Here the nuclei begin to function and induce changes in the cells, which reproduce to form tissues and subsequently mature plants.

Scientists have already successfully transplanted chloroplasts and mitochondria from one plant species into another. There are also reports that mitochondria from one kind of fungus have recombined with mitochondria from another fungus and produced a new type of "hybrid" fungus. Experiments like these offer hope that new and useful types of plants can be produced by transplanting organelles.

TRANSPLANTING DNA FROM ONE CELL TO ANOTHER

A bioengineering technique with important implications for agriculture involves the use of *recombinant DNA*—the hybrid DNA produced by joining pieces of DNA from different sources (parents). Scientists are improving this technique with the long-range goals of building plants that will be immune or resistant to disease, can extract nitrogen from the air, use sunlight more efficiently in photosynthesis, and have "foreign" genes that will improve crop yields and produce more nutritious proteins. Already scientists are constructing specific bacterial mutations at will and routinely producing millions of mutant cells for specialized uses and study. For example, it is now possible to identify and purify some genes, tailor them for insertion into a new bacterial cell, identify the cell, and grow the new cells in quantity to obtain a supply of DNA for future study and use.

Gene Splicing

Recombinant DNA technology (*gene splicing*) is sophisticated and elegant. It is based on the knowledge that an individual chromosome is a double helix (spiral) composed of thousands of *nucleotides* (the fundamental chemical units or building blocks of DNA) attached in a definite sequence (Fig. 8.2). Scientists are using two methods to splice genes into chromosomes. In the gene "tweezers" method, a solution of chromosomes from a desirable host is treated with a specific *restriction enzyme*, which has the

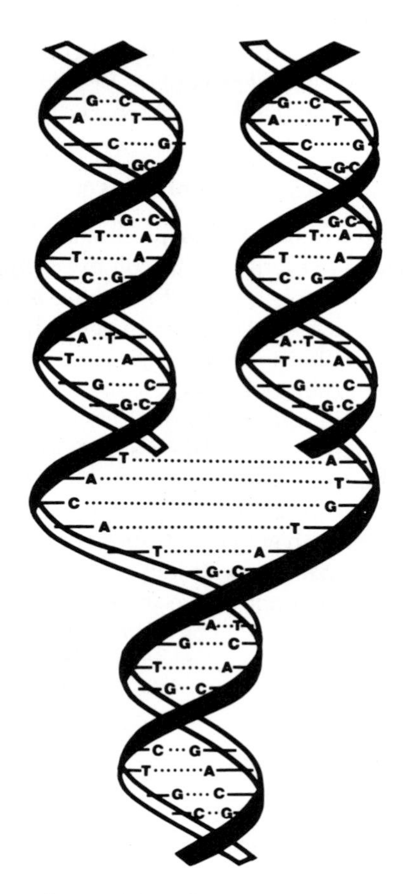

FIGURE 8.2. Deoxyribonucleic acid (DNA) consists of two intertwined strands resembling a twisted ladder. The four kinds of bases linking the stands always pair off in the came combinations: adenine-thymine (A-T) and cytosine-guanine (C-G). When a strand unzips, two new double strands exactly like the original are constructed.

ability to cut double-stranded DNA at a particular point (nucleotide sequence) where the desired DNA is attached. The snipped-out bit of DNA with chemically "sticky" ends is then treated with a *recombinant enzyme,* which helps stick the small piece of DNA or gene to a new chromosome. In other words, a gene from one chromosome is extracted and then inserted into the DNA of another species; a mutation has been engineered. In the "shot-gun" approach, several restriction enzymes are used to sever a large genome of one species into usable segments that can be easily spliced into the chromosome of another species.

Where would the selected genes for gene splicing come from? They might come from a closely related plant, another plant species, or a bacterium—or perhaps even from an animal. Theoretically, any gene that can replicate and function after being spliced into a recipient chromosome can be used, for example, to confer disease resistance. Animal genes might be put into plants to produce crops with meatlike proteins.

Scientists also hope to use recombinant DNA technology to improve the ability of soil bacteria to "fix" nitrogen from the air. Soybeans and other legumes get their nitrogen from bacteria that inhabit nodules on their roots. The infection of soybean plants with super nitrogen-fixing strains could increase crop yields. Another future goal is to build nitrogen-fixing bacteria that will grow on the roots of corn and other crops that are unable to fix nitrogen. This remarkable accomplishment would reduce growers' reliance on expensive commercial sources of nitrogen.

There is evidence, too, that the root cells of plants can be chemically altered so that they are more attractive to certain fungi (mycorrhizae) that grow on or near the roots. In some way these fungi assist plants to increase the uptake of phosphorous from the soil. Scientists would like to develop crop cultivars that encourage beneficial fungi or bacteria to thrive in the vicinity of the plant roots and enable the plants to take up more phosphorous or nitrogen from the soil. Less commercial fertilizer would then be required to maintain or increase crop yields, and crop production costs would be decreased.

Additionally, some of the beneficial fungi that grow on, in, or near roots of certain crop plants make the roots less attractive to root-feeding nematodes that injure or deplete the root system with subsequent decrease in yields. Perhaps superior strains of these beneficial fungi can be identified, isolated, and increased and their spores used to inoculate soils of problem fields. Exudates from the roots of crop plants would stimulate the fungus spores to germinate and produce a web of protective mycelium around and in the root system. Less injury from harmful nematodes would result, and yields would be increased without the cost of additional commercial fertilizer.

Gene Cloning

Another exciting prospect in genetic engineering is the possibility of inserting a useful gene into the cells of another host—say, bacteria, yeasts, or plant protoplasts. Then the transformed cells could be grown in large quantities, the cells harvested, and the useful product extracted. For example, gene coding for antibody production in animals could be implanted into DNA of a specific plant protoplast and cause the plant to synthesize the animal antibodies. The same technique might be used to manufacture other useful products such as antibiotics, pigments, enzymes, and aromatic essences.

The possibilities of forming new kinds of plants and animals are endless. Researchers now are acquiring the knowledge needed to direct or control the evolution of living things, including human beings. However, before genesplicing can be accomplished routinely, scientists must learn much more about gene structure and what genes do. For one thing, very few genetic traits are controlled by one gene. Scientists believe that 15 genes together control the ability of soil bacteria to absorb nitrogen from the air and transform it to a usable form in the roots of legumes. Probably more than 50 genes contribute to the process that makes corn protein different from soybean or tobacco protein. Despite these problems, some scientists expect one day to transfer traits that cannot be transferred by crossbreeding, through the use of recombinant DNA.

Bacterial Vectors of DNA

The pathogenic bacterium *Agrobacterium tumefaciens,* which causes crown-gall disease, attacks many species of plants. It is closely related to the nitrogen-fixing bacteria that grow in nodules on the roots of legume plants. In causing disease, *A. tumefaciens* enters susceptible plants through tiny wounds or injuries. DNA carried in the tumor-inducing (Ti) *plasmid* of the bacteria is transferred to the plant cells by means of the juices bathing the wounded cells, transforming them into tumor cells. These affected cells are greatly stimulated to produce not only the many substances required for host cell division and enlargement, but also compounds that enable the bacteria to reproduce. The host cells begin to grow and divide rapidly, forming soft galls or tumors on any wounded shoot or root; the galls may become many times larger than the root or shoot that bears them. The disease progresses slowly, stunting the plant but rarely killing it.

An important aspect of crown-gall disease is that plasmid DNA of the bacterial pathogen is transferred to the plant cell and actually is incorporated into the host's chromosomal DNA. This bacterial DNA then multiplies and becomes part of tumor tissue, free of bacteria.

There are many strains of the crown-gall bacterium, some of which are only weakly pathogenic. They can invade wounded cells and reproduce, but form only a few small galls or none at all. Apparently they do little harm to the infected plant. Realizing this, scientists have ingeniously used *A. tumefaciens* as a possible vehicle to insert foreign DNA (genes) into plants. Cultures of *A. tumefaciens* are treated with a restriction enzyme that "unlocks" or opens the ring of DNA (Fig. 8.3) in the chromosomal material contained in the plasmid of the bacterial cell. Solutions containing new, desirable DNA, for example, the DNA that controls *phaseolin* (a protein) production in beans, then are added to the bacterial cultures; bits of the new DNA are absorbed into the bacterial chromosomes, where it is replicated. Cultures of the harmless bacteria containing the new desirable DNA (gene) then are inoculated into experimental sun-

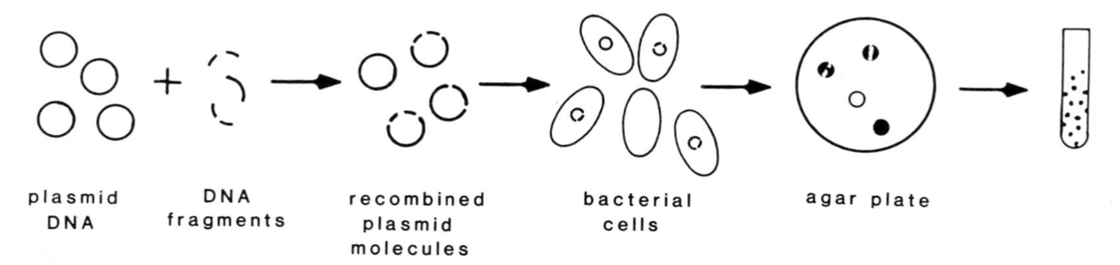

FIGURE 8.3. Transforming bacteria by inserting foreign DNA into a plasmid DNA molecule. Plasmid DNA is treated with a restriction enzyme that opens the plasmid ring. Solutions of desirable DNA fragments (e.g., the DNA, obtained from a bean plant that controls production of a useful protein) are added to the bacterial cultures. A few bacterial cells absorb the donor DNA into their plasmids. These cells are grown on selective media and yield a colony of identical cells or clones. Once the "new" colony has been found, it can be grown in quantity to produce the useful protein. *Courtesy* R. Wetzel, American Scientist 68, *664-675 (1980).*

flower plants. As the bacteria begin to grow, the desirable DNA is transferred to the sunflower cytoplasm, where it becomes part of the normal functioning process of the plant cells, imparting increased protein production to the sunflower.

Scientists also can isolate protoplasts from small tumors of sunflower tissue, regenerate them, and test the new plantlets for increased protein production. Active research is under way using plasmid DNA of nonpathogenic strains of the crown-gall bacterium to insert factors controlling resistance to diseases into commercial crops.

Inserting DNA into Plant Cells by Microprojectile Bombardment

Biotechnology is changing rapidly; almost every month something new is announced. Since 1988, biotechnologists have developed a remarkable method of inserting genes (DNA) into plant cells. The donor DNA is isolated from cell nuclei or from altered plasmid chromosomes of the bacterium *Agrobacterium tumefaciens,* which contain the useful DNA. Microscopic metal particles (gold dust often is used) are coated with DNA, and then are shot at high velocity, by using a bullet or compressed air, into plant cells. The DNA remains biologically active and can be integrated into the chromosomes of the recipient cells to produce stable genetic changes in the plant. This valuable technique allows transformation of plant species such as corn and wheat, which cannot be transformed efficiently by *Agrobacterium.* An important consideration is that DNA can be delivered into intact cells without the removal of the cell walls, a feature that facilitates regeneration of plants from transformed cells.

Viral Vectors of DNA

It is safe to say that all, or almost all, organisms are attacked by viruses (Chapter 17). A plant virus is an infectious agent composed of ribonucleic acid (RNA) wrapped in a protein coat. Viruses can replicate only in living cells. In nature, virus particles are commonly injected into healthy plant cells through tiny wounds made by insects that pick up the virus while feeding on an infected plant. Once inside the cell, the protein coat of the virus dissolves, and the virus particle begins to make copies of itself. Many strains of viruses exist. Some cause severe symptoms on the host plant, others only mild symptoms.

Scientists are exploring the possibility of inserting beneficial, disease-resistance genes into the RNA of mild or harmless strains of plant viruses. These genetically altered viruses then can be injected into plants by an insect vector or with microneedles. Once inside the cell, the virus particle carrying the beneficial gene or group of genes would begin to multiply and make the plant disease-resistant.

CITIZEN CONCERNS AND THE NEED FOR BIOTECHNOLOGY

Many citizens are concerned about the environmental and social dangers involved in breeding and releasing into the environment transgenic plants and genetically engineered microorganisms. These people are afraid that if such organisms escape, they will reproduce unchecked and upset the natural ecosystem or otherwise pollute the environment, causing harm to animals (including humans) and plants.

The scientists who are involved with biotechnology and the production of transgenic organisms also are concerned about the dangers of "creating" new organisms. They always ask (1) how much damage would be done if the new transgenic plant or animal reproduced unchecked, and (2) what the chances are that the new organisms will outcross or breed with other normal species in the environment, posing a threat to the community. If the probabilities are high that the newly created, transgenic organisms will be unduly harmful to humans or the environment, the project usually is abandoned.

In the United States, safeguards are in place against the release of inadequately tested "new" organisms. Scientists must obtain permits from the USDA, Animal and Plant Health Inspection Service, before transgenic organisms can be released or placed in the environment.

Agricultural and medical firms are becoming increasingly interested in the potential products of biotechnology. Pesticide manufacturers realize that many agricultural chemicals in use today may be banned in the future because they contaminate the environment. Also, some pesticides lose their efficacy as new strains of pathogens arise that are resistant to them; for example, some downy mildew fungi have become resistant to the fungicide metalaxyl. In addition, the expenses involved in developing, testing, and releasing a new pesticide are almost prohibitive. Millions of dollars can be required to develop just one new pesticide.

There is another reason why agricultural firms are interested in biotechnology and genetic engineering: transgenic plants can be patented. For example, if a company develops a new corn or soybean cultivar resistant to herbicides, the cultivar can be patented, and only the firm that owns the patent may legally sell the seed. Profits from the sales of such seed or plants could be very high.

9

Nematodes

Ever since an early nomad unknowingly sowed the beginnings of agriculture, a potential enemy has lurked in every acre of farmland and garden soil. It is called the nematode.

There are thousands of kinds of nematodes. Probably every living thing—animal or plant—bigger than a nematode is attacked or parasitized by them. The elephant, the gnat, the whale, the minnow, humans, cattle, crop plants, birds, insects, other nematodes—all carry one to many kinds of these animals within their bodies.

Nematodes that feed on animals, however, will not or cannot feed on plants; likewise nematodes that feed on plants do not attack animals. This is a general rule in biology: Plant parasites do not attack animals; animal parasites will not attack plants.

GENERAL NATURE AND IMPORTANCE

Plant pathogenic species of nematodes are of tremendous importance as disease agents. All plants are parasitized by nematodes, but only a few hundred species of nematodes are known to be plant parasites. Full recognition of the amount of crop damage resulting from nematode diseases and the development of effective management measures have occurred only since the 1940s.

There are many reasons why scientists overlooked the importance of nematodes as a cause of reduced crop yields. Usually, farmers judge the health of plants on the basis of aboveground symptoms; rarely do they examine the roots of unthrifty plants. Moreover, the stunting, yellowing, wilting, and poor growth associated with nematode infections do not produce the same concern caused by the outright death of plants. Most plant parasitic nematodes cause a slow decline in yield from root damage rather than the obvious damage due to low fertility, acid soils, drought, or frost. Often the term "tired land" and "sick soils" are used to describe fields on which plants make poor growth despite proper fertilization, weed control, and good cultivation. The land is not "tired"; it often is infested with plant-parasitic nematodes. Some plant-parasitic nematodes attack seeds (e.g., wheat-gall nematodes), foliage (e.g., foliar nematodes), and even the water-conducting tissues above ground (e.g., pine wood nematodes).

Other reasons these remarkable creatures escape notice are that they are microscopic in size or barely visible to the naked eye, and almost transparent. They spend their lives hidden in the tissues of host plants or in the soil near the roots. Another unusual thing about plant-parasitic nematodes is that they neither feed nor reproduce except on living plants. To survive, each individual nematode, after hatching, must reach a plant on which it can feed before its reserve food supply is exhausted. That means the nematode must swim through the soil in a film of water in search of food.

Geography and climate also played a role in the lack of early recognition of nematode occurrence and importance. Most problems due to plant-parasitic nema-

TABLE 9.1. Estimated annual losses caused by nematodes for selected world crops.

Crop	Estimated production, 1984 (1000 metric tons)*	Estimated yield losses due to nematodes† (%)	Estimated monetary loss due to nematodes (US$)
Banana	2,097	19.7	178,049,979
Barley	171,635	6.3	1,102,926,979
Cassava	129,020	8.4	975,391,200
Citrus	56,100	14.2	4,022,931,000
Cocoa	1,660	10.5	450,391,200
Coffee	5,210	15.0	2,481,261,500
Corn	449,255	10.2	6,736,129,470
Cotton (lint only)	17,794	10.7	4,112,549,280
Field bean	19,508	10.9	1,156,746,300
Oat	43,355	4.2	180,270,090
Peanut	20,611	12.0	1,028,901,120
Potato	312,209	12.2	5,789,403,696
Rice	469,959	10.0	16,072,597,800
Sorghum	71,698	6.9	588,712,270
Soybean	89,893	10.6	2,687,081,500
Sugar beet	293,478	10.9	1,183,596,774
Sugarcane	935,769	15.3	16,464,854,000
Sweet potato	117,337	10.2	2,621,073,906
Tea	2,218	8.2	510,562,300
Tobacco	6,205	14.7	2,732,756,460
Wheat	521,682	7.0	5,806,320,660
		Total	$77,698,508,015

*Figures from 1984 *Food and Agriculture Organization Production Yearbook*.

†Based on estimates from the 1985 Crop Nematode Research and Control Project worldwide nematode survey of nematologists (371 responses).

todes occur in the southeastern United States and California and in countries in tropical and subtropical regions. A large portion of the early work in plant pathology in the United States occurred in such states as New York, Wisconsin, and Minnesota (up to the 1920s and 1930s), where nematodes do not cause significant damage. Thus, diseases caused by fungi, bacteria, and viruses were studied more extensively than nematode diseases by early researchers.

Accurate information on the losses caused by nematodes is difficult to obtain. However, in 1985 a worldwide survey of scientists from 75 countries was conducted by plant nematologists cooperating in the International *Meloidogyne* Project (sponsored by the U.S. Agency for International Development). It was found that yield losses due to plant-parasitic nematodes (principally the root-knot nematode) in important food crops (and several others) exceeded a staggering $75 billion annually (Table 9.1). Nearly 11% of the production of the 21 most important crops is lost worldwide to nematodes each year.

As the world population now approaches 6 billion (and is expected to be 10 billion by 2050) famine is still a very real threat to most of the human populace. Nematodes constitute a significant part of that threat, especially in developing countries, where they are a more serious and complex problem than in the developed countries. Warm, sandy soils are favorable to nematode infection, especially in irrigated areas used for continuous crop production. Perennial crops, and annual crops grown in the same fields year after year, often are so seriously attacked by nematodes that such crops can barely survive, yet the people in the developing countries must depend on perennial crops and nonrotated crops for their food supply.

SIZE, DISTRIBUTION, AND ANATOMY

A nematode is a tiny threadlike animal resembling but not closely related to an earthworm. The nematodes as a group are second only to the insects in their total number. The name nematode comes from a Greek word *nematos,* meaning thread. Most nematodes range from 0.5 mm (1/50 in.) to 6 mm (1/4 in.) in length, but some that live in animals grow much larger. Approximately 80 adult lesion nematodes could lie side by side inside this hyphen: -. About 15,000 lesion nematodes could lie on a thumbnail without overlapping (see Fig. 2.1).

Nematodes are found everywhere and in almost everything: the top of the highest mountain, the bottom of the lowest valley, the most arid desert, and the most fertile farmland. They are found in the icy saltiness of Arctic seas and in the hot springs of Yellowstone Park. Many different kinds live in the soil, but most of these are harmless and some are distinctly beneficial. A single hectare of cultivated soil may contain hundreds of millions; as many as 100,000 nematodes have been recovered from a liter of garden soil.

The basic anatomy of a nematode is diagrammed in Fig. 9.1 and summarized below:

A nematode has	*A nematode does not have*
Digestive organs and glands	Eyes, ears, or a nose
Reproductive structures (male and female)	Blood vessels or blood
	Hair

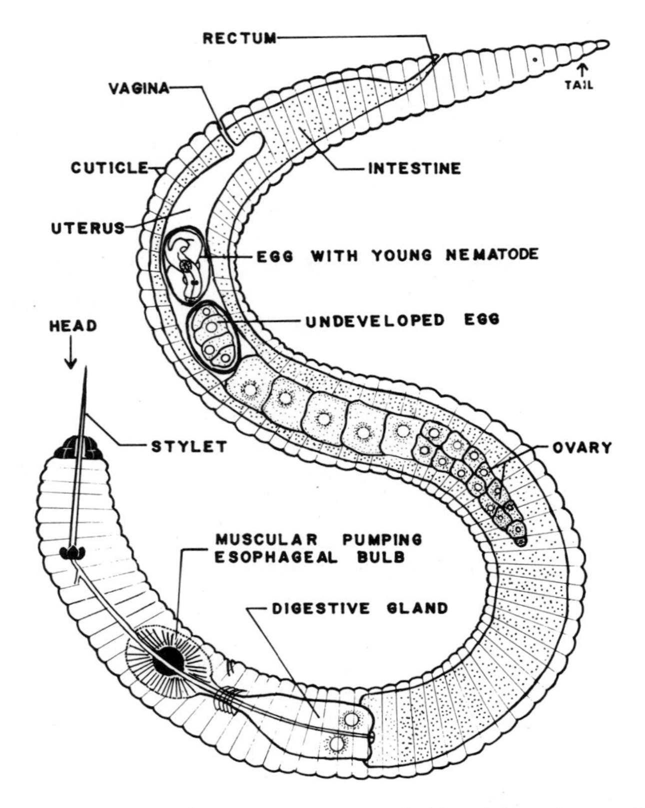

FIGURE 9.1. Anatomy of a plant-parasitic nematode (female). *Courtesy R. P. Esser and the Division of Plant Industry, Florida Department of Agriculture and Consumer Services.*

Excretory structures
Muscles
Nerves
Tough cuticle (nematode skin)

A skeleton
A liver, gall bladder, or lungs
Appendages (arms or legs)

A nematode moves by levering its body up and down, swimming in the water film on the surface of soil particles or surface water like an eel. A nematode moving with its side against the ground would closely resemble a snake gliding across sand.

FEEDING BEHAVIOR AND PLANT INJURY

Nematodes that feed on plants have mouths equipped with a *stylet,* which looks like a miniature hypodermic needle (Fig. 9.2A,B). A nematode feeds on a plant by puncturing the cell wall with this stylet and sucking out the contents of the cell. Suction is induced by contractions of a muscular bulb in the nematode's esophagus. Not only do nematodes suck out juices from plant cells; they also inject liquids into the cells and in this way may introduce viruses into plants. In addition, the tiny wounds or punctures in the tissues provide entry for bacteria or fungi.

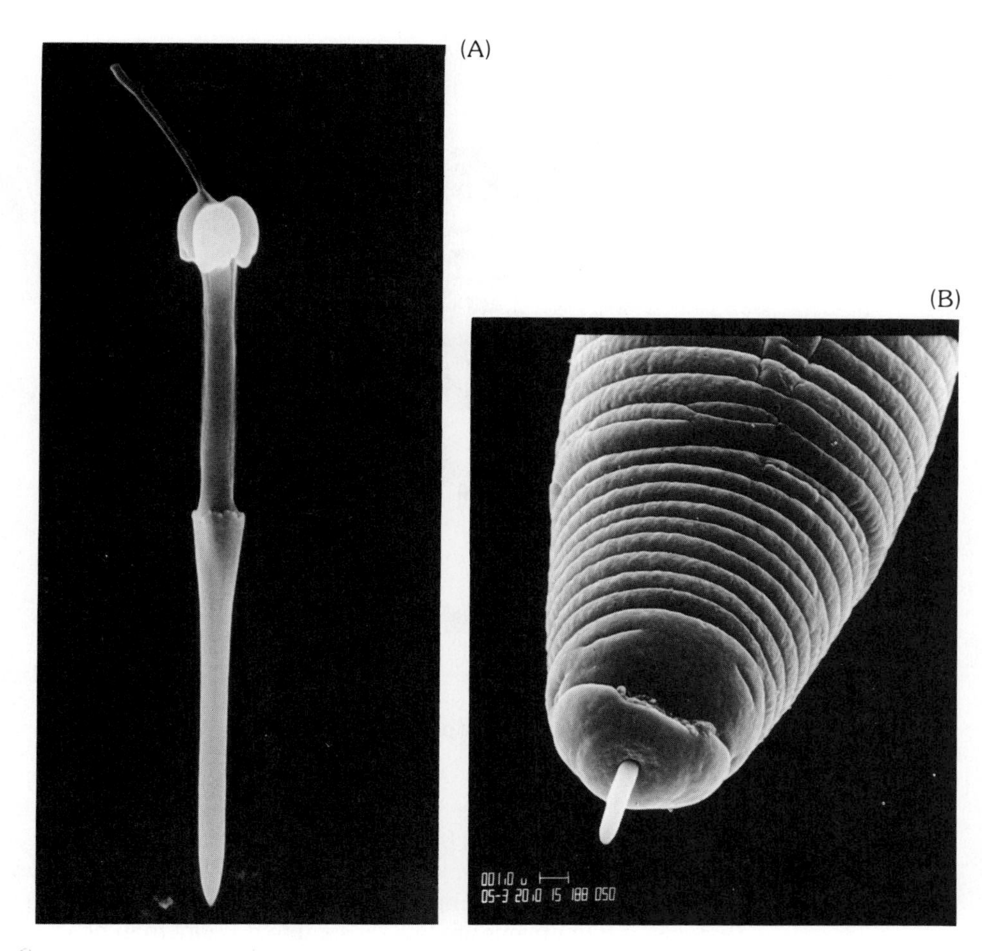

FIGURE 9.2. Stylet and head of a plant-parasitic nematode. (A) Scanning electron micrograph of a nematode stylet, magnified about 5220 times. *Courtesy J. Eisenback.* (B) Scanning electron micrograph of the head of a nematode with protruding stylet, magnified about 7840 times. *Courtesy J. Eisenback and H. H. Hirschmann.*

The injection of the stylet into the cell of a plant may have drastic effects. Sometimes the cells are killed quickly. In other instances (e.g., root-knot nematodes), the cells remain alive and are stimulated to extra growth by the hormones and enzymes contained in the liquids injected by the nematodes. The adjoining cells also increase in size and contribute to the supply of liquids continually sucked out by the feeding nematodes.

Some nematode species (e.g., the citrus nematode) attack only a very few kinds of plants, whereas others (e.g., the root-knot nematode) feed on many different kinds.

The various kinds of nematodes cause damage that interferes with the growth of plants. The weakened roots cannot efficiently absorb water and mineral nutrients from the soil. Reduction in the size of the root system by rotting or galling restricts its

FIGURE 9.3. Peanut plants attacked by sting nematodes are stunted and generally lack vigor. The severity of attack varies within rows. Plants not yet infected by the nematodes (on the right) are larger than the others and fill the rows.

efficiency in obtaining the nutrients and water the plant must get from the soil. Plants whose roots are attacked by nematodes lose vigor, become unhealthy (Fig. 9.3), and may show symptoms of nutrient deficiency such as yellowing, stunting, or burning of the foliage at tips and along the edges. In addition, wilting occurs during hot parts of the day, and the plants are less resistant to drought than healthy plants. Nematodes that attack seed or foliage result in poor seed quality or unsightly and diseased leaves.

On the basis of their feeding habits, nematodes are divided into two main groups:

1. *Endoparasites* enter plant tissues and feed from inside. Some move within the tissues and force their way between cells, injuring the tissue as they feed. Others remain in one spot, feed on only a few cells and become completely encased by host tissue.
2. *Ectoparasites* do not enter the plant. They feed from the outside and rarely penetrate the host tissues.

LIFE HISTORY, REPRODUCTION, AND DISPERSAL

Nematode eggs are found in great numbers in the soil. When the growing season begins, soil water carries substances exuded from plant roots to the dormant eggs, causing them to hatch. From each egg a larva hatches that is capable of wriggling short distances in soil water. Newly hatched larvae, attracted to the plant roots, swim to the roots, enter, and proceed to feed and grow to be adults. This process usually takes place in less than a month if the soil is warm and moist.

The new adults lay more eggs. Several generations of nematodes may appear during the growing season, constantly producing new nematodes to feed on the roots and drain away energy from the plant.

In some species of nematodes males are seldom seen. They are not necessary for reproduction, and the eggs laid by females usually are not fertilized by males. The process by which reproduction occurs without males is called *parthenogenesis*.

Each female nematode lays many eggs inside or outside the plant. The young nematodes, called larvae, develop inside these eggs, which have a tough shell that protects the young enclosed nematode from unfavorable conditions. The larvae come out of the eggs and develop into adults through a series of molts, during which the outer cuticle is shed. The eggs may lie dormant in fallow land, sometimes for several years. Then, when the land is planted to a crop suitable for nematodes, the eggs hatch, and new populations of nematodes build up in the soil. Each succeeding year, if conditions are favorable, many more eggs are present, spreading the nematode population in ever-widening circles.

Nematodes spread slowly by their own efforts. Most nematodes never move more than a few centimeters from the spot where they hatched. Humans have been the principal agent responsible for the spread of nematodes. Nematodes can be unseen contaminants on the parts of plants: corms, rhizomes, roots, tubers, seeds, flowers, leaves, and stems. Nematodes in the soil stick to tools, vehicles, and the feet of animals. Drainage and irrigation water also carry the eggs and larvae long distances to begin new centers of infection. In dry weather, strong winds may blow the eggs or nematodes from one field to another.

ENEMIES OF NEMATODES

All living things have enemies. The fact that a parasite feeds off a living host does not preclude the parasite in turn from being attacked by other parasites and predators.

Nowhere does "survival of the fittest" operate more cruelly than in the soil—a complex microcosm that consists of liquid, gas, and solid; of vegetable, animal, and mineral; of cycles within cycles; of surge and countersurge; of parasite on parasite; of poison juices, traps and nooses, predators, and foes. Plant-parasitic nematodes have many enemies in the soil. At any stage of their life cycle, they may be captured and devoured by some other soil organisms. Indeed, certain types of nematodes eat other nematodes. Some of these predatory free-living nematodes have grasping or cutting teeth or both; others have long hollow stylets that are stabbed into the victim and suck out the body sap quickly.

Certain fungi also prey on nematodes. Some of them have sticky surfaces to which the nematodes adhere; others have loops or rings that close when a nematode starts to crawl through (Fig. 9.4). In either instance, the fungus penetrates the nematode body and kills it.

Even though many plant-parasitic nematodes are parasitized by other soil inhabitants, entire populations seldom are exterminated. As with other parasites, the nematodes manage to reproduce just a little faster than they are exterminated. A single female

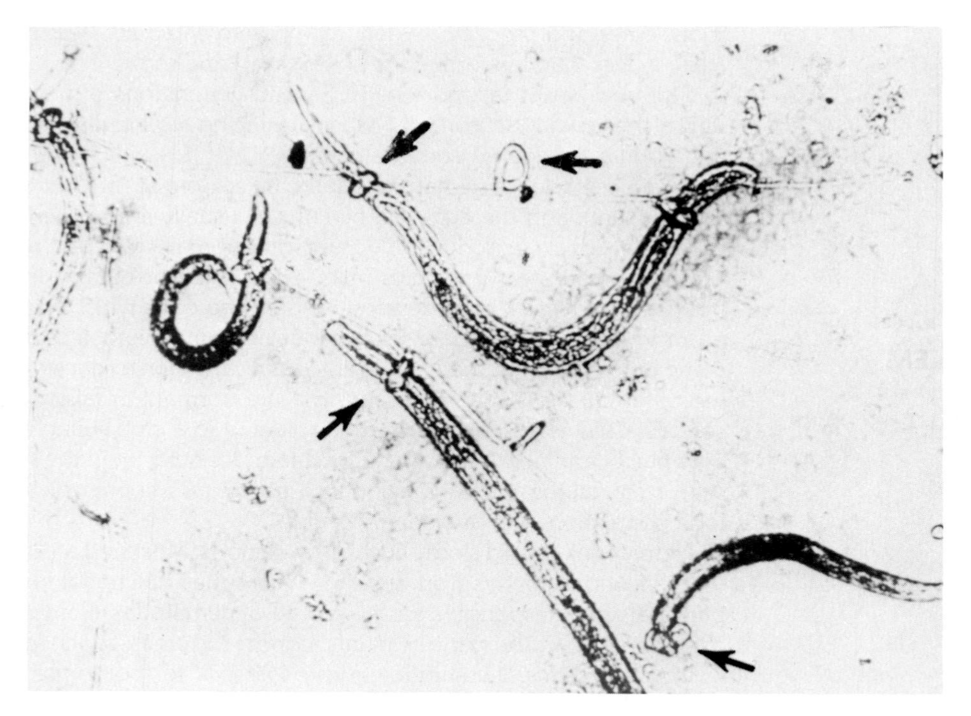

FIGURE 9.4. Nematode-trapping fungi. Some fungi that live in the soil form loops or rings (arrows) that close when an unsuspecting nematode tries to crawl through the noose.
Courtesy M. K. Beute.

root-knot nematode can lay more than 500 eggs. If only a few of them survive and reproduce in turn, an outstanding increase in population can take place during a warm growing season.

ENVIRONMENTAL EFFECTS ON NEMATODE SURVIVAL

Nematode populations do not remain static; they decrease when conditions are unfavorable and increase with favorable conditions. The eggs can survive for a year or more in soil. Nematodes have been known to stay alive in soil flooded for 22 months. Species found in cold climates can easily survive winter and are not killed by freezing of the soil. Some species are killed if subjected to drying, but others enter a dormant state from which they can revive in a short time if moistened. A nematode that parasitizes rye has been revived after dry storage of 39 years.

In warm, damp soils, nematodes are very active; as the temperature cools, they become less active. In warm climates, nematodes multiply at the rate of 5 to 10 generations a year and are also more vigorous in finding and attacking plants than in colder areas. Enormous populations result from this rapid buildup, which helps explain why nematode damage is so severe in warm regions.

The combined effects of sunlight and drying will kill the eggs of many nematodes in 30 min. However, the larvae are the weakest link in the nematode life cycle; they are more active in moist soils, under conditions that favor rapid plant growth, but usually cause more damage during dry weather. However, they die quickly when exposed to direct sunlight, excessive heat or cold, or lack of moisture. As long as the moisture content in the soil is favorable to the growth of plants, nematodes prosper and multiply.

NEMATODES AND OTHER DISEASES

Frequently, plants suffer from more than one disease at the same time, resulting in increased damage to, or death of, the plant. A plant may be able to resist the attack of one parasite; but when two or more attack the same plant, it may be overwhelmed. Nematodes are notorious partners in plant disease complexes. Plants attacked by both nematodes and other pathogens (bacteria, fungi, or viruses) often suffer severe injury, resulting in drastically reduced yields. Nematode-infested plants also may be quite susceptible to chemical injury from improper use of pesticides.

There are four major ways that nematodes combine with other pathogens to cause increased plant damage:

1. *Wounding agents:* The wounds and punctures made by nematodes provide entry points for other pathogens. Nematodes open the doors and make it easy for other pathogens to enter. Once fungi or bacteria get inside, they thrive in the tunnels made by the nematodes.
2. *Host modifiers:* When nematodes feed, not only do they suck fluids out of the host cells, but they also inject digestive juices into the cells. These substances, including both enzymes and hormones, change the cells, increase their nutrient content, and make them a more suitable food source not only for the nematodes but also for other pathogens.
3. *Rhizosphere modifiers:* Many fungi exist in the soil near living roots (the *rhizosphere*) as dormant spores that require a nutrient source before they can germinate and infect roots. Chemicals that leak out of roots are a principal source of nutrients. Nematodes affect this source by (1) making wounds that let increasing amounts of root liquids leak out and by (2) changing the composition of cell sap so that it is a more suitable food source for attacking fungi. The added food source not only allows the fungal spores to germinate, but it also supports growth of the fungi before they enter the root. The nematodes prepare the way for bacterial or fungal infections that would not otherwise occur.
4. *Vectors:* The nematode's stylet is a tiny, hollow spear that can pierce the host cell, suck out cell sap, and also inject salivary secretions into the cell. The stylet diameter of all plant-parasitic nematodes is less than 1 μm. However, the infective particles of plant viruses are smaller than 1 μm; thus, they can be carried in or on the stylet.

Bacteria and fungi sometimes occur in huge numbers in the rhizosphere. Spores of these microorganisms may stick to a nematode's body and be transported by the nematode as it swims through the soil or feeds on plant roots. Because most spores have no means of self-transport, being carried about by nematodes could increase their chance of contacting a host.

MANAGEMENT

Nematodes are so widespread that it is practically impossible to eliminate them from infested fields. We will never get rid of the nematodes completely. Therefore, the primary goal in developing management strategies against nematodes is to learn how to live with them. The aim of management methods must be to keep the total nematode population at a tolerable level and to prevent the buildup of any one kind to a level that could result in great damage. No single method can effectively achieve these goals. Moreover, efficient nematode management requires a continuing year-round program aimed at keeping the crop as healthy as possible from the time the seed is planted until the crop remains are destroyed after the harvest. A combination of practices must be used if crops are to be grown profitably. Moreover, nematode management methods must fit into the overall IPM system in use on any given farm. Finally, any management program must be profitable; that is, an increase in the monetary value of the crop must offset the cost of the management measures. (See Chapter 5.)

Crop Rotation

Crop rotation is the most important cultural practice used in managing nematodes. Nematode management by crop rotation is based on the fact that parasitic nematodes can live and reproduce only when they can feed on suitable crop plants. In the absence of such plants, nematodes will simply starve to death. Therefore, in a crop rotation, susceptible crops are alternated with immune or highly resistant ones. While these crops are growing, the nematode population decreases. At the end of one or more growing seasons, depending on the kind of nematode present, the nematode population will have decreased to a point where the susceptible crop can be grown again with little damage. After one growing season, a new cycle of resistant crops must again be started. Rotations should be designed so that crops in any given field vary as widely as possible. Crops should not be planted in rigid sequence; rather, the farmer should attempt to "rotate the rotation." Also, for some nematodes, such as cyst nematodes, rotation may be only minimally effective because of the longevity of viable cysts in soil.

Small-scale farmers who raise poultry, swine, or cattle can use the animals as part of a rotation for managing nematodes (Fig. 9.5). In any given year the animals are restricted to a single portion of a fenced-in plot. The other sections are planted with various crops that are part of a 3-year rotation—crops for 2 years, animals for 1 year. The animal traffic prevents plants from growing, thus depriving nematodes of a host, and the nematode populations decrease to a low level. The animal manure enriches the soil.

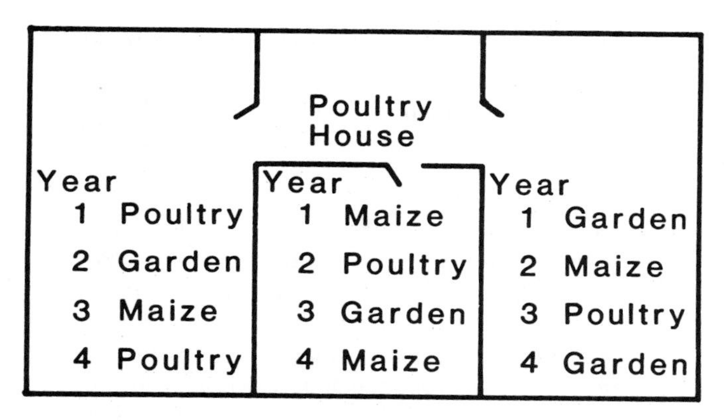

FIGURE 9.5. Crop-animal rotation for managing nematodes.

Chemical Management

Chemicals used for nematode control, called *nematicides,* have been in large-scale use since the 1940s, and their use has increased greatly. It is not possible to kill all the nematodes in the soil even with increasing doses of these chemicals. Therefore, recommended application rates are based on a cost/benefit ratio calculated to give the largest dollar increase in crop value for each dollar invested in the treatment. Usually, farmers expect to receive $3 to $4 for each dollar expended. Nematicides are expensive; so their use usually is justified only on high-value crops such as cotton, golf greens, greenhouse soil, lawns, nurseries, orchards, ornamentals, peanuts, seed beds, strawberries, sweet potatoes, tobacco, and vegetables. (Chemical management of nematodes is discussed in detail in Chapter 7, Soil Treatment section.)

Nematicides also may move into groundwater as pollutants. In an effort to reduce pollution, scientists are increasing their efforts to find additional biological control methods that kill nematodes but do not harm the environment. These "biocontrol" compounds encompass inhibitory chemicals in roots that repel nematodes, nematode parasites or predators, transgenic plants resistant to nematodes, nematode-infecting viruses, and other environmentally safe methods yet to be discovered.

Nematode-Free Transplants

For crops grown from transplants (annual flowers, cabbage, pepper, tobacco, tomato), the seed should be sown in seedbeds previously fumigated with a nematicide that also kills bacteria, fungi, insects, and weeds. Growers should be careful to avoid contaminating beds with infested soil while transplants are being grown.

Other Nematode Management Practices

Disinfecting Plant Materials

Nematode-infected plant material can be disinfected by dipping bare-rooted transplants, corms, or seedlings in solutions or emulsions or nematicides for a few minutes to an hour. The roots are soaked in the prepared solution for the required time, drained, and planted immediately.

Caution: Remember that most nematicides are very poisonous and dangerous. Be sure no liquid touches your skin or clothing. Do not breathe the fumes.

Plowing After Harvest

As soon as possible after a crop is harvested, the roots should be plowed up. Plowing exposes the roots and soil to the drying action of the wind and sun, and prevents further reproduction of nematodes that would occur if the plants were allowed to continue to grow. This destruction of the debris and remains of the preceding crop also helps growers to manage insect pests and other pathogens that would continue to grow on a crop left standing in the field.

After harvest, if labor is available and economical, the plants may be pulled by hand, placed in stacks, allowed to dry, and then burned. One also can do this in home gardens or other small areas where only a few plants must be destroyed.

Using Resistant Cultivars

Considerable progress has been made in breeding cultivars resistant to certain nematodes. To be of value, a cultivar must be resistant to the nematode species and race prevalent in the region concerned and, in addition, have acceptable agronomic qualities, yield well, produce food or fiber of good quality, and, in general, be more profitable for the farmer than cultivars previously grown.

Some nematode-resistant cultivars have been bred and are widely used. However, the job of breeding resistant cultivars never ceases because new races of nematodes that will attack formerly resistant cultivars are continually evolving.

Nematode Assays

Several states now have a nematode advisory service, usually a subdivision of the state department of agriculture. Residents may send in soil samples to these laboratories and have the soil assayed for the number and kinds of nematodes present. With the new requirements for pesticide record keeping included in the 1990 Farm Bill in the United States, one item to be noted is the specific pest or disease targeted by a pesticide application. The results of nematode assays should, therefore, be included in the history and records kept of each field. In some states, computer software programs are available to help growers make management decisions on a field-by-field basis. Nematode assays generally are just one of several services offered, including soil fertility and plant tissue analyses and a plant disease and insect clinic. Reliable services of this kind are not only a valuable asset to farmers, but also a necessity for efficient agriculture.

10

Diseases Caused by Nematodes

The representative nematode diseases discussed in this chapter are common and can be recognized fairly easily. It should be remembered that it is easy to mistake nematode diseases for those caused by some other organisms, and vice versa. Merely finding nematodes in diseased plant tissue or the soil is not conclusive evidence that they are the cause of the trouble. Nonparasitic types of nematodes often are found in great numbers in decaying plants, and the soil always contains a variety of free-living nematodes. One should always obtain positive identification before starting expensive or troublesome management measures. On the other hand, nematodes should always be considered as a possible cause of plant diseases when root systems are galled, shortened, or reduced by rotting; the stems are shortened and thickened, and the leaves do not grow normally; or some other abnormal growth is noted. Important genera of plant-parasitic nematodes are listed in Table 10.1.

TABLE 10.1. Important genera of plant-parasitic nematodes.

Genus	Common name	Genus	Common name
Meloidogyne	Root-knot	*Radopholus*	Burrowing
Globodera and *Heterodera*	Cyst	*Rotylenchulus*	Reniform
Pratylenchus	Lesion	*Helicotylenchus*	Spiral
Ditylenchus	Stem and bulb	*Paratrichodorus*	Stubby root
Tylenchulus	Citrus	*Belonalaimus*	Sting
Xiphinema	Dagger	*Criconemella*	Ring

ROOT-KNOT NEMATODES

Root knot is one of the most important plant diseases in the world. This widespread disease has changed human history by forcing the migration of peoples from one land to another, as root-knot nematodes have caused or contributed to famines. In many warm regions of the world, as the soil has become more and more infested and yields have continued to decline, farmers have had to move to other fields. When new fields become heavily infested, farmers must move again.

Throughout the world, root knot causes an average annual yield loss of about 8 to 11%. This loss would not be of great significance if it were evenly distributed, but it is not. The greatest losses occur to those who can least afford it, namely, the small farmers of underdeveloped countries. Their losses may be as much as 25 to 50% over wide areas of the available farmland.

In the southeastern United States, root knot is a primary threat to many crops. Root-knot nematodes must be managed if crops are to be grown profitably; crop losses from root knot are heavy. In combination with other pathogens, the disease can be disastrous.

Indirect losses associated with root knot are due to the following:

- Secondary attacks by other pathogens.
- Increased cost of weed control because of lack of competition by the dwarfed crops.
- Loss of usefulness of cultivars resistant to other diseases but susceptible to root knot.
- Inefficient utilization of fertilizer and water.
- Cost of chemical treatment.
- Loss of use of desirable fields because of the necessity for crop rotation.

Root-knot nematode management is a good example of the value of agricultural research. Since the 1940s, nematologists and plant pathologists have demonstrated the importance of root knot in crop production and have developed suitable management measures. Research must continue in order to find new and more efficient ways to manage these ever-present parasites.

The root-knot nematode parasitizes more than 3000 plant species. Members of all the main crop families are attacked. Germinating seeds and young seedlings are particularly susceptible to infection, including the seedlings of some plants that are otherwise resistant.

Symptoms

Plants suffering from root-knot nematodes are stunted and yellowed. Severely infected plants occasionally may be killed, especially during dry weather. Within the same field, some areas may be severely affected, whereas plants in other areas may be free of symptoms. Indeed, individually affected plants may be scattered throughout a field. It is unusual to find all plants in a given field affected. Plants with root knot may have symptoms of nutrient deficiencies or drought injury even in the presence of adequate fertilizer and water. During dry periods, affected plants develop severe symptoms. These plants become stunted, and they wilt during the hot part of the day. The plants recover at night and wilt again the next day. Older leaves yellow prematurely, and the tips and edges "burn" or turn brown.

The most distinctive symptoms caused by root-knot nematodes are the *galls* or *knots* on the roots (Figs. 10.1 and 10.2). The galls vary in size from a pin head to compound galls more than 2.5 cm in diameter. The galls are irregular, spherical, or spindle-shaped. Although the knots may be scattered on any part of the main root or its branches, they are most often found on tender rootlets like beads on a string. Small galls contain at least one nematode, whereas larger galls contain numerous females in all stages of development. Galls may be formed on tubers and on parts of the stem in contact with the ground.

Not only are galled roots rough and clublike; the number of feeder roots is decreased. Therefore, the roots can absorb less water and nutrients from the soil than

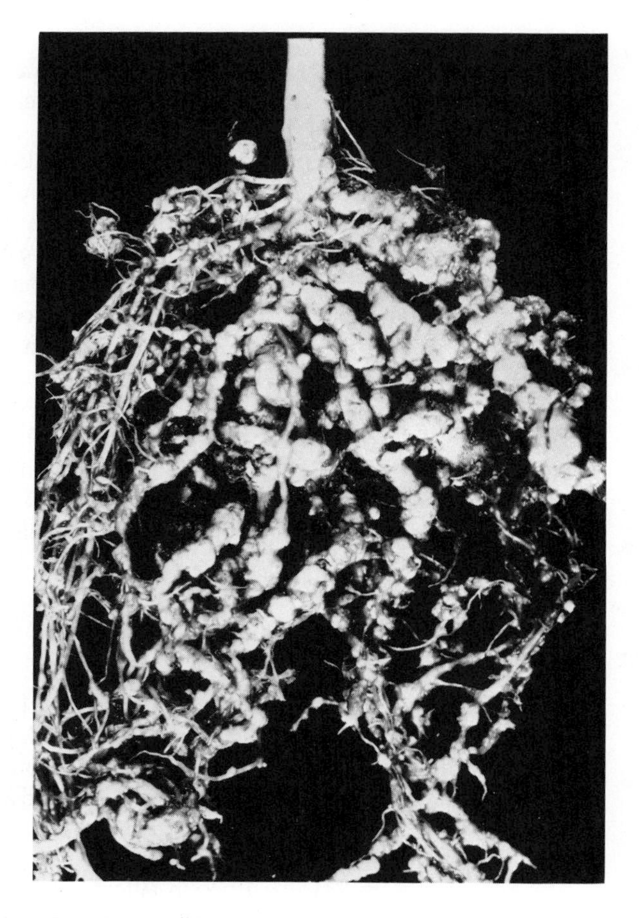

FIGURE 10.1. Root-knot galls on bean roots about one-half natural size. Healthy roots should be long and fibrous without large galls.

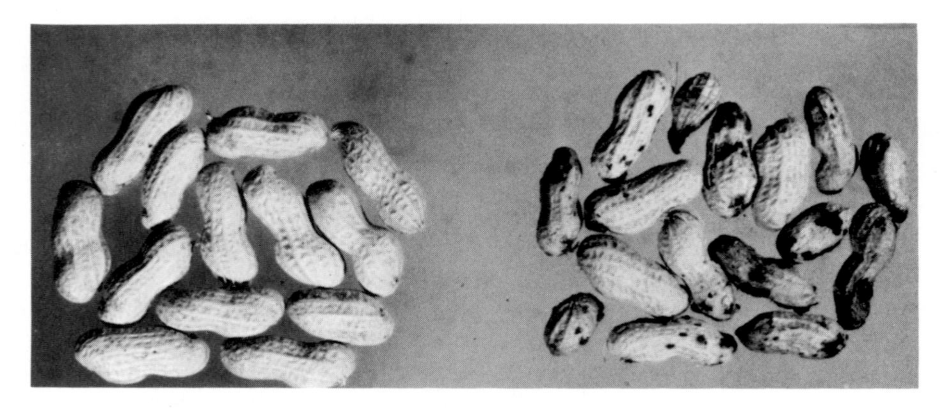

FIGURE 10.2. Root-knot galls on peanut shells (right) compared with undamaged, healthy peanut shells (left).

healthy roots absorb—hence the wilting and nutrient-deficiency symptoms. Another conspicuous symptom is premature rotting and death of the galled tissue, particularly late in the growing season.

Certain insects and nitrogen-fixing bacteria also cause galls on roots. However, positive identification of root-knot galls is made by breaking them open and looking for the pear-shaped, pearly-white females about as big around as the shank of a pin. They are large enough to see with the unaided eye and easy to see with a magnifying glass, particularly in a portion of the root that has begun to rot so that the white females contrast with the brown root tissue. The egg masses of the root-knot nematode also are easy to see. These are brown, often as large as the female itself, and cling to the sides of the roots. When the sac containing the egg mass is lifted off, the female will be found embedded in the root.

Growers who suspect that they may have a nematode problem on their crops should send soil samples to a nematode assay laboratory, which usually charges a small fee. The local extension agent can give instructions about how to collect and mail soil samples. Many agricultural consultants also offer this service as part of their "package" of agricultural advice.

Causal Agent and Disease Cycle

Root knot is caused by species of *Meloidogyne* [from the Greek *melon* (gourd or melon) + *oeides* (resembling) + *gyne* (woman or female) = melon-like female]. There are at least 36 species, but the four most widespread that do the most damage, in order of their importance, are *M. incognita* (Kofoid and White) Chitwood, *M. javanica* (Treub) Chitwood, *M. hapla* Chitwood, and *M. arenaria* (Neal) Chitwood. The white, adult females are 0.4 to 1.3 mm long by 0.27 to 0.75 mm wide. The stylets of mature females range from 12 to 16 μm long. After the female becomes full grown, she begins to produce eggs in a yellow-brown, jelly-like sac fastened to her rear end.

Each female forms on the average 400 to 500 oval or cylindrical, colorless eggs about 80 μm $\times$ 40 μm (Fig. 10.3). The eggs are found in great numbers in the soil.

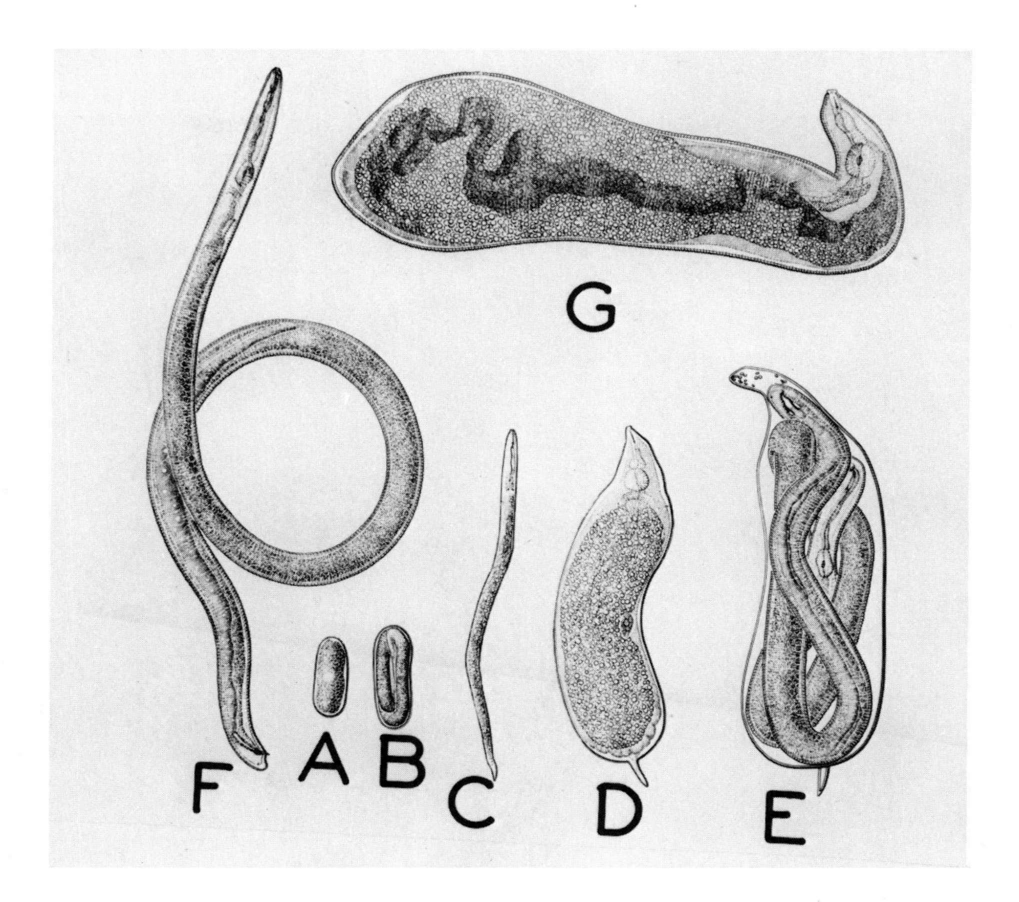

FIGURE 10.3. Development stages of a root-knot nematode. A—unsegmented egg; B—egg containing larva; C—migratory larva as it occurs in the soil; D—sausage-shaped larva as it lives in the plant root; E—larval molt containing fully developed male; F—adult male; G—young female. All magnified about 100 times. *After G. Steiner.*

From each egg, a long, slender, shoestring-like larva, averaging 400 $\mu m \times 15$ μm in size, hatches. The larvae can move or swim short distances in soil water, and are attracted to growing roots by the chemicals that leak out of them. Once a larva touches the root of a susceptible plant, it pushes, forces, and wriggles to the central part of the root containing the water-conducting tissue. Here, under favorable conditions, the larva becomes a female and remains for the rest of her life. Reproduction occurs parthenogenetically in root-knot nematodes. Usually only a few males are seen, but when food is scarce or high populations of larvae are present, greater numbers of males develop.

Root-knot nematodes are called *sedentary endoparasites* because they remain stationary and feed inside the root. A nematode feeds on three to five cells only, puncturing these cells with the stylet and injecting chemicals, enzymes, and hormones that stimulate the plant cells to grow into *giant cells* that remain in a perpetual juvenile

FIGURE 10.4. Adult female of the root-knot nematode with attached egg mass embedded in a protective jelly. Magnified about 100 times. *After G. Steiner.*

state. The enriched cell contents provide an excellent food source for the nematode, which begins to swell rapidly, first becoming a sausage-shaped and then a pear-shaped, whitish body. The plant cells surrounding the nematode are stimulated to divide, forming the galls. After the female becomes full grown, she begins to produce eggs (Fig. 10.4). When the root decays, the galls rot, and the nematode eggs are released in the soil.

Environmental Effects

Root-knot nematodes are quickly killed in areas where the soil freezes, and infection is rare at 10 to 12°C. Above 25°C, root knot is severe, and the nematode's life cycle is completed in about 20 days. In warm climates, nematodes multiply at the rate of 5 to 10 generations a year and also are more active in finding and attacking plants than in colder regions.

As long as the moisture content of soil is favorable for the growth of the plants, soil moisture seems to play only a small part in the development of root knot. Lighter, sandy soils generally are regarded as most favorable for increase of the nematode. These soils have large pore spaces that favor nematode movement and support relatively few organisms that feed on nematodes.

Root Knot and Other Diseases

Although the root-knot nematode acting alone can cause serious damage to plants, it often teams up with other pathogens. When two or more pathogens act together, they may cause greater damage than the sum of their individual efforts. Three well-known examples of pathogens acting in concert with the root-knot nematode are the fungi *Fusarium* and *Phytophthora* and the bacterial-wilt organism. Not only do the nematodes provide the wounds that permit the other pathogens to enter the roots, but they also modify the host cells to provide a more suitable food source for the fungi and bacteria. Under the combined attack of two or more pathogens, the host plant quickly dies, and great losses occur.

Management

The general procedures discussed in Chapter 9 (Management section) and Chapter 7 (Soil Treatment section) are effective in managing root-knot nematodes.

CYST NEMATODES

Cyst nematodes, so called because the female's body becomes a brown leather-like sac (cyst) containing large numbers of eggs, occur worldwide on many crops. Cyst nematodes cause severe diseases on alfalfa, clover, soybeans, sugar beet, tobacco, tomato, and many weeds. The potato cyst nematode is also called the golden nematode. In addition to stunting some hosts, cyst nematodes suppress the reproduction of mycorrhizal fungi and inhibit the activity of nitrogen-fixing bacteria. The soil type does not appear to be a limiting factor because the cyst nematode occurs in fine as well as coarse-textured soils.

Symptoms

Symptoms include severe stunting, wilting, and a greatly reduced root system. Dark brown, oval cysts, about 0.5 mm in diameter, attached to the roots, confirm the presence of cyst nematodes.

Causal Agent and Disease Cycle

Globodera and *Heterodera* are the two principal genera of cyst nematodes. *Globodera rostochiensis* (Wollenweber) Behrens attacks potato. *Heterodera glycine* Ichinohe

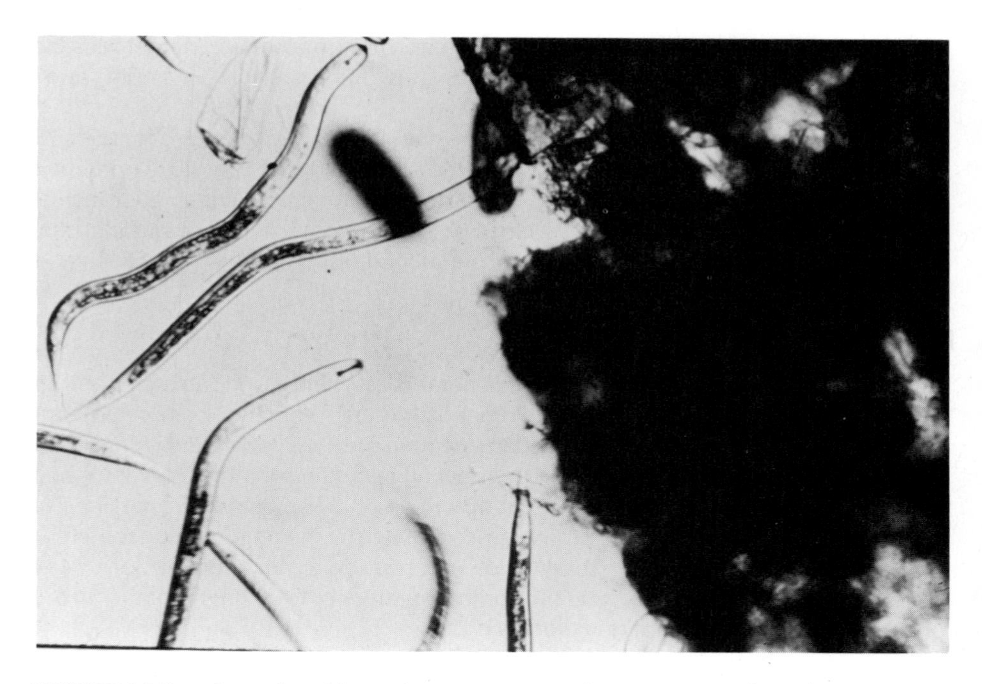

FIGURE 10.5. Juveniles of the soybean cyst nematode moving toward a soybean root. Note the dark stylet that the nematodes will use to feed on the root. *Courtesy D. P. Schmitt.*

attacks soybeans, and *Heterodera schachtii* Schmidt attacks sugar beets. Living females are pearly white. Eggs are retained within the female's body, and when the female dies, her body becomes a brown leather-like sac (cyst) containing hundreds of eggs. The cysts are resistant to degradation by soil organisms, drying, and the chemicals employed to manage other types of nematodes. Eggs contained in the cysts may survive in the soil for several years in the absence of a host plant. When stimulated by root exudates of host plants, the eggs hatch, and the shoestring-like juveniles emerge, migrate to plant roots (Fig. 10.5), and feed on host cells in the stele of the rootlet. During development, the females become spherical and break through the root surface. With adults, only the head and neck portions remain in the root. Mature males emerge from the root and fertilize the females. Approximately 20 days are required for the female to produce eggs and complete its life cycle. Many of the eggs will hatch within a few days and invade rootlets. Four to five nematode generations may be produced on a single crop. Apparently most field populations are a mixture of different strains or races that differ in their ability to attack different cultivars.

Management

The general procedures used to control root-knot nematodes also are effective for cyst nematodes. Successful management requires the integration of long-term crop rotation with nonhosts (corn), resistant cultivars, weed control, nematicides, and efficient land management. Continuous or frequent use of resistant cultivars permits the development of pathogenic races that eventually render the resistant cultivars useless.

Computer software programs are available that are designed to help growers make control decisions for individual fields based on assays telling kinds and numbers of nematodes present, crop and cultivars to be grown, nematicide specificity and costs, and kinds and numbers of weeds present.

LESION NEMATODES

In contrast to root-knot nematodes, which stimulate host cells to grow and swell, lesion nematodes (sometimes called meadow nematodes because they were first found in meadows) kill the cells on which they feed. Lesion nematodes are one of the primary factors involved in the general root destruction of many plants. These nematodes occur all over the world and feed on many kinds of crops. Different species of lesion nematode have broad and overlapping host ranges. They do much damage to crops such as alfalfa, apple, carrot, cherry, corn, cotton, grasses, peach, potato, tomato, tobacco, and many ornamentals. However, the direct damage caused by these pests is only part of the story. All root-invading nematodes provide sites for entry of secondary invaders and alter the host cells, thus causing compound damage. It is difficult to assess the damage caused by lesion nematodes, as the disease often is associated with and leads to increases in other root diseases.

Symptoms

In a given field, lesion nematodes usually occur in well-defined areas, which vary greatly in size and shape. Diseased plants may occur individually or in small groups. Plants attacked early in the season grow poorly and become worthless. Aboveground symptoms resemble those caused by root-knot nematodes; that is, the plants are stunted and yellowed, and wilt during the hot part of the day. The edges of the leaves turn yellow, then brown and die ("rim fire").

Lesion nematodes do not form galls on the roots. The nematode larvae feed in the *cortex* of roots and kill the cells on which they feed. Fungi and bacteria then colonize the dead tissue. In the early stages, the only visible symptom is a small red-brown lesion on the root. As the nematodes feed, the lesion enlarges, often girdles the root, and eventually severs it. Affected roots often shrivel and die; extensive root pruning results; feeding roots are destroyed, and the entire root system is reduced (Fig. 10.6). Diseased roots show various degrees of mutilation. There may be small yellow to brown to black lesions on the feeder roots. The lesions break open, and the cortex sloughs off like a sleeve, leaving only the vascular cylinder.

Often secondary roots appear above the nematode lesions, resulting in the formation of "blind" roots and a stubby, discolored root system. Surviving roots are grouped near the soil surface. Other nematodes and soil organisms also cause the same sort of damage. Positive identification of the disease, therefore, requires identifying the lesion nematode whose eggs, larvae, and adults are often seen in the cortex of infected roots.

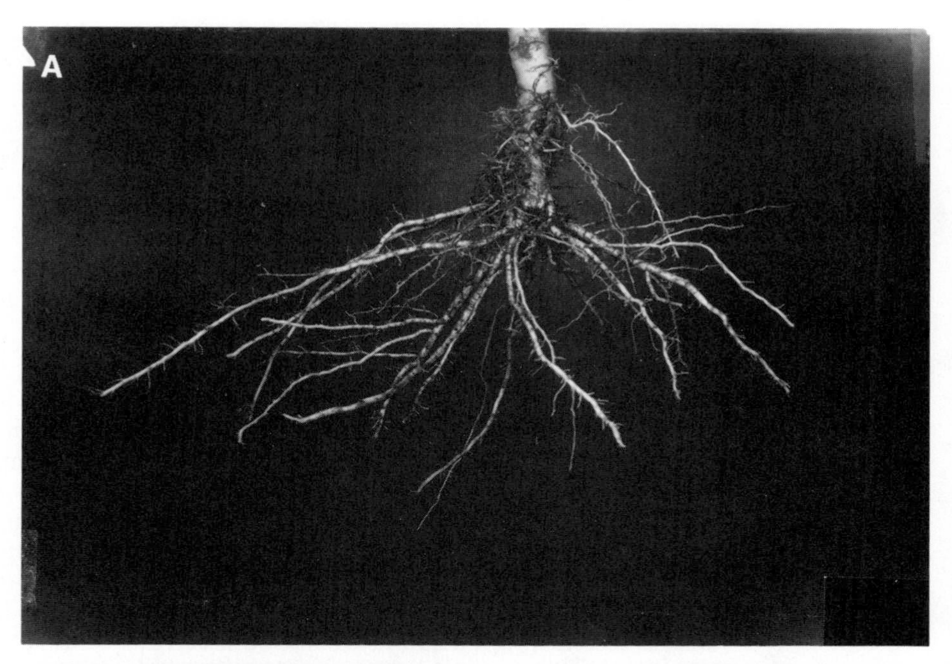

FIGURE 10.6. Roots of plants grown in a field with a high population level of lesion nematode. (A) Soil untreated; most of the feeder roots have been destroyed by nematodes. (B) Soil fumigated with methyl bromide; there is an extensive root system with numerous fibrous roots. *Courtesy C. J. Nusbaum.*

Causal Agent and Disease Cycle

Lesion nematodes belong to the genus *Pratylenchus*. *Pratylenchus brachyurus* Godfrey, *P. scribneri* Steiner, and *P. zeae* Graham are important species. Different species are adapted to different crops and climates. Some thrive in warm, tropical regions; others do more damage in cooler regions. Lesion nematodes (Fig. 10.7) are stout and cylindrical, with a blunt head, strong short stylet, and rounded tail. They range from 0.4 to 0.7 mm long and 20 to 25 μm in diameter. In contrast to the root-knot nematodes some lesion-nematode species reproduce sexually. The eggs are laid

FIGURE 10.7. Lesion nematode eggs, larvae, and adults in a feeder root about 0.8 mm in diameter. *Courtesy C. J. Nusbaum.*

singly, each gravid female averaging one egg a day for about 30 days. Eggs deposited in root tissue hatch in 6 to 17 days, depending on environmental conditions. These endoparasitic, migratory nematodes wriggle through the cortex of infected roots, feed on, and destroy the cells. Eggs, larvae, and adults may overwinter in soil and in live or dead roots.

Management

Lesion nematodes can be managed by the procedures discussed in Chapter 9 (Management section) and Chapter 7 (Soil Treatment section).

STEM AND BULB NEMATODES

Although it occurs worldwide, the stem and bulb nematode is most destructive in temperate climates. It is a serious pest of many crops, including alfalfa, bean, clover, corn, flower bulbs, onion, pea, small grains, strawberry, and sugar beets. More than 400 species of plants are hosts to this nematode, which causes heavy losses by killing seedlings and dwarfing plants, making bulbs unfit for propagation, and reducing yields.

Symptoms

Retarded germination of seedlings and reduced stands often occur in fields infested with stem and bulb nematodes. Many seedlings are twisted, deformed, stunted, and yellowed. Swellings or galls develop on the shortened, thickened stems. The leaves become twisted, dwarfed, and distorted. Bulb tissue becomes soft and spongy and begins to decay. Nematodes can be found in large numbers in the affected parts but are difficult to see without a microscope.

Causal Agent and Disease Cycle

Ditylenchus dipsaci (Kühn) Filipjev, the stem and bulb nematode, is a long slender nematode (1.0-1.3 mm long; 30 μm in diameter) with a short stylet. The preadult larvae can withstand long periods of dormancy; they have been revived after storage in dry plant material for 23 years. During this dormant period they show no sign of life, and their metabolic activity is practically at a standstill. They can withstand long periods of freezing or extreme drying, and with the return of suitable temperature (15-20°C) and adequate moisture, the larvae become active again.

When the aboveground parts of plants are covered with a film of moisture, the nematodes move upwards to the new leaves and stems and enter through stomates or wounds. The nematodes feed on young succulent tissue and stimulate gall formation. Unlike root-knot nematodes, stem and bulb nematodes must mate to reproduce, and the females lay fertilized eggs. Eggs and larvae overwinter in dried, infected host material.

Disease incidence is most severe in cool climates in moist loam or clay soils. Damage is usually slight in dry years.

There are many pathogenic races of *D. dipsaci*, which vary in their ability to infect different hosts.

Management

As stem and bulb nematodes can stay alive in infested seed or bulbs for many months, the need for nematode-free seed is obvious. One should use seed, bulbs, or other planting material from nematode-free plants. If bulbs from plants grown in infested soil are used, they can be freed of nematodes by heating in hot water at 46°C for 1 hr or soaking in a 0.5% formaldehyde solution at 43°C for 4 hr. Long rotations with resistant crops such as carrots, lettuce, spinach, and potatoes are helpful in reducing infestations. It helps to pull out and destroy diseased plants. Large fields can be treated with nematicides; small areas or greenhouse soils can be fumigated or sterilized with steam (see Chapter 7, Soil Treatment section, and Chapter 6, Reduction of Inoculum section).

STUBBY-ROOT NEMATODES

Root-knot and lesion nematodes are classified as *endoparasitic* because they enter the roots and feed on the inside. In addition to these nematodes, there are a large number of *ectoparasitic* species, including the stubby-root nematodes. This group lives in the soil and feeds on the roots and other underground parts largely from the outside (Fig. 10.8). For many years the damage caused by stubby-root nematodes was unknown or overlooked because of their external feeding habits.

Unfortunately, it is hard to prove that a given ectoparasite is the cause of a certain disease. Even though a suspected nematode is found in large numbers in the soil

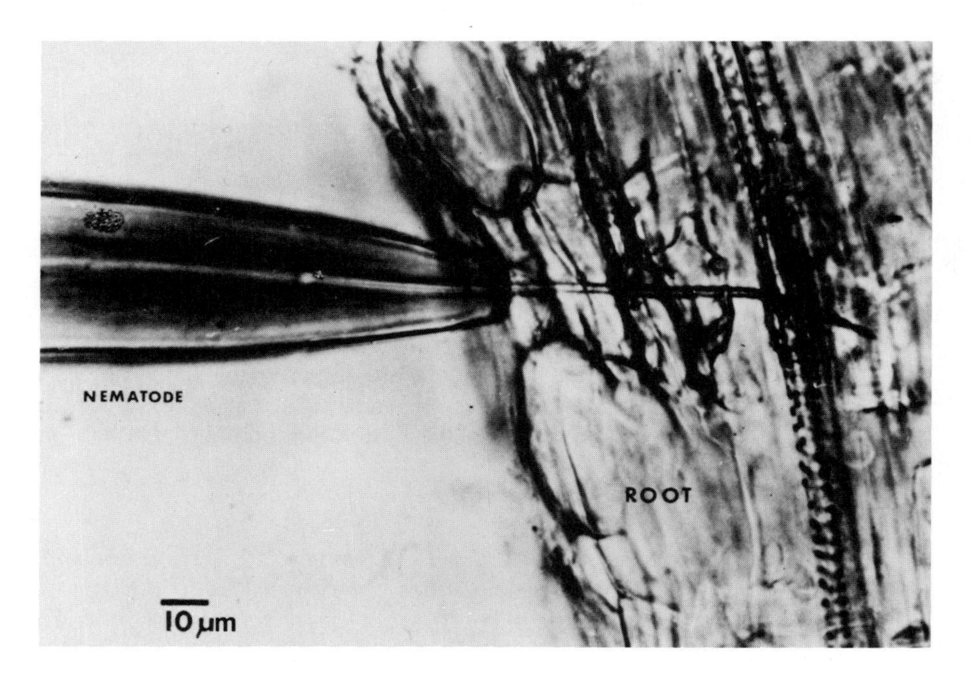

FIGURE 10.8. Ectoparasitic nematode feeding on a petunia rootlet from the outside. Note the collapsed cells punctured by the stylet. *Courtesy R. S. Pitcher.*

surrounding roots of affected plants, many other nematodes also may be found, several of which may have caused the damage. Many of the ectoparasitic forms do not cause distinctive symptoms; any condition or agent that kills or injures the roots may cause similar symptoms. Furthermore, the life history of many of the ectoparasitic nematodes is not completely known; at certain times during the growing season they are found in large numbers in the soil surrounding the roots, but at other times the populations are low.

However, we now know that several genera of ectoparasitic nematodes are important plant parasites. These include the *dagger, lance, ring, stubby-root, spiral, sting,* and *stunt* nematodes. In addition to the damage caused by the feeding of these nematodes, some ectoparasites such as *Xiphinema,* the dagger nematode, also transmit plant viruses. If a nematode pierces the root cells of a virus-infected plant, some of the virus particles may be sucked up into the stylet. Then, if the same nematode moves to the root of a healthy plant nearby and injects salivary secretions containing virus particles into cells of a healthy root, it may become infected with the virus. The realization that nematodes can transmit plant viruses opens new avenues of investigation that may help in managing both nematodes and viruses.

Stubby-root nematodes are typical examples of ectoparasitic nematodes. They have been found in many parts of the world attacking a wide range of crops including beans, cabbage, clover, corn, grape, grasses, peach, tobacco, and tomato. They are found most frequently in sandy or porous soils. Stubby-root nematodes seldom kill the host plant, but they weaken the root system. Crops grown in fields with high populations of stubby-root nematodes cannot withstand drought conditions. The plants are stunted and produce reduced yields of low quality.

Symptoms

Poor stands and reduced growth are typical symptoms of stubby root early in the growing season (Fig. 10.9). Plants attacked by stubby-root nematodes remain stunted and *chlorotic* (yellowed). They have fewer and smaller leaves and branches than healthy plants. They wilt easily on sunny, hot days.

Stubby-root nematodes feed at or near the root tips. The injured roots are weakened and may stop growing. There is abnormal growth of lateral roots and branch roots, which in turn are attacked by the nematodes. The result is a root system composed of short stubby roots often arranged in clusters. Parasitized roots are not killed, although they usually are darker in color than normal roots. The fine feeder roots typical of healthy root systems are conspicuously absent. Undoubtedly, such root systems cannot efficiently take up water and nutrients.

Causal Agent

Two closely related genera of nematodes, *Trichodorus* and *Paratrichodorus,* are called stubby-root nematodes. They vary in length from 0.5 to 1.5 mm and from 30 to 50 μm in thickness. These nematodes never enter the roots. They lay their eggs in the soil, and the life cycle is completed in about 20 days during favorable conditions. Both

FIGURE 10.9. Stunted corn growing in soil with a high population level of stubby-root nematodes. The farmer got no yield and lost the money invested in preparing the land and planting the crop. *Courtesy K. R. Barker.*

females and males are found. Populations vary considerably, depending on the kind and age of the plant, soil type, temperature, season of the year, and competition from other nematodes and pathogens.

Management

The general measures outlined in Chapter 9 (Management section) and Chapter 7 (Soil Treatment section) manage stubby-root nematodes. However, scientists need to know much more about these parasites before they can develop resistant cultivars and determine the most efficient nematicides to use on various crops.

11

Fungi

Fungi are microorganisms that contain no chlorophyll and usually reproduce by producing spores. Although they have sometimes been classified as simple plants, they have no roots, stems or leaves. Fungi are now placed in a kingdom of their own, Mycetae, and are considered to be a unique group of organisms.

The fungi are a very old group of microorganisms that have lived on earth for more than two billion years. Fossils of fungi have been found. Also, we know that fungi have plagued humans by attacking plants for thousands of years. Spores of plant pathogenic fungi have been found in burial chambers in the pyramids of ancient Egypt on grains that were left to aid the Egyptian Pharaohs on their journey to the next world.

Approximately 100,000 species of fungi have been described, and more are being found each year. Mushrooms, toadstools, molds, and mildews are fungi known by most everyone, but many fungi are so small that it takes an expert to find and see them through a microscope.

GENERAL NATURE AND IMPORTANCE

The body or vegetative stage of a fungus consists of branching, delicate, threadlike structures called *hyphae.* A single strand is known as a *hypha.* Many hyphae together form an interlacing tangle known as a *mycelium* (plural *mycelia*). The hyphae of many fungi are divided by crosswalls or *septae* (singular *septum*) into definite cells full of protoplasm. The protoplasm is held within a semipermeable cytoplasmic membrane containing one or more nuclei. Hyphae of other fungi possess no septae or crosswalls and are essentially long tubes containing protoplasm; the streaming back and forth of the active liquid protoplasm is often visible under the microscope. Mycelia may form on or inside the host plant or on decaying organic matter and have different branching habits and structures, which help identify the fungus.

Most fungi are beneficial. They are used for food, in medicine, and in industrial processes. Most grocery stores carry at least one type of edible mushroom in the produce section. Also, cream of mushroom soup is commonly found among the canned soups on supermarket shelves. The baking of bread, the brewing of beer, the making of wine, and even the production of soy sauce all require the use of specific fungi. In medicine, the life-saving antibiotic penicillin was first derived in 1929 from a fungus that had contaminated a culture in Alexander Fleming's laboratory.

Together with bacteria, fungi are principal agents of decay, and through the decomposition of organic matter they play an essential role in recycling nutrients and in the nutrition of plants. All fungi are capable of breaking down complex organic foods into simple substances and using these substances as sources of energy. The fungi that can live exclusively on decaying organic matter are called *saprophytes*. Fungi decompose all sorts of materials, including books, cloth, food, fruits, leather, meat, nuts, paper, stored seed, twine, vegetables, and wood.

Some fungi form a *symbiotic* association (i.e., an association beneficial to both organisms involved) with plant roots. These fungi form *mycorrhizae* and live primarily in (*endomycorrhizae*) or on (*ectomycorrhizae*) roots. The fungi obtain needed nutrients from the plant roots, but also aid the roots in the absorption of certain nutrients such as phosphorus. Some mycorrhizae also may help protect roots against invasion by harmful or pathogenic fungi.

Harmful fungi cause diseases of humans, other animals, and especially plants. About 8000 species of fungi can cause plant diseases, and all plants are attacked by some fungi. Some plant-parasitic fungi can attack many species of plants; others attack only one type of plant. Some fungi can grow and multiply only on a living host and are called *obligate parasites*. Others can grow and multiply on dead organic matter as well as on living plants and are called *facultative* or *nonobligate parasites*. In general, fungi that cause plant diseases will not attack humans, and fungi that attack animals also do not cause plant diseases.

Some substances produced by plant-parasitic fungi in food, called *mycotoxins*, are toxic to animals. When cereal grains such as wheat, barley, or maize (corn) or foods such as peanuts are placed in storage, certain fungi can grow on the stored seeds if adequate moisture is present (Fig. 11.1). Some infections of the seeds occur in the field, some in storage. On corn, growth of the fungus *Aspergillus flavus* (Fig. 11.2) results in the production of aflatoxin. If the level of aflatoxin is sufficiently high, it is poisonous to poultry, swine, cattle, and, in some cases, humans. Certain *Fusarium* spp. also can grow on cereal grains and produce a range of dangerous toxic substances. Ergot of rye (caused by the fungus *Claviceps purpurea*) also results in the contamination of the grain with sclerotia (survival structures of the fungus), which can contain several highly toxic and often hallucinogenic chemicals that can be quite harmful to livestock and humans.

In addition to the problems caused by mycotoxins, plant diseases caused by fungi increase the grocery bill of each U.S. household by more than $600 each year. Billions of dollars of crop losses occur yearly due to plant diseases and postharvest spoilage of food caused by fungi throughout the world. In the United States forests lose more

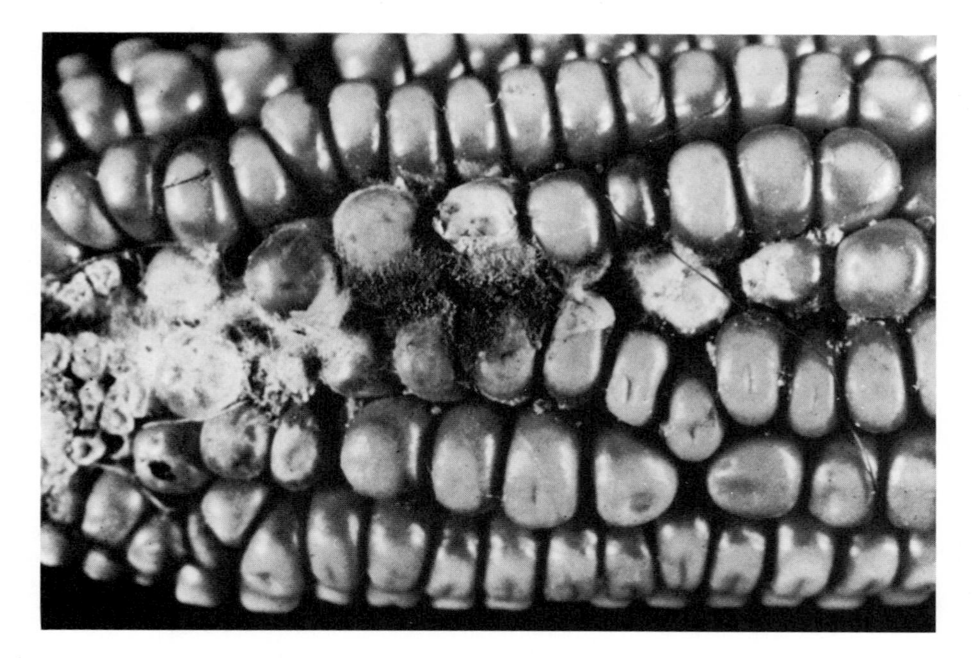

FIGURE 11.1. A mycotoxin-producing fungus, *Aspergillus flavus*, growing on an ear of maize (corn).

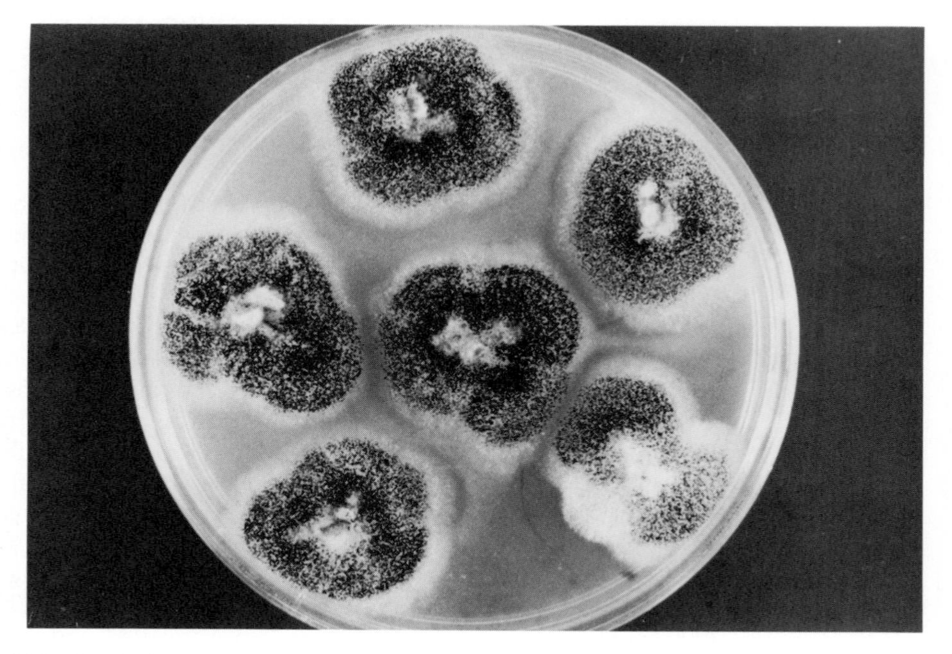

FIGURE 11.2. The fungus *Aspergillus flavus* growing in culture on an agar medium. The dark masses are conidia or spores; the white growth at the edges of each colony is mycelium.

than 20 billion board feet of lumber each year to diseases caused primarily by fungi. Hundreds of millions of dollars also are spent annually for chemicals (*fungicides*) to combat pathogenic fungi. More millions are spent in breeding resistant plants throughout the world.

GROWTH, REPRODUCTION, AND CLASSIFICATION

A hypha usually originates through the germination of a *spore*. The simplest spores are one-celled bodies of microscopic size, containing a nucleus and cytoplasm. They may or may not be able to swim. A spore germinates by pushing out, from a thin place in its wall, a tubelike or threadlike structure known as a *germ tube*. The germ tube develops into a hypha, which branches into many hyphae to form the mycelium. Growth is largely localized at the tips of the hyphae and is made possible by the absorption of liquid materials directly through the cell walls. The mycelium usually is substantial but hidden in the substance on which the fungus grows. The growing hyphal tips are able to penetrate directly through the cell walls of many plant tissues, even the hardest wood. They penetrate by means of organic substances known as *enzymes*, which they secrete. Enzymes have the power to dissolve or digest substances composing cell walls and other cell parts.

In some fungi the mycelium may pass into a dormant or resting stage by the formation of closely compacted hyphae, known as *sclerotia*. These vary in size from a few cells (*microsclerotia*) up to many thousands of cells. Some sclerotia may weigh up to 4 kg, but most are quite small. Sclerotia also vary in color (colorless, pale yellow, brown to black) and shape (irregular, spherical, or elongate). In the form of sclerotia, fungi can hibernate or survive through conditions unfavorable for growth, such as cold, drought, heat, or the absence of a host. On the return of favorable conditions, the sclerotia may form hyphae that infect a host, form a new mycelium, or produce a reproductive body of some sort.

Many fungi reproduce by forming spores at the ends of, inside, or on specialized hyphae and structures. These spores are tiny, microscopic bodies the size of dust, and are produced in huge numbers; their function is similar to that of seeds in higher plants. They are carried by animals, diseased plants, insects, machinery, humans, soil, wind, or water.

Fungi that live and grow in wet conditions often produce *zoospores* that can swim. The zoospores are produced in a *zoosporangium*, and after a brief swimming period, they encyst or come to rest on the host and germinate. However, most fungi produce nonmotile spores that are blown by the wind, splashed by rain, or moved with soil. These spores are formed in several different ways, and often a given fungus may produce more than one kind of spore. Many species have a spore stage in which they pass the winter or unfavorable conditions, and thus the spore becomes a *resting spore*. The different kinds of spores may be quite different in color, size, structure, and wall thickness.

To germinate, spores must be exposed to certain temperature and moisture conditions. Some spores require an actual film of water, whereas others are able to germinate in a moist atmosphere. Few spores germinate when it is very cold; the optimum temperature for germination of most spores is between 15°C and 25°C.

There are five main groups of plant-pathogenic fungi, differentiated for the most part by their hyphal characteristics and how they reproduce. The major groups or subdivisions of fungi, under the division Eumycota (or true fungi) (Fig. 11.3), are the Mastigomycotina, Zygomycotina, Ascomycotina, Basidiomycotina, and Deuteromycotina. Within these subdivisions the fungi often are referred to as Oomycetes and Chytridiomycetes (Mastigomycotina), Zygomycetes (Zygomycotina), Ascomycetes (Ascomycotina), Basidiomycetes (Basiciomycotina), and imperfect fungi (Deuteromycotina). Several plant-pathogenic fungi also occur in the division of fungi known as the Myxomycota. We will discuss only the five major subdivisions of the Eumycota or true fungi here.

Fungi in the Mastigomycotina have filamentous hyphae that usually lack cross walls or septa and are multinucleate. The vegetative hyphae are generally believed to be *diploid* (2n), and meiosis occurs in the reproductive structures just before fertilization, as it does in animals. The sexual spores in the Oomycetes are produced by the union of two morphologically different gametes (*oogonium* and *antheridium*). Most of these fungi produce spores called zoospores that have flagella and can swim in water. The zoospores are formed in sporangia.

Fungi in the Zygomycotina have filamentous hyphae that also lack cross walls or septa. The sexual spores, called *zygospores*, are thick-walled and are produced by the union of two morphologically similar gametes. Nonmotile spores are produced in sporangia, and no zoospores are formed.

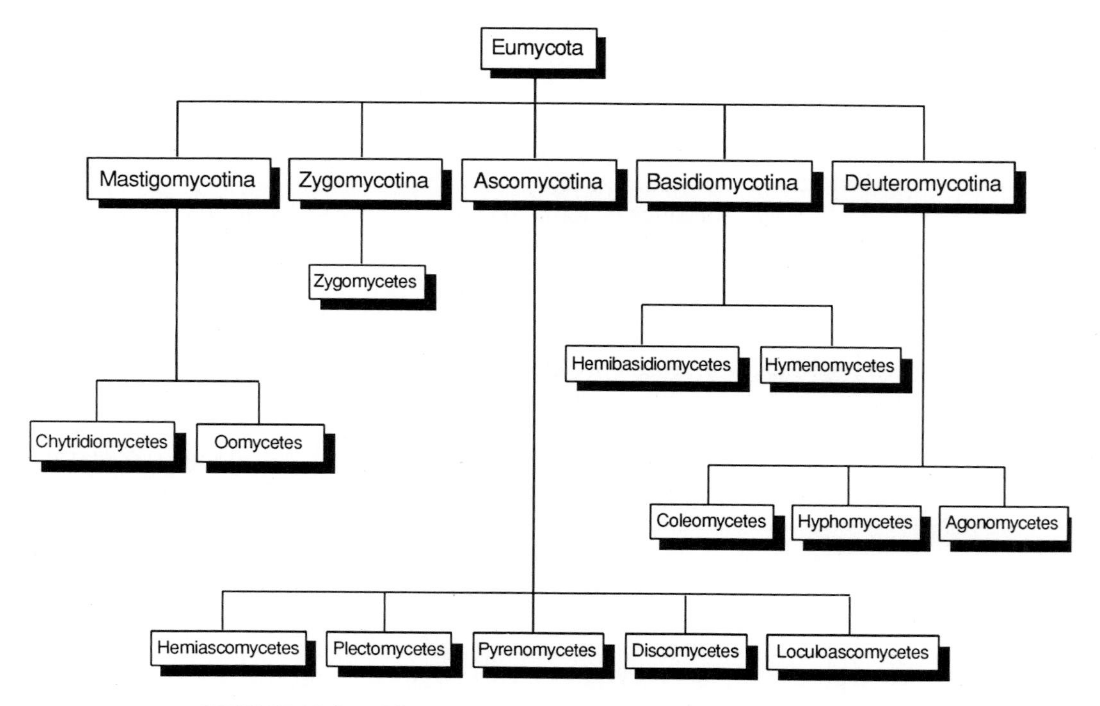

FIGURE 11.3. The major groups or subdivisions and classes of the fungi in the division Eumycota or true fungi.

Fungi in the Ascomycotina or the ascomycetes have vegetative hyphae that have cross walls or septa, and the nuclei are haploid (n). Ascomycetes can have both sexually and asexually formed spores. The asexual spores, called *conidia*, are haploid and can be formed in structures called *pycnidia* (singular *pycnidium*). The diploid stage (2n) is formed during sexual reproduction and occurs only briefly before meiosis occurs. The *haploid* (n) spores that are formed after sexual reproduction are called *ascospores*. Generally eight ascospores are produced as a result of sexual reproduction. The ascospores are contained within a sac-like structure called an *ascus* (plural *asci*). Asci may be contained in a larger body called an *ascocarp*. The way the ascocarp is formed and the way the asci are borne within the ascocarp form the basis for further classification of fungi in the Ascomycotina. The Hemiascomycetes have naked asci (i.e., no ascocarp). The Plectomycetes have asci borne in *cleistothecia*. The Pyrenomycetes have asci borne in *perithecia*. The Discomycetes, also known as the cup fungi, produce asci at the surface of a fleshy cup- or saucer-shaped structure called an *apothecium*. The Loculoascomycetes produce *pseudothecia*, which are perithecia-like structures with asci in separate or single large cavities formed in host tissue.

Fungi in the Basidiomycotina have vegetative hyphae with septa or cross walls, and the nuclei are generally haploid. For some basidiomycetes, extensive mycelial development and infection of plants can occur only after one strain of haploid mycelium fuses with another genetically different haploid mycelium to form vegetative cells with two unique nuclei. This nuclear condition is called *dikaryotic* (n + n). During sexual reproduction, fusion of the haploid nuclei (i.e., *karyogamy*) occurs, and the diploid nucleus then undergoes meiosis to produce haploid nuclei that will be contained within *basidiospores*. The basiodiospores are produced externally on a spore-producing structure called a *basidium* (plural *basidia*). In the Hemibasidiomycetes, the basidium has cross walls or is the *promycelium* from a special type of spore known as a *teliospore*. Rusts and smuts are hemibasidiomycetes. In the Hymenomycetes basidia do not have cross walls and are produced in definite layers (i.e., *hymenia*). Some of these fungi have the basidiospores formed in specific structure such as a mushroom. The wood decay and some root rot fungi are in the Hymenomycetes.

Fungi in the Deuteromycotina have mycelium with septa and haploid nuclei. Some cells may have more than one nucleus and are called *multinucleate*. Fungi were originally placed in the Deuteromycotina or the imperfect fungi because sexual reproduction and sexual structures had not been observed. For some fungi, the sexual stage still has not been found; for others, the sexual stage is known to be an ascomycete or a basidiomycete that can be produced in culture but occurs only very rarely in nature. As a result some fungi have two scientific names, one for the asexual stage or the *anamorph* and the other for the *teleomorph* or sexual stage. The sexual stage is also known as the perfect stage, which means that the asexual stage is the imperfect stage—thus the name for the imperfect fungi. For example, the fungus *Rhizoctonia solani* is the anamorph for the corresponding teleomorph *Thanetephorus cucumeris* (which is a basidiomycete). Because the imperfect stage often is the one found in nature, some fungi that have teleomorphs are known more widely by the name of the corresponding imperfect or asexual stage.

The imperfect fungi are further classified on the basis of the presence or absence of spores and the type of structure in which the spores are formed. Fungi in the Coleomycetes have spores or conidia borne in pycnidia or acervuli (Fig. 11.4). Fungi

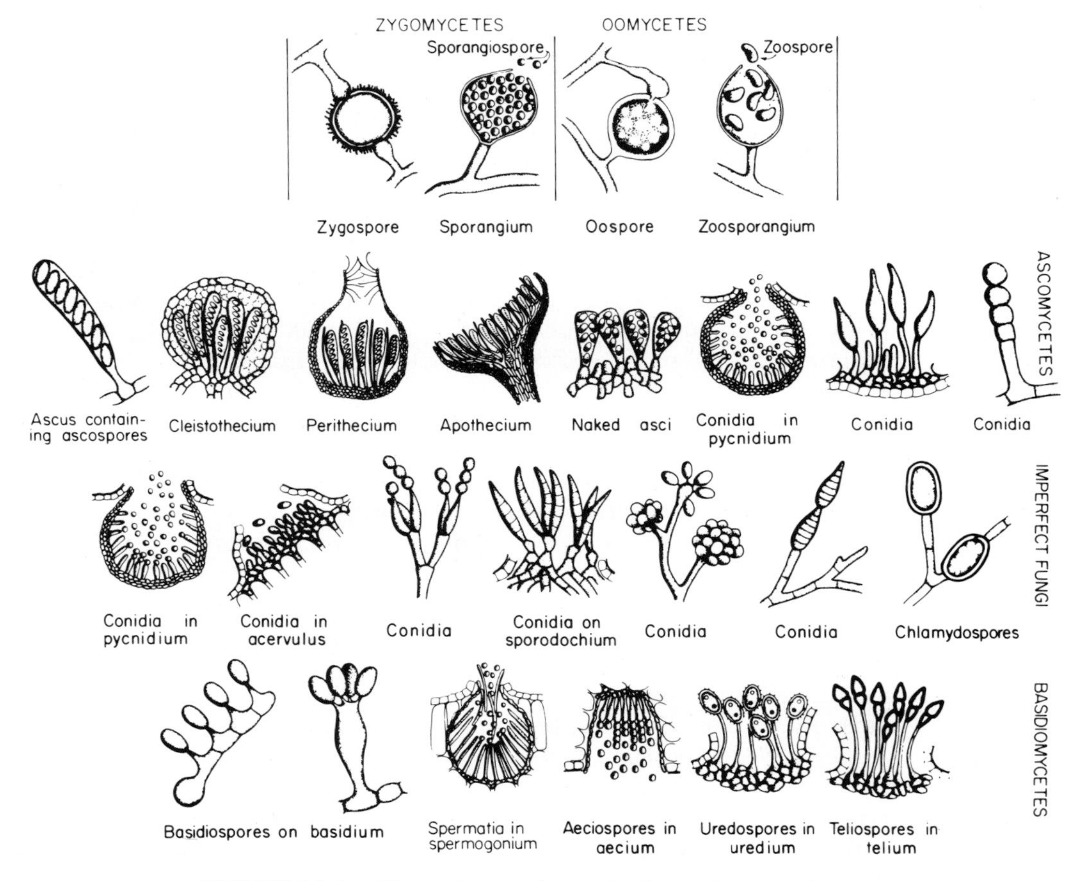

FIGURE 11.4. Representative fruiting bodies and spores of the main groups of fungi within the Eumycota. *Courtesy G. N. Agrios and Academic Press.*

in the Hyphomycetes have asexual spores produced on or within hyphae that are freely exposed to the air. Fungi in the Agonomycetes or Mycelia Sterilia produce no spores and exist in the asexual stage only as mycelium or structures formed from mycelium such as sclerotia.

DISTRIBUTION AND DISPERSAL

Fungi grow almost everywhere. They thrive in the soil, in living and dead plants and animals, and in all kinds of organic matter. They grow in both fresh and salt water; the dustlike spores may be blown thousands of kilometers. Fungi can use many substances for food. Provided that sufficient moisture is present, they will grow on most organic matter. Some even grow on paint and oil, and others attack some synthetic materials and insulating materials.

In general, plant diseases caused by fungi can be classified into two major types: those caused by airborne fungi and those caused by soilborne fungi. As the names imply, the means by which each type of pathogen is spread differ. Many airborne fungi are spread by wind and blowing rain or may be splashed from one plant to the next by the impact of a falling raindrop. Soilborne fungi can be moved about in blowing dust, in soil washed between fields, or on farm equipment. The spread of airborne pathogens usually is more rapid and over longer distances than the spread of soilborne pathogens. Sometimes the forces of pathogen dispersal are equalized because either type of pathogen can be moved on infected plant material.

Several diseases caused by soilborne fungi on agronomic and horticultural plants are discussed in Chapter 12; those caused by airborne fungi are considered in Chapter 13. The strategies and techniques for the management of specific diseases caused by fungi are considered in these chapters. Diseases of shade and forest trees caused by fungi also cause severe losses. The management of tree diseases is particularly difficult because the host plants are widely dispersed and live for many years. Several important fungal diseases of trees and decay in wood are discussed in Chapter 14.

12

Diseases Caused by Soilborne Fungi

Soilborne fungi cause some of the most widespread and serious plant diseases. Spores or mycelium of many of these fungi can overwinter or survive adverse conditions in soil or on plant debris, and once an area has become infested with soilborne fungi, it is generally difficult to get rid of them. Often root diseases caused by soilborne fungi may not be noticed until extensive damage has been done, because many of the early symptoms occur only on roots. The representative diseases discussed in this chapter are summarized in Table 12.1.

TABLE 12.1. Some important diseases caused by soilborne fungi.

Disease	Pathogen	Hosts
Root rots	*Phytophthora* spp.	Soybean, alfalfa, tobacco, ornamentals, and others
Damping-off	*Pythium* spp.	Many crops
Root rots, leaf blight, and damping-off	*Rhizoctonia solani*	Beans, tomato, cereal grains, turfgrasses, and many others
Root and stem rots and wilts	*Fusarium* spp.	Many crops
Wilts	*Verticillium* spp.	Cotton, potato, tomato, alfalfa, some flowers, fruit trees, shade trees, and others
Southern blight	*Sclerotium rolfsii*	Many vegetable crops, peanut, soybean, forage crops, and others
Crown and stem rots, watery soft rot	*Sclerotinia* spp.	Many vegetable crops, forage legumes, peanut, soybean and others
Black rot	*Cylindrocladium crotalariae*	Peanut, soybean, other legumes

PHYTOPHTHORA ROOT ROTS

Among the fungi called "water molds," because they thrive in wet soils, are the *Phytophthora* spp., which occur almost everywhere crops are grown. This genus contains several species of soil-inhabiting fungi that cause serious, widespread, and difficult-to-control root and stalk rots in warmer regions throughout the world.

Phytophthora root and stem rot of soybeans caused by *Phytophthora megasperma* f. sp. *glycinea* is prevalent throughout the soybean-growing regions of the United States. The disease can attack soybeans at any stage of development and may reduce yields by more than 50% in susceptible soybean cultivars. Phytophthora root rot of alfalfa, caused by *P. megasperma* f. sp. *medicaginis,* occurs in nearly every area of the world where alfalfa is grown. It is devastating to stands of seedling alfalfa and also can cause severe stand loss in established fields. Black shank of tobacco, caused by *P. nicotianae,* is prevalent throughout the southeastern United States, and can destroy a large part of the crop in a short time. *Phytophthora cinnamomi* causes an important root disease on many ornamental plants, including azalea, rhododendron, camellia, mountain laurel, and yew. This pathogen also causes a devastating disease on Fraser fir, a popular Christmas tree species. Another species, *P. fragariae,* causes red stele of strawberry, which destroys the feeder roots of the plant.

Phytophthora Root and Stem Rot of Soybean

Symptoms

Phytophthora root rot can occur on soybeans of any age. If the fungus attacks seedlings early, damping-off may result. If seedlings are slightly older, root and stem rot can cause wilting, yellowing, and seedling death. Older plants are killed gradually, and plant vigor is reduced throughout the season. Upper leaves turn yellow, and plants wilt. Affected plants die, but withered leaves often remain attached to the plant for a week or more. Taproots turn dark brown, and this discoloration may extend up to the fourth or fifth node. Infected lateral and branch roots usually are destroyed. Inside the plant, the cortex and vascular tissues also are darkened.

Causal Agent and Disease Cycle

Phytophthora megasperma Drechs. f. sp. *glycinea* T. Kuan and D. C. Erwin is the causal agent. Anywhere that plant roots are growing in waterlogged soils provides good conditions for *Phytophthora* spp. to attack. The hyphae of *Phytophthora* spp. have no cross walls or septa and branch profusely. *Phytophthora megasperma* survives in soil primarily as sexually produced *oospores,* which form abundantly in roots when an *antheridium* fertilizes an *oogonium* (Fig. 12.1). The oospores are generally round, and dormant oospores have thick, smooth inner and outer walls. New pathogenic strains of *Phytophthora* spp. can be produced as a result of genetic recombination during sexual reproduction. When water is present, sporangia arise from germinating oospores on infested plant debris. Sporangia (42-65 μm $\times$ 32-53 μm) are colorless and lemon-shaped. Each sporangium is attached to a fragile, slender stalk called a *sporangiophore.* When the sporangia are ripe and temperature and moisture are suitable, the protoplasm inside each sporangium divides into about

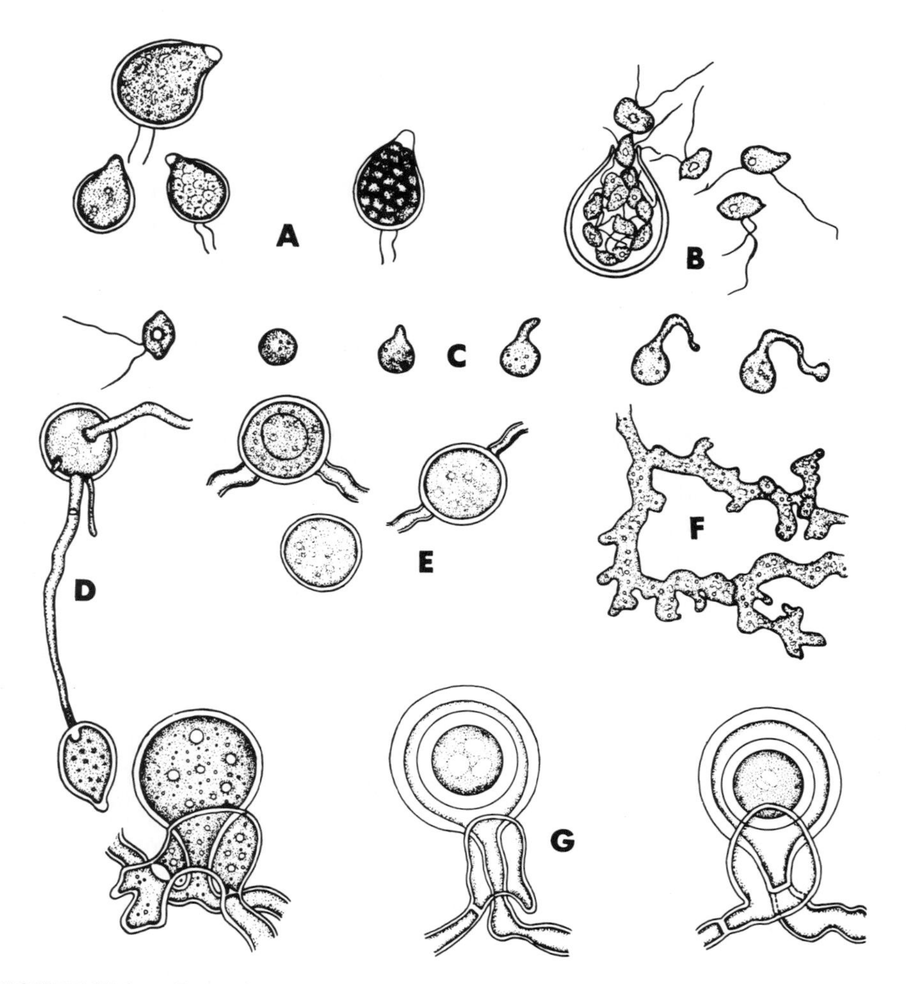

FIGURE 12.1. Reproductive and growth structures of a *Phytophthora* spp. magnified about 400 times. A—sporangia; B—zoospores emerging from sporangium; C—stages in zoospore germination; D—chlamydospore germinating to form sporangium; E—chlamydospores; F—hyphae; G—oospores with antheridia. *Courtesy P. D. Dukes.*

30 colorless, bean-shaped *zoospores.* These remarkable spores can swim (most spores cannot) by propelling themselves with two whiplike structures (*flagella*). Zoospores are attracted to roots by the leachates or juices that leak out of the roots. When a zoospore encounters a root, it attaches to the root surface by encysting, and it germinates. Within 2 hr the germ tube has penetrated the root, and infection has begun. The mycelium penetrates all through and kills the root tissues.

When the food supply is all used up or the root tissues begin to die, oospores are formed as a survival or resting structure. The life cycle of the fungus is now completed. Many *Phytophthora* spp. also form a type of resting, asexual spore called a *chlamydospore.* These chlamydospores often arise from a hyphal swelling and may survive for long periods in soil.

Phytophthora root rot occurs most commonly in heavy, tightly compacted, clay soils. High soil moisture (soggy soils) and rainfall favor disease development. If a dry period occurs before or after planting, the disease will be less severe. Cool temperatures also favor disease development. The greatest danger of root damage occurs in infested soils when temperatures are about 15°C (60°F). Little or no disease develops at 35°C (95°F).

Management

Several physiological races of *P. megasperma* f. sp. *glycinea* can attack soybeans. Thus, it is important to identify the fungal races present in an area. Once this is known, an appropriate high-yielding Phytophthora-resistant cultivar can be selected from the numerous ones available. Additionally, growers should always plant in warm [18°C (65°F) or warmer], well-drained soil. Growers should avoid planting susceptible cultivars in low-lying areas in poorly drained soil or in areas where the disease has been observed in the past. Rotation with nonhost crops such as wheat or corn helps reduce the amount of fungus in the soil.

Phytophthora Root Rot of Alfalfa

Symptoms

Mature alfalfa plants that are infected wilt, and the leaves, particularly on the lower portions of stems, become yellow or reddish brown. Plants are slow to regrow after harvest. On taproots, lesions have diffuse margins and are tan to brown in color. Lesions usually begin where lateral roots have emerged. A diagnostic symptom is the yellowish discoloration that extends through the root cortex and into the xylem. Taproots can be affected at any depth in soil and usually will rot off at the level where drainage of water is slowed or halted by a compacted soil layer. New roots may grow if environmental conditions do not continue to favor the pathogen.

Causal Agent and Disease Cycle

The causal agent of this disease is *Phytophthora megasperma* Drechs. f. sp. *medicaginis* T. Kuan and D. C. Erwin. Isolates from alfalfa often are host-specific. The fungus survives in soil as mycelium in infected plant tissues or as oospores. The general life cycle of the fungus is similar to that of *P. megasperma* f. sp. *glycinea,* the causal agent of Phytophthora root rot of soybeans. The fungus can infect plants only when free water is present. Thus, the disease is important only where soil moisture is excessive. If alfalfa is irrigated, the irrigation water can carry zoospores that are released from sporangia that have formed on infected roots.

Management

Resistant cultivars are available and should be used wherever Phytophthora root rot is a potential problem. Management of soil and water also is important in managing the disease. Deep tillage of soil to break up compacted soil layers improves the movement of water through it and reduces the time that the soil remains saturated. Alfalfa should not be planted on heavy, poorly drained soils. If alfalfa is irrigated, the length of time that water is applied should be regulated to prevent long periods of soil saturation.

Black Shank

Symptoms

Black shank causes a damping-off of young seedlings in seedbeds. When vigorously growing plants 30 cm or more high are attacked, the first indication of the disease is the sudden wilting or drooping of the leaves. After a few days the leaves begin to turn yellow and hang down the stalk. The entire root system and the base of the stalk turn black, decay, and die. When the stem of a diseased plant is split in half lengthwise through the black lesion, the pith appears dry and brown to black and usually is separated into platelike disks (Fig. 12.2). This is one of the most characteristic symptoms of this disease.

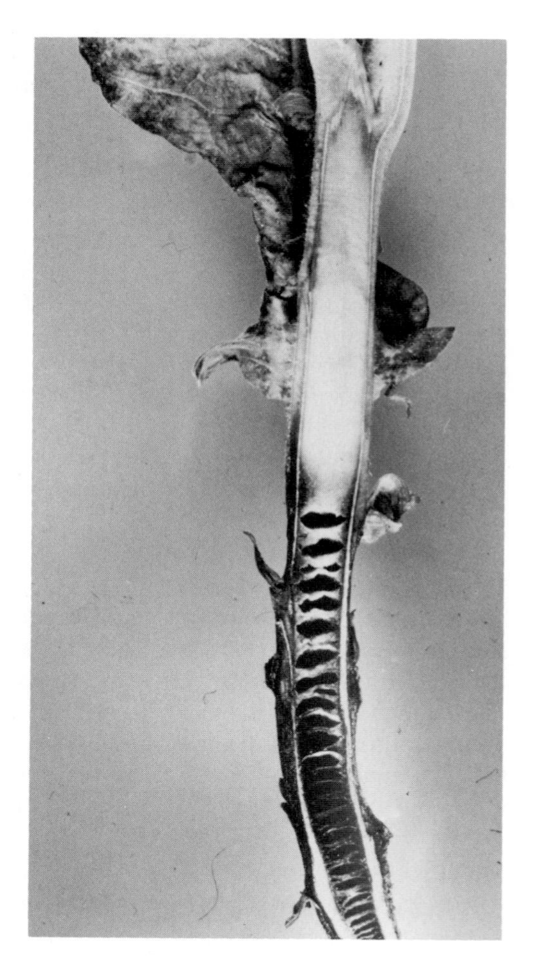

FIGURE 12.2. Split lower stem of a tobacco plant with black shank. The stem is split lengthwise to show the typical decay, "disking" of the pith, and darkening of the infected tissue.

Causal Agent and Disease Cycle

This disease is caused by the fungus *Phytophthora nicotianae* Breda de Haan. Any place where plant roots are growing in waterlogged soils provides good conditions for *Phytophthora* spp. to attack. The characteristics of the fungus and the general disease cycle are similar to those for *P. megasperma.*

Black shank is a warm-weather disease. The fungus thrives in warm, wet soils. When temperatures range from 20°C to 35°C with frequent rains, resulting in high soil moisture, zoospores are produced in great numbers. These spores are washed and/or swim from one place to another, infecting any tobacco plants they contact.

Management

To effectively manage black shank, farmers must coordinate several practices, all of which are necessary. The backbone of any management program is the use of resistant cultivars.

Farmers in an area infested with black shank should try to grow their own disease-free transplants. Seedbed soil should be fumigated to kill bacteria, fungi, nematodes, and weeds (see Chapter 7, Soil Treatment section). Plant beds should not be watered from streams or ponds that drain infested fields.

Root knot (caused by the nematode, *Meloidogyne* spp.) and black shank each increase the damage done by the other. Therefore, root knot also must be managed. Fields that are to be planted in tobacco should be treated with appropriate multipurpose chemicals that kill nematodes, fungi, and bacteria.

Rotations with nonhost crops (corn, cotton, grasses, peanuts, small grains, sweet potatoes) also are valuable in managing black shank, by helping to reduce the amount of fungus in the soil. The same rotations should not be used year after year. Change the cropping sequence; that is, "rotate the rotation" to prevent root-knot nematodes or other pathogens from building up to dangerous levels.

After harvest, stalks should be cut and the roots plowed to expose them to the drying action of sun and wind. This reduces carryover of nematodes to the next crop.

Phytophthora Root Rot of Ornamentals

Symptoms

The symptoms of Phytophthora root rot on ornamentals vary with the cultivar. On azaleas new leaves are smaller than normal with interveinal chlorosis, purple discoloration, and defoliation. This chlorosis often is confused with a deficiency of iron or other nutrients. Applications of iron and fertilizer usually improve the green color of the leaves only for a short time. Usually, plants slowly decline in vigor and die branch by branch over a period of several months to years, but sometimes they can die rapidly. Roots are dark-colored, brittle, and often limited to the upper part of a container of soil. The reddish-brown discoloration advances to the larger roots and eventually to the main stem.

On rhododendron the primary symptoms of Phytophthora root rot are a rapid wilting and death of leaves (Fig. 12.3). The leaves droop but remain attached to the limb. Root symptoms are similar to those on azalea.

FIGURE 12.3. The rhododendron plant on the right, with drooping, rolled leaves, is dying from Phytophthora root rot. Roots of the apparently healthy plant on the left intertwine with those of the diseased plant. Therefore, the healthy plant also probably will become infected and die. *Courtesy R. K. Jones.*

On *Camellia japonica* the main symptoms are a gradual decline in vigor and loss of older leaves. Root symptoms are similar to those on azalea. *Carmellia sasanqua* is tolerant to the pathogen. Yew trees die rather suddenly, with the foliage turning reddish brown, when infected by *Phytophthora* species. Roots also are reddish brown, and the discoloration may extend into the main stem. Fraser fir also is susceptible. Needles on diseased trees turn reddish brown, but remain attached to the branches.

Causal Agent and Disease Cycle

The pathogen that causes this disease is usually *Phytophthora cinnamomi* Rands. The fungal characteristics are similar to those of *P. megasperma.*

Phytophthora root rot of ornamentals is favored by high soil moisture and warm soil temperatures. The frequency of root rot increases in areas where surface water collects, and in shallow soils with underlying rock or compacted hard pans. Heavy clay soils; overwatering; tight, poorly drained soil mixes in containers; and containers with a few small holes in the bottom also are predisposing factors that favor disease development. Settling of containers into sawdust or other similar material used as a ground cover for weed control may also increase the disease.

Management

The most practical management of Phytophthora root rot in nurseries and Fraser fir plantations involves avoiding the disease, either by the production and use of disease-free plants or by the prevention of disease development on established plants.

PRODUCING DISEASE-FREE PLANTS

To assure proper sanitation of planting areas, the following practices are recommended:

- Remove all old rooting media, old roots, and plant debris from propagating benches and beds and the surrounding area.
- Treat all unpainted wood, flats, frames, and benches with 2% copper naphthenate. Treatment with a 1:10 dilution of a commercial chlorine bleach or fumigation with recommended chemicals under plastic will eradicate the fungus from wood surfaces, pots, tools, and equipment.
- Remove all plants from the greenhouse at the beginning of the propagating season, and disinfect the interior.
- Fumigate rooting media or potting media under a plastic tarp or with steam [82°C (180°F) for 30 min] before use. Store rooting and potting media on higher ground than the rest of the nursery to prevent contamination by surface runoff water from irrigation and rain. A paved area is suitable for this purpose.

Except in areas with deep, well-drained, sandy soils, it is not possible to grow disease-free, *Phytophthora*-susceptible plants in the field. Planting on raised beds with improved drainage will reduce the symptoms and nursery losses due to Phytophthora root rot, but the plants may have the fungus in the soil or roots and die after transplanting. Planting in raised beds with bark (e.g., pine bark) incorporated in the planting site is important, however, in landscape plantings to prevent root-rot development.

Much cleaner plants can be grown in pathogen-free, well-drained potting mixes in containers than are possible in the field. This potting mix should contain little or no peat moss or soil and no sawdust. Potting mixtures containing large amounts of composted tree bark (without many fine particles) and sand have a high air volume (20-25%) and good water drainage; such mixtures do not favor the growth and development of *Phytophthora* species. Containers should be placed on a well-drained site on crushed rock, gravel, or coarse bark so that water can drain away from the container, and surface water contaminated with the fungus cannot enter the drainage holes in the bottom of the container. Some fungicides specific for oomycetous fungi are available for use.

Do not overwater plants susceptible to Phytophthora root rot. Even in well-drained potting mixes, *Phytophthora* spp. can develop if plants are constantly overwatered. All *Phytophthora*-susceptible plants should be grouped together in a nursery so that they can be given the special attention necessary to keep them free of disease. If at all possible, water *Phytophthora*-susceptible plants with well water. Lake, pond, or stream water can be contaminated with spores of *Phytophthora* species by surface runoff water. Drainage ditches should be constructed so that surface water does not drain into irrigation water sources other than wells. Also, do not overfertilize or set plants too deep because these practices can increase the chance of *Phytophthora* root rot development.

PREVENTING ROOT ROT ON ESTABLISHED PLANTS

Phytophthora root rot on established ornamentals and in Fraser fir plantations must be prevented because few chemicals are effective in controlling this disease after aboveground symptoms become obvious. The following suggestions may aid in the prevention of root rot:

- Purchase disease-free plants from a reputable nursery. Avoid plants that lack normal green color or appear wilted in the morning, and evergreen plants that have excessive defoliation or dark discolored roots.
- Plant *Phytophthora*-susceptible plants in well-drained areas. If excess water from any source collects in the planting site, avoid planting susceptible plants. If soil is heavy clay or does not have good internal drainage, set susceptible plants in raised beds and thoroughly mix a porous material such as bark (not sawdust or peat) into the bed. The material should be incorporated to a depth of 20 to 30 cm. In some areas drain tile plus gravel, placed 15 to 30 cm below the surface, may also help reduce excessive soil moisture.
- Do not set a new plant any deeper than the soil level in the container or the soil line in the nursery.
- In areas where susceptible plants have died, replant with plants that are not susceptible to root rot.
- The spread of *Phytophthora* spp. into and among plants also can be reduced through the use of specific fungicides, but these chemicals may not kill the fungus in infected plants.

DAMPING-OFF

Damping-off disease of seedlings occurs all over the world. Indeed, the soil in every cultivated field probably is inhabited by fungi that cause this disease. Greenhouse and nursery growers are constantly threatened by outbreaks.

Losses from damping-off vary with the crop, season, soil type, and weather. The disease is responsible for many seedbed failures and poor stands. Sometimes during wet, cold planting seasons, it is almost impossible to get a crop off to a good start because of damage from damping-off fungi. Damping-off is a common disease of flower and vegetable seedlings and causes extensive losses through death of seedlings or production of low-quality plants that perform poorly in the field.

Symptoms

Damping-off fungi may rot seeds soon after they are planted, or kill young seedlings before they emerge from the soil (Fig. 12.4) (*preemergence damping-off*). Later these fungi may attack young seedlings above ground or near the soil line (*postemergence damping-off*). A brown, watery, soft-rot develops, the young stems are girdled, and the seedlings fall over. In advanced stages of the disease, affected tissues collapse, and the stems either disintegrate or become shrivelled and limp. Affected areas in a seedbed, containing shrunken, collapsed, or dying seedlings, may be 30 cm or more in diameter.

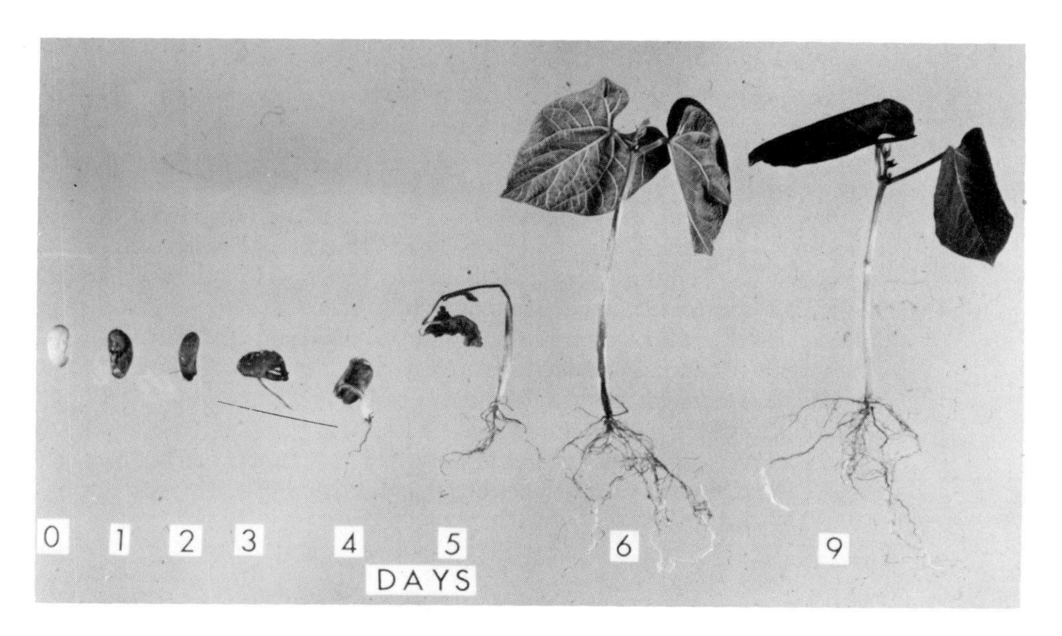

FIGURE 12.4. Pythium damping-off of snap bean on successive days after planting. Day 0—a healthy seed; 1 and 2—seed rot; 3—preemergence damping-off; 4, 5, and 6—postemergence damping-off; 9—a healthy seedling that has escaped infection. *Courtesy G. Abawi.*

Causal Agent

Many *Pythium* spp. cause damping-off diseases. These fungi, which are oomycetes, live on dead plant and animal material as *saprophytes*. They also can act as low-grade parasites, which attack any young succulent tissue, particularly fibrous roots and tender stems. The slender, profusely branched mycelium has few, if any, cross walls. The mycelium grows rapidly and produces many sporangia, each one containing 30 to 100 zoospores which are capable of swimming in water. When the zoospores are released, they swim about in the soil water for a few minutes. If the roots of susceptible plants are growing nearby, the zoospores are attracted to the roots by exudates that serve as nutrients or stimulants. Once the zoospore touches the host tissue, it stops swimming and within 2 hr produces a germ tube, which quickly penetrates the root tissue to start a new infection. Food from the root tissue enables the fungus mycelium to grow rapidly and produce a new crop of sporangia in a few days. The fungus also produces oospores or sexual spores. Because oospores require a resting period before they germinate, they also are called resting spores. Oospores germinate by germ tubes that develop into mycelium or by producing zoospores in sporangia. The type of germination of both sporangia and oospores is controlled by temperature. At temperatures above 18°C (64°F), germination is by germ tubes; temperatures between 10° (50°F) and 18°C (64°F) induce zoospore production.

Another fungus, *Rhizoctonia solani* Kühn, also causes damping-off. This fungus causes other diseases in addition to damping-off and will be discussed in the next section.

Disease Cycle

Pythium spp. enter seeds by direct penetration of the moistened, swollen seed coat, or through cracks in the seed coat or wounds in roots. The fungus is attracted by the chemicals leaking from seed and seedlings during germination and from sites where newly formed roots have emerged. The fungus penetrates the embryo or emerging seedling tissues through mechanical pressure and enzymes, which dissolve those parts of the walls that hold the cells together. Further breakdown results as the fungus grows between and through cells. Other enzymes digest the proteins of the cell, and the fungus consumes the cell contents. Thus, infected seeds and tissues are killed and turn into a rotten mass consisting primarily of mycelium and undigested host parts. The severity of damping-off depends upon several factors, including soil moisture and pH. *Pythium*-induced damping-off is most severe in wet soils and at high pH.

Pythium species also can infect fleshy vegetable fruits or leaves in the field, in storage, in transit, or at the market. For instance, an entire cucumber fruit on wet soil may be rotten and soft 3 days after being infected by *Pythium* species.

Management

The most efficient management approach to damping-off is to avoid it altogether. Commercial cultivars or varieties of most plants are susceptible to damping-off pathogens. Significant progress has been made in reducing Pythium damping-off by treating seeds of some plants with biological control agents. Once damping-off has started in a plant bed, a seedling flat, or the field, it may be difficult to contain. Several general practices are employed to prevent damping-off:

- Proper soil preparation and management to provide good aeration, drainage, structure, and water-holding capacity; fertilization based on soil test results to provide adequate plant nutrition.
- Proper soil treatment to reduce the level of fungi that cause damping-off (in the greenhouse).
- The use of certified, fungicide-treated seed with high germination.
- Proper seeding rates to avoid dense plant stands, poor air movement, and low light intensity.
- Strict sanitation to avoid reinfesting treated soil with disease-causing organisms.

Many outbreaks of damping-off in greenhouses can be attributed to poor sanitation practices after treating the soil. For small bedding-plant operations, it may be practical to buy a sterilized soil mix and eliminate the soil preparation, fertilization, and treatment practices described above.

Once damping-off has started in a bed or flat, it may be managed by providing drier conditions for seedling growth. This can be done by ditching inside and outside the plant-bed structure to give efficient water drainage and by removing dirty covers, overhanging branches, or other obstructions that shade the bed, to increase the amount of light.

If these preventive management measures fail, several fungicides are available that may be effective if applied as a drench or heavy spray as soon as the first symptoms of damping-off are observed. Also, several fungicides are available as seed treatments that help prevent damping-off in the field. The fungicides in combination with proper growing conditions may stop the spread of damping-off. Rapid identification of the causal fungus should be obtained so that proper chemicals can be applied. Several applications of the fungicide may be necessary. Check the label carefully to be sure the proper fungicide and application rate are used on a particular crop.

RHIZOCTONIA ROTS

The pathogen *Rhizoctonia solani* Kühn is one of the most important plant-pathogenic soil-inhabiting fungi. It is often placed in the Deuteromycotina (or Fungi Imperfecti), but does have a sexual stage, *Thanatephorus cucumeris* (A. B. Frank) Donk, which is in the Basidiomycotina. The sexual stage occurs infrequently in the field in temperate zones, but occurs frequently in moist, tropical areas. It causes more different types of diseases to a wider variety of plants than any other plant-pathogenic fungus known. It even parasitizes other fungi. This destructive fungus occurs worldwide in crop plants and also is common in forests and noncultivated desert land.

Rhizoctonia Diseases and Their Symptoms

One of the most common diseases caused by *R. solani* is seed rot. The fungus invades the planted seed from infested soil; the seed decays and never germinates. Often the rotted seed provides food for the fungus, which then grows through the soil and infects adjacent seed. In other instances the seed germinates, but the fungus attacks and destroys the seedling before it emerges from the ground. These seed rots and preemergence damping-off result in poor stands, particularly in seed flats and nursery beds. Any condition, such as cold soil, that delays germination may increase the chances of damping-off. Young seedlings also are attacked after they emerge from the soil. Young, succulent tissues are easily invaded, and decay of the young stem at the soil level causes the plant to fall over. Damping-off tends to occur in circular or irregular patches that represent foci of infection.

As plants mature, they become increasingly resistant to the seedling diseases caused by *R. solani*. Some species, however, including cabbage, forest tree seedlings, snap bean, and tomato, are attacked through the transplant stage. The dark brown discoloration of the stalk at or near the soil line, extending upward for several centimeters, is a characteristic symptom of diseases known as sore shin or wire stem (Fig. 12.5). Affected stems are smaller than normal, discolored, tough, and woody. When diseased plants are set in the field during conditions favorable for disease, a poor stand often results.

Cereal grains in some countries sustain serious losses from root and crown rot caused by *R. solani* and another species, *Rhizoctonia cerealis* Van der Hoeven. Underground fleshy roots often are affected by cankers as well as by crown rot. Carrots, celery, potato tubers, radishes, sugar beets, tulip bulbs, and turnips become discolored and rotted by the fungus.

During warm, wet weather, *R. solani* may spread through the top of the plants; this disease is called web, leaf, or thread blight. The closer plants are to each other, the

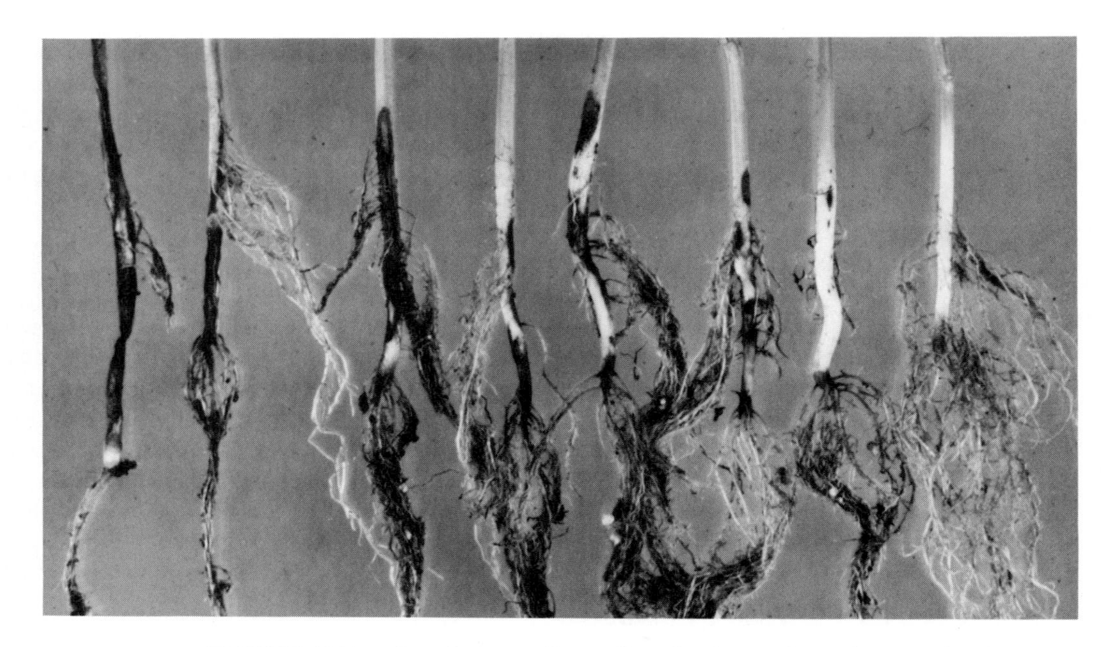

FIGURE 12.5. Snap bean seedlings affected with sore shin or hypocotyl and root rot, caused by *Rhizoctonia solani*. The seedling on the right is healthier than the others. *Courtesy G. Abawi.*

greater the spread of aerial blights. Beans, tall fescue, and other turfgrasses, lespedeza, rice, and soybeans are attacked this way.

Brown patch of turfgrasses is an important aerial blight (Fig. 12.6). The diseased areas, often 1 to 2 m in diameter, have a dark purple-green advancing margin in which the webbed mycelium is visible in the early mornings. At first the grass appears water-soaked; soon it turns dark; finally the leaves wilt, then die, and turn light brown. During hot, humid weather a smoky gray to black border of wilted, "webbed" grass, called a smoke ring, may be seen in the early morning, bordering the light brown diseased area. The crowns and roots are rarely invaded. Leaves are infected through stomates and mowing wounds. Sclerotia are formed near the base of the plants. Because the grass grows close to the soil and when there is much dew, high humidity enables the fungus to row rapidly. "Poling" (using a long bamboo pole as a brush) the grass in the early morning to knock off the water droplets reduces the rate of fungus spread. Isolates of *Rhizoctonia solani* that cause brown patch require high temperatures and high relative humidity to thrive and produce the disease symptoms.

Although most foliar blights caused by *R. solani* occur on tropical crops, an interesting foliar disease called target spot has occurred on tobacco in the southeastern United States since 1984. The disease is caused by the teleomorph or sexual stage of *R. solani*, which is known as *T. cucumeris*. The disease is initiated by basidiospores that form under conditions of high relative humidity and moderate temperatures [20-30°C (68-86° F)]. The basidiospores are blown from plant to plant by wind, but

FIGURE 12.6. Brown patch, caused by *Rhizoctonia solani*, on Bermudagrass. Some diseased areas may be 2 m (6 ft) or more in diameter.

the fungus survives as sclerotia or mycelia in organic matter in soil. Losses were high in 1989 when weather conditions favored the disease, but were negligible in 1990 with unfavorable weather.

Apparently, any aboveground plant part, especially fruits, that comes into contact with soil infested with *R. solani* may become rotted. Infection occurs through wounds or the intact skin. Tomato fruits, either green or ripe, develop small, firm, brown spots where they touch the soil. These later enlarge, become soft, and have concentric zones of dark tissue. Mycelium and sclerotia may develop on the surface.

Belly rot of cucumber is caused by *R. solani;* similar spots occur on bean, eggplant, pea, and peppers. Small infections at harvest time continue to develop in storage; infected seeds help start the disease in the next crop.

On Irish potatoes, one of the best-known symptoms of *R. solani* is the appearance of black sclerotia on potato tubers. It is called "dirt that will not wash off," black scurf, black speck, and black scab. The dirty, ugly black mycelium really does not reduce the value of the potatoes; it just makes them unattractive. However, if potatoes are stored under warm, wet conditions, the fungus will begin growing, and the potatoes will rot.

Causal Agent and Disease Cycle

Diseases caused by *R. solani* may be confused with diseases caused by *Pythium* or *Phytophthora* species; however, the characteristic mycelium of *R. solani* readily distinguishes it. The hyphae are broad and coarse (8-12 μm in diameter), colorless

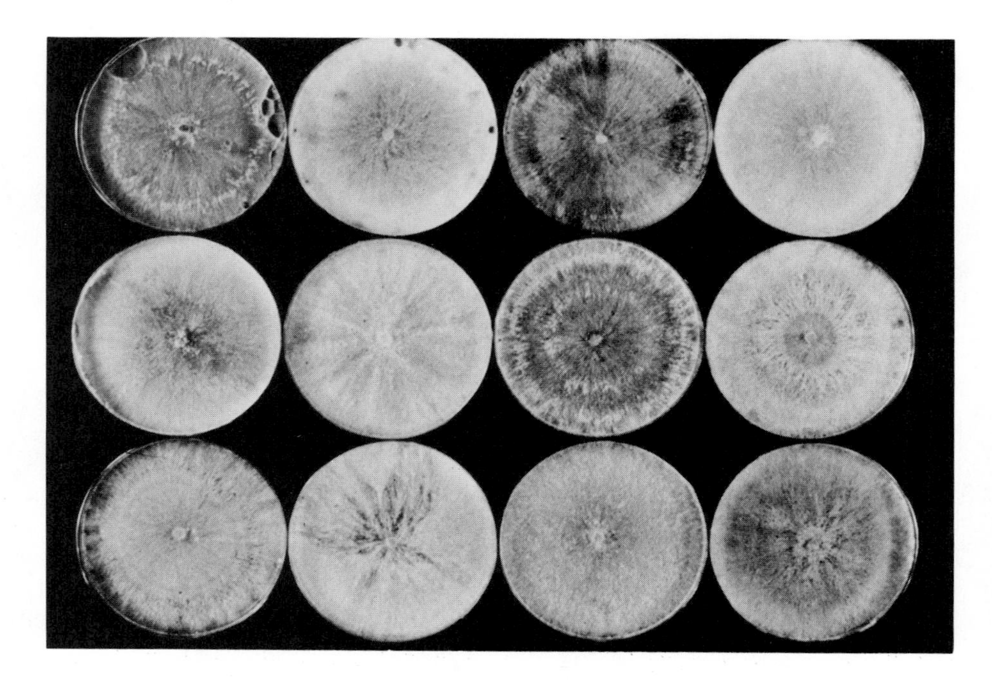

FIGURE 12.7. Twelve isolates of *R. solani* growing on an agar medium in petri dishes. The isolates vary in color, abundance of aerial mycelium, sclerotial formation, and pathogenicity.

when young, but turning dark brown with age. Young branches are inclined at angles of 45° to 90° from the direction of growth of the parent hyphae and usually are constricted at the point of origin. A septum or cross wall is always present near the base of the branch. Sometimes tan to black sclerotia ranging up to 6 mm in diameter form on decaying plants, and these help identify the pathogen.

There are many different strains of *R. solani* (Fig. 12.7). Some grow more rapidly above ground; others are adapted to life in the soil. Strains also differ in host range, pathogenicity, pH response, and temperature requirements.

R. solani is a primitive parasite with simple food requirements. It grows rapidly, and as a saprophyte it can live on dead or decaying plant debris for many months. It persists in the soil as hyphae or sclerotia, which may contain large amounts of stored food and persist for long times in soil. Infections often start from sclerotia (Fig. 12.8), which can be carried on roots, seed, and tubers and in soil dry enough to blow as dust. Even in moist soil, sclerotia can survive 5 years.

Exudates that leak out of germinating seedlings and actively growing roots stimulate the growth of *R. solani* from sclerotia. The sclerotia germinate by hyphal growth. Once contact with host tissue is made, the hyphae penetrate directly or form "infection cushions," composed of hyphal clusters on the susceptible tissues, which cause discoloration and death. The hyphae penetrate living cells through this dead tissue or invade directly through cracks and wounds. The invading hyphae grow rapidly

FIGURE 12.8. Top: Infection hyphae of *R. solani* attached to the outside of a bean stem showing typical branching and cross walls, magnified about 400 times. Bottom: Section of a bean stem cut through the center of the dome-shaped infection of *R. solani* with numerous hyphae already growing in the stem tissue. *Courtesy T. Christou.*

through the host tissue, causing it to turn brown and collapse. The fungus produces numerous enzymes that digest the host tissue and make it suitable for fungus use. Under favorable conditions, symptoms appear within 3 to 7 days after infection.

Diseases caused by *Rhizoctonia* species can occur in relatively dry soils and over wide pH and temperature ranges. They occur at cool temperatures in seedbeds and in the field when temperatures are high; they usually are most severe at temperatures that are unfavorable for the host. Excess solutes in the soil, resulting from overfertilization, increase the incidence of these diseases. Increased disease severity results from the interaction of *R. solani* on plants already infected with root-knot nematodes.

Management

Several basic guidelines, if followed, will aid in managing diseases caused by *Rhizoctonia* species:

- Do not use seed contaminated with sclerotia of *R. solani.*
- Do not use infected corms, roots, or tubers for planting stocks.
- Use certified seed treated with fungicides whenever possible.
- Use disease-free transplants grown in soil treated with aerated steam, dry heat, or methyl bromide.
- Plant seed at the proper depth or practice shallow seeding so that seedlings emerge as quickly as possible.
- Plant at temperatures favorable for the host.
- Use resistant cultivars when available.

To manage brown patch of grasses, avoid seeding in hot weather and overfertilization with nitrogen; mow frequently to prevent matting of foliage and to permit more light penetration and aeration. Rake and remove clippings if there are heavy accumulations. Water deeply and infrequently instead of lightly at frequent intervals. Spray with recommended chemicals when the disease is first noticed and at regular intervals during warm, moist weather.

FUSARIUM ROOT ROTS, YELLOWS, AND WILTS

Fungi in the genus *Fusarium* contain many species that attack a large number of crops, causing seed rots, root rots, foot rots, stalk rots, wilts, yellows, and ear and kernel rots. Most *Fusarium* spp. were originally placed in the Deuteromycotina, but many have sexual or teleomorph stages that are in the Ascomycotina. This variable fungus, composed of many pathogenic species and strains, lives in the soil and attacks all cultivated crops and many wild plants. The fungus is capable of attacking all parts of the plant. Large amounts of loss may result, especially on susceptible cultivars when weather conditions are favorable for disease development.

Fusarium Diseases and Their Symptoms

Fusarium seed rots and root rots occur when spores carried on seed or present in soil infect germinating seedlings. Infected seed often rot and die before they emerge from the soil. Small lateral roots usually are killed soon after infection (Fig. 12.9).

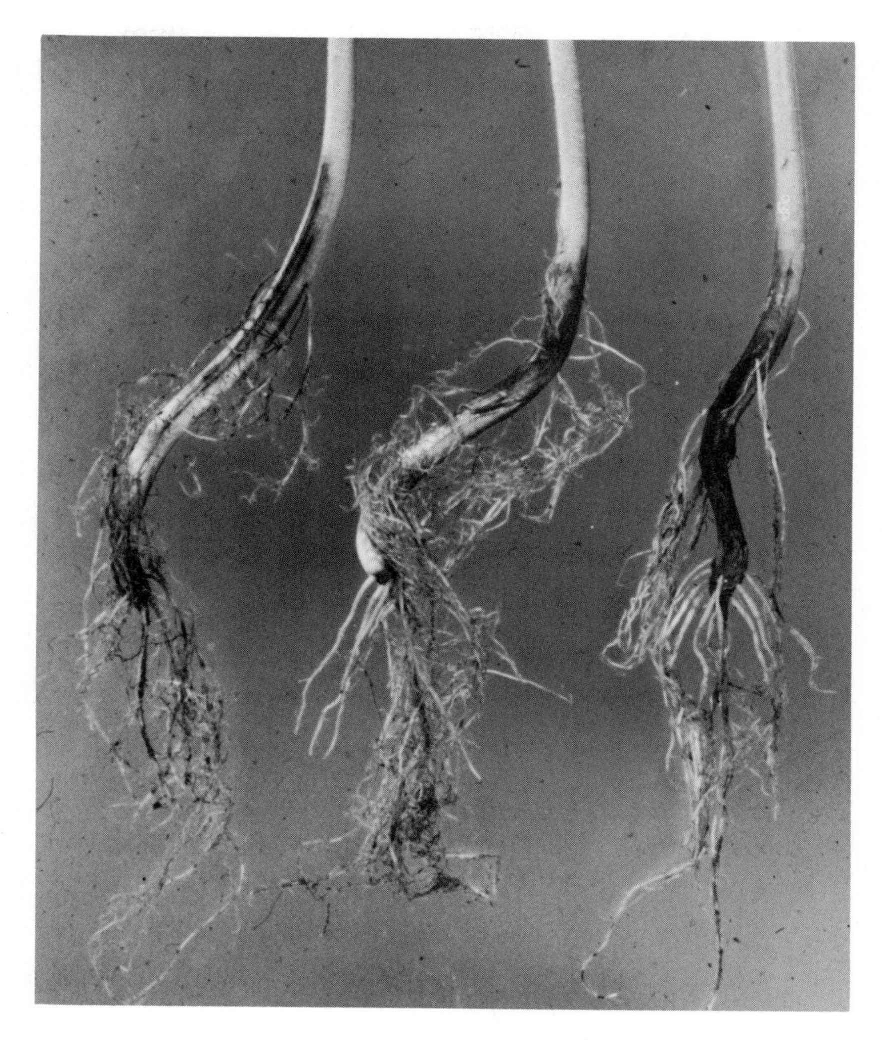

FIGURE 12.9. Fusarium hypocotyl and root rot of snap beans. *Courtesy G. Abawi.*

Fusarium stalk rot of corn is present each year and may cause considerable damage, particularly if abundant rainfall occurs during the latter part of the growing season. Stalks previously injured by cold, leaf diseases, or chinch bugs are especially susceptible to attack. Diseased stalks ripen prematurely and are subject to excessive stalk breaking. Not only do stalk rots add to the cost of harvesting, but the ears on broken stalks may touch the ground, increasing their chances of rotting.

Ear and kernel rots of corn may cause serious losses when conditions are favorable for their development. Severe infection not only reduces yield but also lowers the quality and grade of the grain produced. In addition to these losses, the infected kernels may contain substances toxic to animals and are worthless as seed. The fungus typically causes a pink reddish rot of the ear. Apparently healthy looking kernels may be heavily contaminated with spores during harvesting or shelling. If storage conditions for the corn are not proper, particularly if too much moisture is present, the fungus may grow on the stored grain and produce toxic substances. Spores carried on the seed or in the soil will infect the new corn crop to repeat the infection cycle.

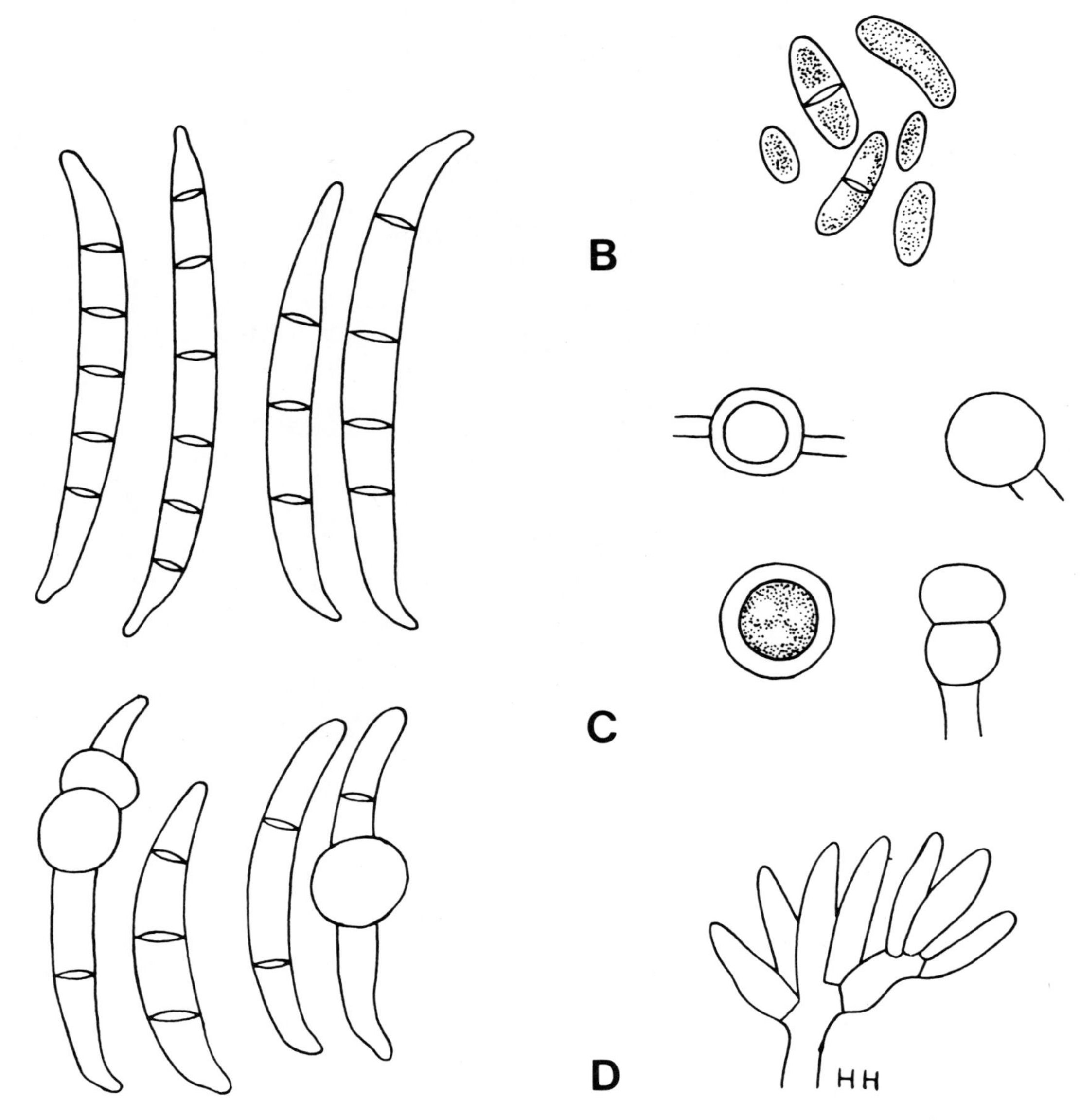

FIGURE 12.10. Reproductive structures of *Fusarium solani*. A—macroconidia; B—microconidia; C—chlamydospores; D—conidiophore. A, B, and C magnified about 1000 times; D magnified about 380 times. *Drawing by H. Hirschmann after S. Johnson.*

Fusarium root rot is a limiting factor in snap bean and pea production. It is particularly severe during cool, wet soil conditions. Depending on soil and environmental conditions, *Fusarium* spp. may act alone as a rotting agent; but more often they act in combination with species of *Pythium, Rhizoctonia,* and other soilborne pathogens. Often such plant disease interactions make the growing of legume crops unprofitable in certain years.

Other *Fusarium* spp. penetrate roots principally through wounds made by nematodes. The fungus grows through the water-conducting tissues, producing toxins that kill the cells, stunt the plant, and yellow the leaves. Often the plant wilts and dies from lack of water. Symptoms usually appear only on one side of the stem and progress upward until the foliage is killed and the stem dies. The water-conducting tissues just under the bark turn brown; this discoloration is visible in cross sections of stems near the base of infected plants. The fungus mycelium can be seen, with the aid of a microscope, growing in these tissues.

Causal Agent

Fusarium spp. produce three kinds of spores (Fig. 12.10). *Microconidia* are colorless, single-celled, spherical spores about 6 to 15 μm in length and 3 to 5 μm in diameter. The sickled-shaped, colorless *macroconidia* have three to five cross walls and average 30 to 50 μm in length and 2 to 5 μm in diameter. The smooth, spherical, single cells called chlamydospores are produced on older mycelium and average about 10 μm in diameter. All three spore types are produced in soil or on infected plants. After infected plants die, the fungus and its spores are returned to the soil, where they may persist indefinitely.

Contaminated seed, infested soil, and infected transplants help spread the pathogen from one place to another.

Management

Seed treatment with the proper fungicides helps prevent seed rots and seedling diseases caused by *Fusarium* spp. *Fusarium*-free seedlings of transplant crops should be grown by treating seedbed soils with methyl bromide (see Chapter 7, Soil Treatment section). Resistant cultivars (cabbage, cantaloupe, tomato) should be used when available; rotation should be practiced, and root-knot nematodes controlled.

VERTICILLIUM WILTS

Verticillium wilts have a worldwide distribution but are most important in temperate regions. Species of *Verticillium* live in soil and plant residue and can attack over 200 species of plants. Many economically important crops such as cotton, potato, peanuts, alfalfa, tomato, pepper, watermelon, cantaloupe, apricots, cherry, peach, and mint can be attacked.

Verticillium Diseases and Their Symptoms

The symptoms of Verticillium wilts are nearly identical to those of Fusarium wilts. In fact, on plants attacked by both pathogens it often is difficult without laboratory examinations to determine the actual causal agent. In many hosts, infection by *Verticillium* spp. results in defoliation, gradual wilting, and death of branches beginning at ground level and proceeding up the stem, or, in some cases, a sudden overall wilt and plant death. Older plants that become infected may be stunted. The rate at which symptoms develop depends on the susceptibility of the host plant.

Verticillium wilt generally is most severe during cool, wet weather and may not be evident during warm, dry weather. Verticillium wilts occur at lower temperatures than Fusarium wilts. Also, symptom development is usually slower for Verticillium wilts than for Fusarium wilts.

In cotton, symptoms often appear first about blossom time following rainy weather. Chlorotic or yellowed areas appear on leaf margins and between the veins of lower leaves. These areas eventually turn brown and become necrotic. The chlorosis and necrosis of leaves moves up the plant, and defoliation may occur. Fruiting branches and bolls may be dropped. Light to dark brown vascular discoloration often is present in the stem and branches of the lower half of the plant. In susceptible plants, the discoloration may extend to the top of the plant.

On potatoes, Verticillium wilt causes early senescence of plants. Leaves become pale green or yellow and die earlier than normal. Thus, in potatoes the disease is often called "early dying." Plants may wilt, especially on sunny days. Single stems or leaves on one side of a plant may wilt first. If the stems are cut at ground level and a long slanting cut is made toward the top of the stem, the vascular tissue can be seen and will be light brown. Tubers from infected plants may develop a light brown discoloration of the vascular ring.

Causal Agents

Two species of *Verticillium* cause wilt diseases: *Verticillium albo-atrum* Reinke & Berth. and *V. dahliae* Kleb. Both species produce spores that are short-lived. *V. dahliae* also produces microsclerotia, whereas *V. albo-atrum* produces mycelium with dark, thick-walled cells. Some strains of *Verticillium* attack only certain hosts, but most strains attack a wide range of host plants. The fungus can enter the young roots of plants directly or through wounds. After infected plants die, the fungus and its spores are returned to the soil. Microsclerotia of *V. dahliae* can survive in soil for at least 15 years, and mycelium of both *Verticillium* spp. can survive in perennial weed hosts. These pathogens can spread among fields on farm equipment or in infested soil that is moved by wind or water. The seed of some crops also can be infected by *Verticillium* spp.

Management

The use of resistant or tolerant plant cultivars is essential. Susceptible plants should not be grown in infested fields. Rotation with nonhost crops and effective weed management programs are helpful; however, inoculum of the pathogens can survive long periods in soil. Nematode management also is important because certain nematodes can increase the incidence and severity of Verticillium wilts, particularly in potatoes. In irrigated fields, soil should not be kept too wet because excessive moisture favors the development of Verticillium wilts. Plant parts should be removed from equipment moving between fields, particularly when moving between farms. In some cases, soil fumigation can be profitable if it is used to protect high-value crops, but fumigation is too expensive to use on large areas. Soil solarization, where soil is heated to high temperatures by energy provided by the sun, may be useful for managing *Verticillium* spp. in regions with high summer temperatures and low rainfall, such as California and the countries of the Middle East.

SOUTHERN BLIGHT

Southern blight, also known as southern wilt and southern stem rot, is a serious and frequent disease of many crops in the southeastern United States and in semitropical and tropical areas of South America and Africa. It attacks a number of vegetable crops, including bean, cantaloupe, carrot, pepper, potato, sweet potato, tomato, and watermelon. In addition, several field crops (cotton, peanut, soybean, sugar beets, tobacco) and forage crops (clover, alfalfa) are affected.

Southern blight occurs to some extent every year and at times causes severe crop losses. Losses are more severe in fields where the same crop has been grown for several years and may be particularly severe on plants in fields with minimum till or no-till practices. Affected plants usually are scattered, but occasionally large areas ("hot spots") of diseased plants are found in a field. The fungus may also decay harvested produce, particularly carrots.

Symptoms and Signs

The disease is recognized by wilting (Fig. 12.11) and yellowing of leaves; when the plant is pulled up, it is apparent that the lower stem and upper roots are infected. In watermelon, one or two runners may be affected; in cantaloupe, the melons usually are affected first. The edible roots of sweet potatoes have circular, sunken, dark gray surface spots 5 to 10 mm in diameter. The stems of erect plants such as green bean, pepper, potato, tobacco, and tomato usually are rotted at the soil line. The fungus causes a watery fruit rot on cantaloupe and tomato; in carrot, the whole root becomes decayed.

The rots are not associated with an offensive odor, at least initially. A white, moldy growth is evident on affected stem tissues and adjoining surface soil; later smooth, light tan to dark brown "mustard seed" sclerotia (about 2 mm in diameter) are evident in the mycelium. The sclerotia are diagnostic for the pathogen (Fig. 12.12).

Causal Agent and Disease Cycle

Southern blight is caused by the soilborne fungus *Sclerotium rolfsii* Sacc. The fungus is in the Basidiomycotina, but the sexual stage is seen only rarely in nature. The white mycelium grows in strands. The sclerotia begin to form on stems, soil, etc. after about a

FIGURE 12.11. Healthy tomato plant (left) and tomato plant (right) with typical symptoms of southern blight caused by *Sclerotium rolfsii*. Sclerotia will form at the base of the diseased plant.

week as small intertwined, white tufts of mycelium, but become yellow to brown with age. Although the sclerotia may survive 5 years in dry soil, they live for only about 2 years or less in moist soil. Once the food source is used up, the mycelium dies after a few weeks.

The southern-blight fungus overwinters as sclerotia and mycelia in host debris in the soil. The fungus generally is restricted to the upper 5 to 8 cm of soil and will not survive at greater depths. In most soils the fungus does not survive in significant numbers when a host is absent for 2 years or more. The fungus is more active in warm, wet weather than in cool, dry weather; it requires the presence of undecomposed crop residue to initiate infection.

Southern blight is favored by warm weather, high soil moisture, and aeration. Other factors influencing outbreaks are not fully understood. Both cropping history and cultivation practices during late summer, as well as climatic factors, influence disease severity. Outbreaks usually are more severe following excessive hilling or bedding of plants by cultivation or when there is an accumulation of fallen leaves around the base of plants. The fungus is capable of growing on undecomposed litter from a previous crop or other slowly decomposing organic matter, using this material as a food base. Once established and making vigorous saprophytic growth, the fungus can infect uninjured host tissue although injuries make penetration easier.

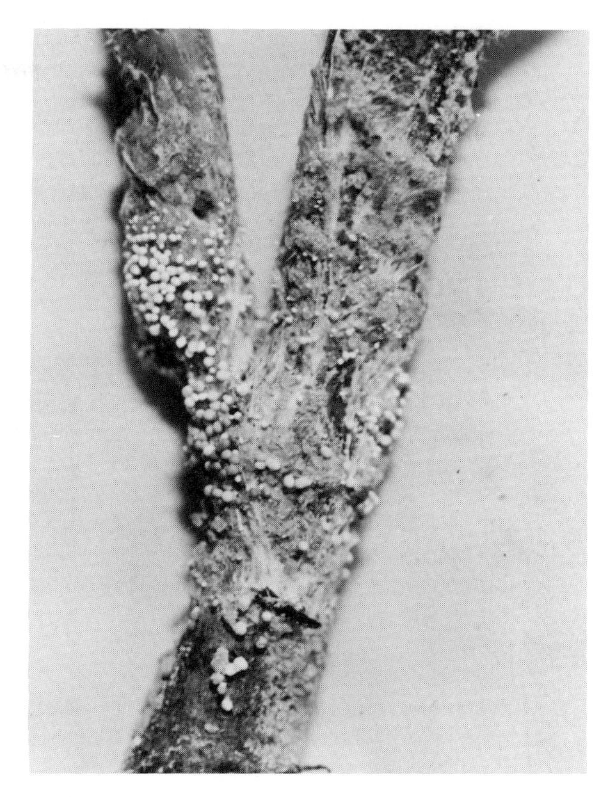

FIGURE 12.12. Sclerotia (white to brown in color in nature) of the *Sclerotium rolfsii* forming on a diseased soybean stem. *Courtesy D. P. Schmitt.*

Management

The control of southern blight is difficult, but losses can be reduced by following an integrated program of disease management over a period of several years.

Proper land preparation is vital in managing southern blight. The previous crop must be well decomposed before planting, which may require disking the field several times in the fall and in the spring. The previous crop litter should be buried with a moldboard plow equipped with heavy-duty concave disk-type coulters to a depth of 20 to 30 cm; this plow depth must be below the depth of later cultivation equipment. The crop litter should be covered with 8 to 14 cm of soil. None of the buried litter should ever be brought back near the soil surface during the current season by cultivation.

Other important management practices include the following:

- Avoid problem fields.
- Rotate susceptible crops with corn, small grain, or other grass crops.

- Do not throw soil with debris against plant parts during the growing season.
- Control foliar diseases because dead leaves on the ground may trigger infection. Weeds should be controlled early in the season for the same reason.
- Use a fungicide in the transplant water or as a drench spray for tomatoes and peppers. Dust formulations also may be applied over V trenches prior to transplanting. Follow all label directions carefully.

CROWN AND STEM ROTS AND WATERY SOFT ROTS

Several species of the soilborne fungus *Sclerotinia* are causal agents for crown and stem rots and watery soft rots, which occur in more than 170 species of plants worldwide. Especially susceptible vegetables include bean, cabbage and other crucifers, carrot, cucumber, eggplant, Irish potato, lettuce, pepper, and squash. Many other crops, such as alfalfa, clover, soybean, peanut, tobacco, and turfgrasses, are susceptible. Some *Sclerotinia* species also attack ornamentals. On many vegetable crops, diseased plants wilt and flop in the field; but of equal importance is a watery soft rot that continues to damage the harvested vegetables during transit and storage.

Symptoms and Signs

On vegetable crops, diseases caused by *Sclerotinia* species can be recognized by a soft, watery rot with white, moldy growth on stems, petioles, and leaves. Infection may start on leaves that touch the soil or on flowers (as in snap beans). The fungus grows through the petiole to the stem, turning it black and often girdling it. Then the plant flops. If conditions remain moist, a large amount of cottony, moldy growth can be seen on the dead tissue. As this growth progresses, hard, black, irregularly shaped sclerotia form on the surface or in the pith of the stems (Fig. 12.13). The sclerotia look like rat dung and are diagnostic signs for the disease. Sclerotia range from 2 to 10 mm in length, tend to be about two to three times longer than they are thick, and are white to pinkish inside. After the progress of the disease apparently has stopped the line of demarcation between healthy and diseased tissue is very sharp. Often the diseased tissue is a light straw color.

On alfalfa, clovers, and other forage legumes, the disease is called crown and stem rot, but the fungus attacks all parts of the plant. Symptoms first appear in the fall as small, brown spots on leaves and petioles. Frost-damaged leaves are most susceptible to attack by the fungus. Heavily infected leaves turn gray-brown, wither, and become covered with the white, cottony mycelium, which spreads to the crowns and roots. By late winter or early spring, the crowns and basal parts of infected stems show a brown soft rot that extends downward into the roots. Consequently, part or all of the new growth of the infected plant wilts and dies. Diseased stolons become soft and rotten. Black sclerotia form on the white mycelium. On peanuts, symptoms are similar, but the disease occurs in late spring and summer.

In newly seeded crops of forage legumes, crown and stem rots cause severe reduction in stands. The damage often is confused with winter killing. The disease

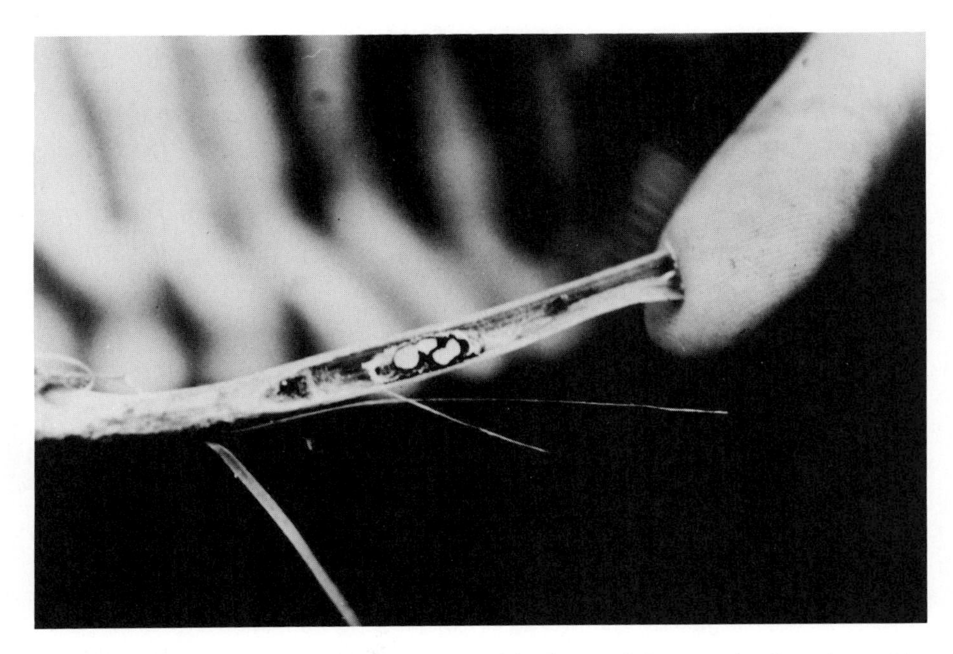

FIGURE 12.13. Irregularly shaped sclerotia of the fungus *Sclerotinia* that have formed in the pith of a plant stem. The sclerotia have a white interior (medulla) and dark exterior (rind). *Courtesy NCSU Plant Pathology Photo Collection.*

occurs in patches throughout a field, but when conditions become favorable, the patches merge, and all plants in large areas of a field may be killed.

On closely mowed grasses such as turf grasses on golf greens, the disease (called dollar spot) appears as light yellow or tan circular spots 5 to 6 cm in diameter, about the size of a silver dollar, hence the name. As the disease progresses, these spots merge into large areas of dead, straw-colored grass.

Sclerotinia species also cause a petal blight or flower spot of azaleas and camellias. Infected petals turn brown and lifeless before they fall to the ground. Often the diseased flowers cling to the plant and present an unsightly appearance. Sclerotia form on the fallen petals beneath the plant.

Causal Agent and Disease Cycle

Sclerotinia spp. are in the Ascomycotina and overwinter (or oversummer) in the soil as black sclerotia that measure 5-10 mm × 3-6 mm in diameter. Sclerotia can survive up to 7 years in dry soil. However, if the soil stays warm and moist, the sclerotia will be killed in less than a year. Following a dormant period, if moisture and temperature are suitable, the sclerotia germinate by resuming vegetative growth or by forming as many as 35 small (5-mm-diameter) mushroom-like bodies, called *apothecia* (Fig. 12.14). The apothecia produce enormous numbers of spores, which are blown about and cause primary infections. They are the only infective spores in the disease cycle.

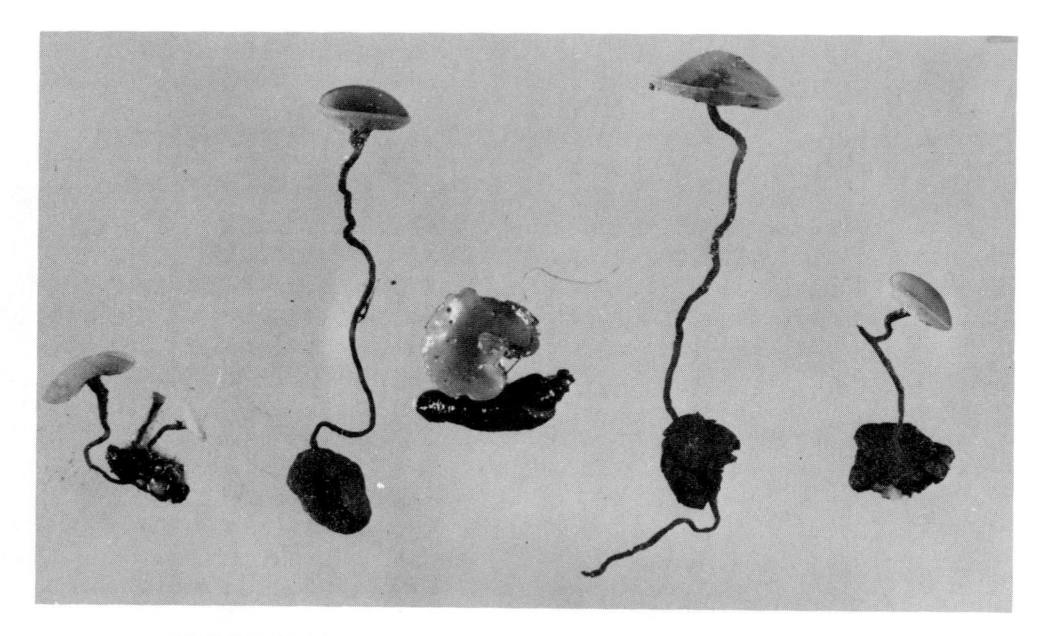

FIGURE 12.14. Sclerotia of *Sclerotinia* from which apothecia have grown (shown natural size). Ascospores will be formed in asci within the apothecia. *Courtesy D. E. Ellis.*

Once the fungus is established, it continues vegetative growth as long as there is sufficient moisture and host tissue available. The disease develops when moisture is high and the temperature cool. Temperatures between 15°C (59°F) and 20°C (68°F) are ideal, but the disease will progress slowly at 3°C (32°F) and at temperatures as high as 25°C (77°F). For the sclerotia to germinate and form apothecia, the temperature must be below 20°C (68°F); for spores to cause infection, the relative humidity must be above 90%.

Management

Management of diseases caused by *Sclerotinia* spp. is difficult. Clean cultivation, deep plowing to bury sclerotia, and long rotations are helpful. Spraying with fungicides may reduce the damage on field crops, vegetables, and ornamentals. The most promising method of management for pasture legumes appears to be the breeding of resistant and/or adapted cultivars.

Watery soft rot can be controlled on plant materials in storage and transit by harvesting and shipping disease-free stock only. The disease can be held in check by precooling before loading for shipment and maintaining storage and transit temperatures as near 0°C (32°F) as feasible.

Azalea and camellia petal blight can be managed by fungicidal sprays. Winter applications of fungicidal dusts or drenches to the soil underneath plants severely infected the preceding spring will help kill overwintering sclerotia.

Management of *Sclerotinia* species on greenhouse crops requires continuous efficient management. Greenhouse soils should be fumigated with methyl bromide or heated to kill the sclerotia. At the end of the season, all plants should be removed from greenhouses. Livestock manure and plant mulches should not be used unless disinfested by heat or methyl bromide. During the growing season, the surface soil should be kept as dry as possible and the air at as low a humidity as possible. Fungicides should be used as needed.

The most effective management of dollar spot on golf greens is to prevent outbreaks by maintaining healthy vigorous turf with proper cultural practices, such as the following:

- Providing adequate fertility based on soil test recommendations (nitrogen-deficient turf is more susceptible to dollar spot than that with adequate nitrogen).
- Reducing thatch more than 2 cm thick by proper mowing.
- Maintaining adequate soil moisture. Avoid light, frequent watering; deep, infrequent watering is more effective during drought.
- Removing grass clippings where disease is active.
- Cleaning mowers and other equipment of infected grass before moving to uninfected areas.
- Using recommended fungicides. Repeated use of some chemicals may result in the development of tolerant strains of the fungus; therefore, use chemicals only where necessary, and alternate the use of different fungicides.

CYLINDROCLADIUM BLACK ROT

Cylindrocladium black rot of peanuts, caused by the fungus *Cylindrocladium crotalariae* (Loos) Bell & Sobers, first was identified on peanuts in Georgia in 1965. Since then it has spread to most of the peanut-growing states in the southeastern United States, where it now causes serious losses every year.

Black rot has also been found on soybeans. Symptoms appear as clumps of three or four dead plants in a pattern similar to that which occurs with southern blight. Examination of stems near the soil line, however, often reveals the characteristic reddish-orange fruiting bodies of the sexual stage or teleomorph *Calonectria crotalariae* (C. A. Loos) D. K. Bell & Sobers called *perithecia*. Root systems are heavily rotted. The fungus also attacks other legumes, but does not appear to harm corn, cotton, or cereals.

Other species of *Cylindrocladium* cause a serious damping-off and seedling blight of both hardwoods and conifers. Before the discovery of adequate management measures, Cylindrocladium blight seriously hampered the production of healthy forest tree seedlings.

Symptoms and Causal Agent

On peanut, cylindrocladium black rot causes an overall stunting and yellowing of the plant and leads to death, in some cases, as the season advances. Previously healthy plants show signs of wilting on warm days and may eventually collapse. Infected plants have severely blackened and somewhat shredded tap roots almost completely stripped of lateral roots. The blackened area extends up to the crown of the plant and, in some cases, on the stems above the ground. Immature pods and pegs also show various stages of blackening. Within infected roots the fungus forms small, black, survival structures about the size of ground pepper. These structures, called *microsclerotia*, are released into the soil as the surrounding tissues decay and account for the ability of the fungus to survive in the soil for long periods (as least 3 years), even in the absence of susceptible host plants.

A certain diagnostic sign of black rot is the appearance of small, reddish-orange, spherical *perithecia* (0.4 mm in diameter) (Fig. 12.15). These reproductive structures of the fungus appear in dense clusters on stems, pegs, and pods just above or beneath the soil surface. They are especially abundant in moist areas under dense foliage. *Ascospores* are exuded from each perithecium as a viscous, yellow ooze, 2 to 3 weeks after the perithecium appears. These spores may play a role in the spread of the disease within a field but are not capable of long-distance movement under normal circumstances. Cylindrical colorless conidia (75 μm × 6 μm) also are produced abundantly and may contribute to local spread of the pathogen.

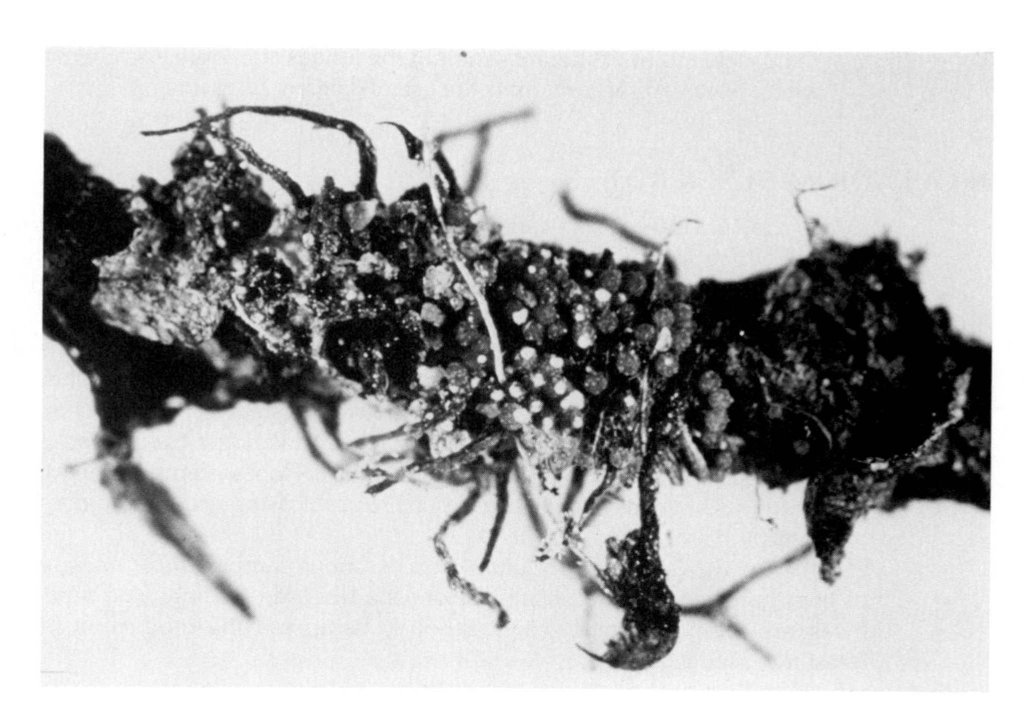

FIGURE 12.15. Perithecia of *Calonectria crotalariae* (imperfect stage *Cylindrocladium crotalariae*) on a plant stem. The white material at the top of some perithecia is a mass of exuded ascospores. *Courtesy NCSU Plant Pathology Photo Collection.*

Management

The most important means by which *C. crotalariae* spreads from farm to farm and within farms is by small microsclerotia (seed), which are present in diseased peanut roots and vines and are scattered in plant debris during digging and harvesting operations. Management of the disease is aimed primarily at curtailing the spread of the fungus in peanut fields.

In fields where only small localized areas of newly diseased plants occur, all affected plants should be removed from the field before harvest. In fields with larger infested areas, farmers have two options:

- Disking under all diseased plants before the digging and harvesting operation, making sure to avoid these areas with equipment during harvest.
- Digging out and removing all diseased plants where feasible and either burning them or burying them in a sanitary landfill.

Soil and plant trash should be removed from all equipment used in infested fields since the microsclerotia can be moved on it. Equipment should be washed with a strong stream of water immediately before it is moved from infested areas.

Small grain crops and corn are not susceptible to Cylindrocladium black rot and, therefore, do not increase inoculum carryover in infested fields. Consequently, rotations of 3 to 5 years, using corn, small grain, or perennial grasses such as fescue, are recommended for infested fields. In addition, cotton is a possible alternative crops on infested land. As a preventive measure, farmers should plant resistant cultivars that are now available. Until additional effective means for disease management are found, growers are urged to avoid planting peanuts or soybeans in fields where cylindrocladium black rot has occurred.

Effective management of the species causing Cylindrocladium seedling blight is possible by fumigating forest nursery seedbeds with methyl bromide. There is evidence, however, that strains of the fungus resistant to methyl bromide have developed. Therefore, in some areas it has been necessary to apply increasing amounts of the gas to nursery soils to eliminate the fungus so that disease-free seedlings can be produced.

13

Diseases Caused by Airborne Fungi

Because airborne fungi can be spread long distances on the wind, they can cause devastating losses, particularly on crops that are planted over large regions. In a given year, for example, the entire North American wheat crop from Mexico to Canada may be exposed to stem rust. These pathogens have the potential of causing, and in the past have caused, widespread epidemics on a variety of crops. Several of the most important diseases caused by airborne fungi are discussed in this chapter (Table 13.1).

RUSTS

Some people say that rusts are the most serious plant diseases in the world. They occur everywhere, reducing the yields and quality of the important grain crops on which all human civilization rests.

Thousands of years ago when our ancestors found they could use seeds of wheatlike grasses as a reliable and palatable food source, they ceased to be nomads. No longer did family groups have to wander from place to place to find food in the form of cereals; humans became tillers of the soil. Civilization changed as people began to cluster together in villages and cities. Jacob Bronowski says in his book *The Ascent of Man* that the single largest step in the ascent of humans was the change from nomadic life to that of settled village agriculture. Yet, down through the centuries rusts have threatened principal food crops.

Wheat is one of the most important food crop in the world today (the other is rice), providing 20% of the world's food calories and occupying approximately 20% of the world's cultivated land. Most is grown in the Northern Hemisphere, principally in North America and Europe. Wheat is an annual grass that is adapted to clay-loam soils and semiarid climates. Most wheat is eaten as bread, but it also furnishes livestock feed and various products for brewing and other industries.

TABLE 13.1. Some important diseases caused by airborne fungi.

Disease	Pathogen	Host(s)
Stem rust	*Puccinia graminis*	Wheat, many grasses
Leaf rust	*Puccinia recondita*	Wheat, many grasses
Cedar-apple rust	*Gymnosporangium juniperi—virginianae*	Cedar/apple
Coffee rust	*Hemileia vastatrix*	Coffee
Corn smut	*Ustilago zeae*	Corn
Covered smut	*Tilletia* spp.	Barley, oats, wheat
Loose smut	*Ustilago tritici*	Barley, oats, wheat
Late blight	*Phytophthora infestans*	Potato, tomato
Downy mildew	*Plasmopora viticola*	Grape
Downy mildew	*Pseudoperonospora cubensis*	Cucurbits
Powdery mildew	*Erysiphe* spp., *Podosphaera* spp., *Spaerotheca* spp.	Many plants
Blast	*Magnaporthe grisea*	Rice
Ergot	*Claviceps purpurea*	Rye, many other grasses
Gray mold	*Botrytis* spp.	Many crops
Leaf spot	*Cercospora* spp.	Bananas, beets, celery, forage legumes, peanuts, soybeans
Leaf blight	*Bipolaris maydis*	Corn
Brown spot	*Bipolaris oryzae*	Rice
Leaf spot	*Alternaria* spp.	Many crops
Anthracnose	*Colletotrichum* spp.	Many crops
Scab	*Venturia inaequalis*	Apple
Soft rot	*Rhizopus* spp.	Crops with fleshy organs

Wheat falls prey to several rust diseases (stem rust, leaf rust, and stripe rust). Sometimes, annual crop losses to wheat rusts amount to more than one million metric tons worth billions of dollars. In years when wheat rust is severe, there is not enough grain to make bread. Famines result. So far the United States has never had a great famine. Most Americans have always had their "daily bread," and reports of widespread starvation are seldom heard. But in other countries famine is a way of life.

In addition to the rusts of cereal grains, rusts attack many other grasses, reducing their forage value. Cattle grazing on rusted grasses find them less palatable and sometimes develop allergies after eating infected forage.

Another rust, the cedar-apple rust, is an annual threat to apple crops. Many vegetables and ornamental crops also are infected by their own special rusts. Forest trees also are attacked. Fusiform rust of pine, one of the most important diseases of this valuable pulp and timber crop, is discussed in Chapter 14.

Stem Rust of Wheat

Symptoms and Disease Cycle

Wheat steam rust is caused by *Puccinia graminis* Pers. f. sp. *tritici* Eriks. & E. Henn. In the course of the disease cycle, five types of spores are produced. These are red spores (urediniospores), black spores (teliospores), basidiospores, pycniospores, and yellow spores (aeciospores). Figure 13.1 diagrams the life cycle of stem rust.

The first symptoms of stem rust on wheat are small yellow flecks that develop into long, narrow, yellow blisters or pustules on the stem, leaves, and leaf sheaths of young seedlings or plants ar any stage of growth. Even the glumes and beards become infected. The *pustules*, 3 to 10 mm in size, break open, revealing a powdery mass of brick-red or rust-colored spores. These are called summer spores, *urediniospores* (synonym: uredospores), or repeating spores. Each spore is about 25 μm long. A single pustule may produce more than 350,000 spores. The red spores escape from the pustules and are blown by the wind to other plants where they germinate, invade the plant, and produce more pustules.

The fungus does great damage to the host plant by using water and nutrients needed by the developing kernels. As a result, kernels are badly shriveled and wrinkled; many of them are so light they are blown out with the chaff in thrashing. Stems of infected plants are weakened and lodge or break over easily. Sometimes losses range up to 90%, and the crop is not worth harvesting.

Later in the season as the wheat plant approaches maturity, the red pustules turn

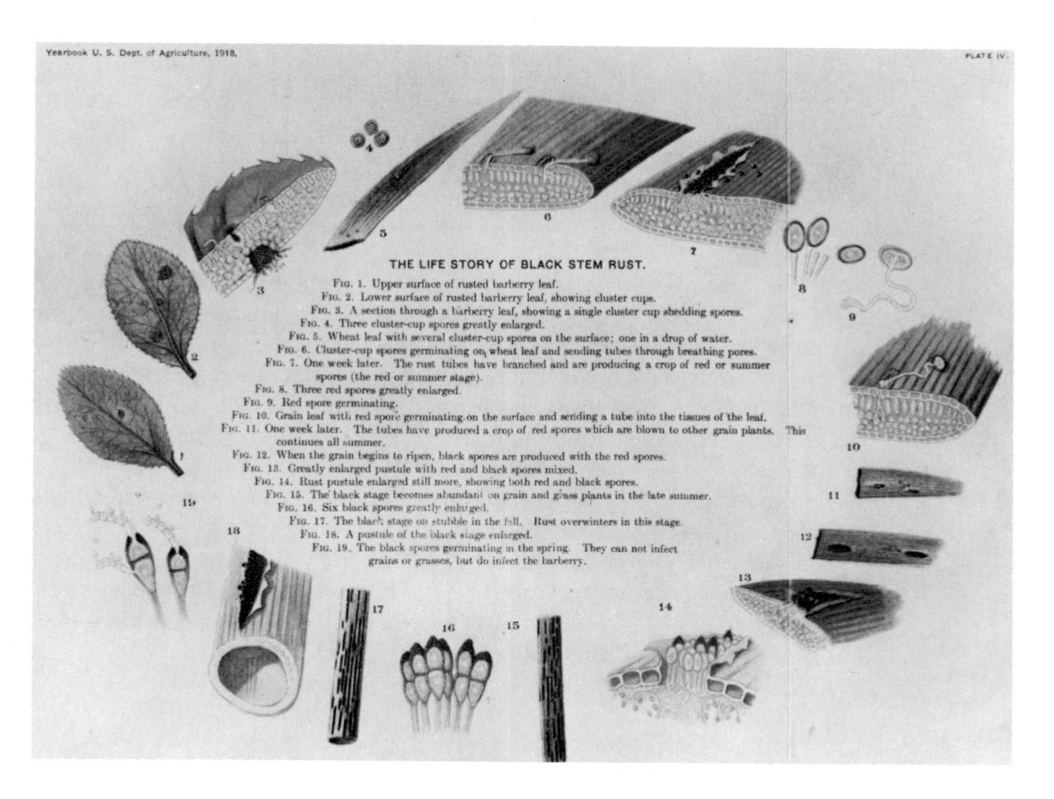

FIGURE 13.1. The life cycle of stem rust of wheat, published by E. C. Stakman in 1918. *Courtesy USDA.*

black because the fungus forms black, winter spores (*teliospores*) instead of red spores. The teliospores remain dormant all winter on the straw and chaff. In the spring, with the onset of warm weather, the teliospores germinate to form *basidiospores* (*sporidia*) of two different mating types (+ and −). These tiny, colorless spores can infect only the leaves of native barberry. If a few basidospores from an infested field are blown onto a wet barberry leaf and germinate, the spores infect the leaf and form yellow pustules (*pycnia*), sometimes up to 10 mm in diameter, within a week or 10 days.

The *pycniospores* formed in pycnia of one mating type (+ or −), when splashed by rain or carried by insects, can fertilize the *receptive hyphae* in the pycnia of an opposite mating type. Cluster cups (*aecia*) form on the underside of the barberry leaf and produce myriads of yellow spores (*aeciospores*), which are carried by the wind to infect any wheat plants growing nearby. As many as 70 billion aeciospores may be produced on one large barberry bush. The aeciospores can infect only wheat. If these spores land on a wet wheat leaf, they germinate quickly and form a new crop of pustules containing red spores in 7 to 10 days.

Dispersal

The fabled winds of the Great Plains of North America, heralded in song and story, are both a blessing and a source of potential disaster to crops. Wind is necessary for pollination of some cereal grasses because it carries tiny pollen grains from one flower to another, resulting in cross-pollination; fertilization follows within a few hours, and the developing seed enlarges rapidly as food is transported to and stored in the seed heads.

However, the wind is also the principal agent for moving the invisible rust spores from one location to another. Local breezes spread urediniospores from one field to another to start new infections; of more importance are those spores wafted upward 3 km or more where they may be blown hundreds or thousands of kilometers before they descend with the falling rain.

If weather conditions are favorable in the spring, as warm weather advances, the rust multiplies, and the spores sweep northward on the prevailing winds. A heavy rust epidemic in Mexico and Texas threatens the wheat crop in Oklahoma and Kansas. These rusts in turn supply the spores for infection in the Dakotas, Minnesota, and intermountain states, and Canada.

Spores blown southward in summer attack volunteer wheat, early-sown wheat, and certain grasses. The fungus overwinters as mycelium in such hosts and the next spring produces more red spores to begin another epidemic.

Management

RUST-RESISTANT CULTIVARS

The principal means of managing or containing wheat stem rust is the use of rust-resistant cultivars. Yet a resistant cultivar is useful for only about 5 years after its introduction before it becomes susceptible to the rust, and yields go down. This happens because there are several hundred parasitic strains or physiologic races of stem rust that differ in their ability to attack different cultivars of wheat; and more of these pathogenic races arise every year. As rapidly as scientists breed a new resistant cultivar, new fungus races arise that can attack it. There are three principal ways that new pathogenic races arise:

1. *Sexual matings:* When teliospores germinate, new combinations of genetic characters occur in the basidiospores. Upon fertilization in the pycnia, the different genetic characters are combined in aeciospores and result in the appearance of new races of the fungus. These new races may be able to infect new or old wheat cultivars that were previously immune to existing races.
2. *Hyphal fusions:* When urediniospores of different pathogenic races happen—by chance—to infect the same wheat leaf, some of the hyphae may fuse and exchange nuclei. As these hyphae grow, new pathogenic races result from the combined influence of the exchanged nuclei.
3. *Mutations:* Unexplained genetic changes occurring in the nuclei of a given race may change its parasitic capability, forming a new race.

Although new rust-resistant cultivars must be developed and introduced on a continuous basis, they are responsible in large part for the consistently high wheat yields obtained in the United States. Experts agree that without rust-resistant cultivars annual wheat yields in the United States would be much lower. Moreover, resistant cultivars provide insurance against disastrous losses, which could seriously disrupt the milling industry, lead to possible shortages of wheat supplies, and result in higher prices for bread and other wheat products. Resistant cultivars, however, are not foolproof; other methods to help manage the stem-rust fungus must be sought.

FUNGICIDES

Before 1960 there was little or no use of fungicides to manage wheat rust. Fungicides were expensive; they did not give good enough management; and income per hectare (1 ha = 2.4 acres) from wheat was low. Therefore, the cost/benefit ratio for fungicides was not high enough to justify their use in managing stem rust on wheat.

The use of chemical sprays, however, became economically feasible for some growers with the introduction of ULV spraying and new fungicides, which permitted effective treatment of wheat fields with as little as 8 liters of fungicide per hectare applied by helicopter or airplane. Two applications of appropriate fungicide mixtures, coordinated with forecasts that predict weather favorable for rust epidemics, may reduce stem-rust damage as much as 75% compared with untreated fields. These chemicals both protect the plant and eradicate the fungus. Therefore, two sprays, the first when rust is just barely detectable and the second 10 to 14 days later, can give control and are feasible, for example, in Europe, where the primary emphasis is on maximizing yield rather than maximizing profits.

Systemic fungicides that are applied to the soil as granules just before sowing and again in midseason as a spray are now available in some areas of the United States. They are used particularly when wheat is raised for certification (see Chapter 6), and thus has a greater than usual value.

Leaf Rust of Wheat

Leaf rust of wheat is found in every place where wheat is grown. Almost every year it causes damage in the southeastern United States and the Ohio and Mississippi valleys, where weather conditions are most favorable for its development.

Leaf rust usually attacks the leaf blades (Fig. 13.2) and sheaths—hence the name. Sometimes it occurs on the stem proper, just below the heads, and also on the chaff and beards. Leaf rust seldom shrivels the kernels, but it does reduce the size, number, and quality of grain. Furthermore, heavily rusted plants are weakened and dwarfed. In such cases, the yield may be reduced by as much as 90%.

There are many strains of the leaf-rust fungus (*Puccinia recondita* Rob. ex Desm.). The urediniospores are the only important ones in the United States. They are found all year on wheat in southeastern states. There the spores remain alive over the winter on volunteer wheat plants and fall-sown wheat. In the spring the spores are carried by wind to infect plants where the fungus may not have survived the winter. During warm moist weather a new generation of urediniospores is produced every week or 10 days. In some other countries, the alternate host (meadow rue, *Thalictricum* spp.) of *P. recondita* is present, and thus the sexual cycle does play a role in fungus reproduction.

To date, leaf-rust-resistant cultivars have been the principal means of reducing leaf-rust losses. However, for such cultivars to be of use, they must also be resistant to powdery mildew, another disease that is prevalent in the eastern United States.

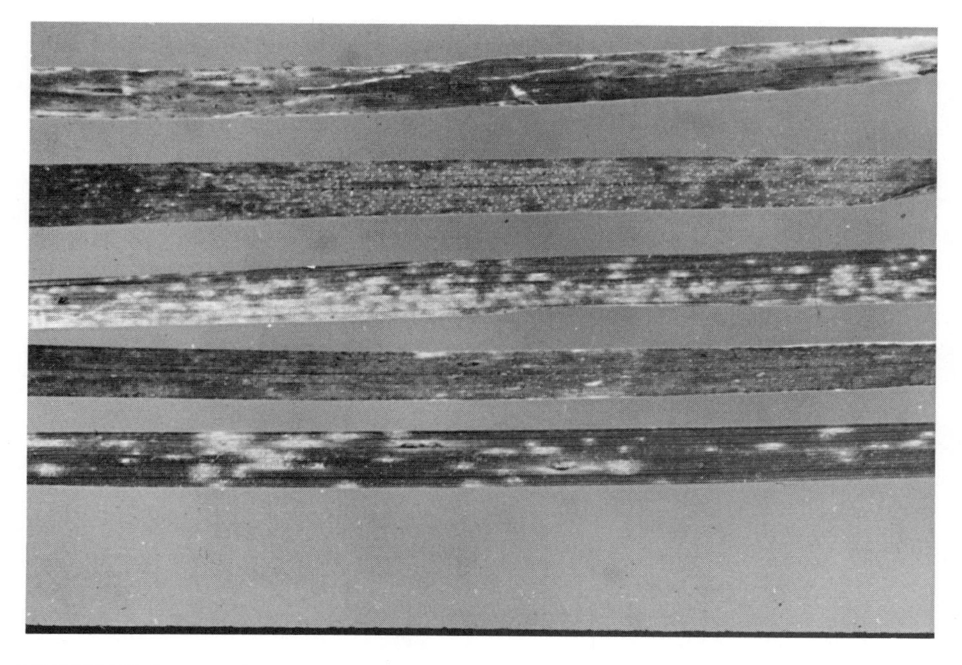

FIGURE 13.2. Leaf rust of wheat. The pustules are filled with masses of orange urediniospores.

Resistant cultivars must be replaced periodically because new fungus strains arise that can attack them, as described in the section on stem rust.

Cedar-Apple Rust

Apples and red cedars both grow well in the eastern United States. In the Midwest, where cedars do not grow naturally, the pioneers often planted them for windbreaks near orchards. Unfortunately, these early settlers did not know that both kinds of trees suffered from the same disease, cedar-apple rust. Where cedar trees are numerous and close to apple trees, rusts can cause serious losses of apples.

Symptoms

Rust diseases injure apples in several ways. Fruits of infected trees are deformed and ugly, and so reduced in grade that most of them are classified as culls. Considerable loss from secondary rots may follow rust infection. Fruit size is reduced by heavy foliage infection. Many infected leaves fall off by midsummer. The trees are weakened and may suffer winter injury. Young trees may die from infection and defoliation.

Disease Cycle

The fungus that causes cedar-apple rust (*Gymnosporangium juniperi-virginianae* Schw.) spends almost 2 years of its life cycle on cedar trees. Cedar leaves are infected in late summer or fall by aeciospores blown from lesions on apple leaves. Sometimes the aeciospores lie dormant in the leaf axils before infecting cedar trees the next spring. From these infections small brown galls, called cedar apples, appear during the summer, but they do not mature until the following spring, when they may grow up to 5 cm in diameter. After a few warm spring rains, about the time that apple blossoms are pink the galls absorb water, increase in size, and squeeze out long, think, bright orange, gelatinous tendrils that form a mass several times that of the original brown galls (Fig. 13.3A). These tendrils, or horns, produce two-celled teliospores that germinate by forming four basidiospores. All teliospores do not germinate at the same time. With each rain the horns push out farther and expose more spores. When the supply of spores is gone, the galls dry and may drop. Often they remain hanging on the cedars through the next year.

As the air becomes dry, mature basidiospores are discharged forcibly into the air where they can be blown several kilometers. If they happen to fall on a wet apple leaf when the temperature is 15 to 20°C (59-68°F), they germinate quickly and penetrate the leaf within an hour or so. Young developing apples also may be infected, usually near the blossom end (Fig. 13.3B).

Yellow rust spots (up to 10 mm in diameter) appear on the upper leaf surface within 1 to 3 weeks, depending on the temperature and the susceptibility of the apple. As many as 300 spots may occur on a single apple leaf. As the spots increase in size, a sticky gum containing pycniospores appears. Insects are attracted to this gum; they carry spores from one rust spot to another and thus fertilizer the fungus. After fertilization the fungus grows through the leaf and produces fruiting bodies, called aecia, on the lower surface of the leaf. The aeciospores ripen in July and August and

(A)

(B)

FIGURE 13.3. Cedar-apple rust. (A) On cedar, swollen galls with orange tendrils or horns on which teliospores will be formed. (B) Symptoms of infection by cedar-apple rust fungus on apple fruit. *Courtesy NCSU Plant Pathology Photo Collection.*

escape from the aecia. Errant winds blow the spores to nearby cedar trees to complete the 2-year life cycle of this rust fungus.

Management

In neighborhoods where apples are an important backyard crop, a simple, practical way to control cedar-apple rust is to cut down any cedar trees nearby. If apple orchards are in valleys fringed by cedar-bearing hills or if the aesthetic value of cedars on some estates outweighs the value of nearby orchards, however, it may not be practical to cut down the cedars.

In the eastern United States it is difficult to grow apples profitably unless a rigid spray program for containing diseases and insects is used. Cedar-apple rust is one of the diseases that make this spraying necessary.

Apple cultivars differ in their susceptibility to cedar-apple rust. Whenever possible, resistant cultivars should be planted where cedar-apple rust is a problem.

Coffee Rust

Coffee rust is the most destructive disease of coffee in the world. It has caused large losses in coffee-producing countries in Asia and Africa, and in 1970 it was discovered in Brazil. Since that time it has been spreading throughout the world's most important coffee-growing countries in South America and Central America. The fungus attacks all species of coffee. The disease is particularly severe, however, on *Coffea arabica*, which produces the best-quality coffee beans.

Symptoms

The first noticeable symptom is the appearance of orange-yellow powdery spots on the lower side of coffee leaves. The rust pustule is initially small (about 5 mm in diameter) and circular. Pustules often grow together and form larger patches. Eventually the center of the pustule turns brown, and the leaf falls off. Early or premature leaf drop is a characteristic symptom of diseased coffee trees.

Causal Agent and Disease Cycle

The disease is caused by *Hemileia vastatrix* Berk and Broome. The fungus exists largely as mycelium in leaves and as urediniospores. Teliospores may be formed and will germinate to form basidiospores; but basidiospores do not infect coffee, and no alternate host has been found.

The disease cycle then is a simple one. Urediniospores from infected and fallen coffee leaves are easily spread by wind, rain, and perhaps insects. When a spore lands on a susceptible leaf, it germinates and quickly infects the leaf through the stomata on the lower leaf surface. Young leaves generally are more susceptible to attack than older leaves. New urediniospores can be produced on the lower leaf surface within 10 to 20 days of infection. The cycle then is repeated continuously as long as environmental conditions (high humidity and the occurrence of free moisture) are favorable for infection. Because infected leaves fall off prematurely, the coffee is weakened, branches may die back, yields are reduced, and eventually trees may die.

Management

Management of coffee rust first was attempted through quarantines designed to keep the pathogen from spreading into disease-free areas. The quarantines slowed the spread for many years, but now alternative strategies must be used. Management of the disease is very difficult once it is established in a planting of coffee trees. Copper-based fungicides offer some protection, and newly developed systemic fungicides offer some hope for managing coffee rust. Fungicides need to be applied frequently (every 2 to 3 weeks) before and during the rainy season—the time when most disease develops. Fungicide application is difficult and expensive because most coffee plantations are planted on hillsides and are not accessible to tractors and other machinery for pesticide applications. Most spraying is done, therefore, with backpack sprayers. Some resistant cultivars are available, but in general the coffee beans of these cultivars are of a lower quality. It also is not known how long disease resistance will be effective before new strains of the fungus develop.

SMUTS

Until the twentieth century, smut diseases throughout the world were second only to rusts as the cause of serious grain losses in cereals. Today, total losses to smuts are low, but serious localized losses can occur, particularly in developing countries. Unlike the rusts, which attack the stems and leaves of cereal crops (and many other plants) smuts do damage by attacking the grain kernels. The reduction in yield brought about by these diseases is immediately observable because kernels of grain are replaced by the black, dusty spore masses of the smut fungi. Kernels that do develop may be covered with these dark, oily spores, and thus are reduced in quality.

There are three basic types of smuts, which are distinguished by the way in which they infect a host plant. Corn-smut spores can penetrate the plant and cause disease at any growing point (*local infection*), including the tassel, the leaves, and most important the developing corn kernels. Covered smut (also known as stinking smut or bunt) is caused by a fungus that infects young barley, oats, or wheat seedlings (*seedling infection*). Loose smut is caused by a fungus that infects the ovary of a cereal plant when the plant is in flower (*floral infection*).

Corn Smut

Symptoms

Minute galls form on leaves and stems of young seedlings infected with corn smut; infected seedlings remain stunted and eventually may die. When older plants are infected, localized areas such as the axillary buds, flowers of the ear (silks) or tassels, leaves, or stalks will develop symptoms. However, most damage occurs when ears are infected (Fig. 13.4).

The fungus penetrates growing tissue and grows throughout a section of the plant. Fungus invasion causes cells of the host to enlarge and to divide at an accelerated rate. The result is a swollen growth (*gall*), which is, at first, covered with a greenish-white layer of fungal tissue. As the gall becomes older and matures, the greenish-white

FIGURE 13.4. Massive smut galls on corn ears. Sweet corn is especially susceptible to corn smut. *Courtesy NCSU Plant Pathology Photo Collection.*

membrane turns silvery white, and the gall is filled with a mass of black, oily spores. When mature, a gall may range from 1 to 20 cm in diameter and contain millions of spores. When the gall ruptures, this mass of black spores escapes into the air; spores often can be seen as black dust on plant parts near the ruptured gall.

When the galls are young and no hint of the black, sooty spores is found, the natives of Mexico and Central America eat the young galls as a great delicacy. As with any fungus, however, one should aways make sure precisely what the proper identity of the fungus is before eating it!

Causal Agent and Disease Cycle

Corn smut is caused by *Ustilago zeae* (Beckm.) Unger, which produces two kinds of spores. Black spores (teliospores) are produced in the galls and overwinter in debris in the soil. Basidiospores are produced when the teliospores germinate in the spring.

After the fusion of mycelium growing from two basidiospores, the resulting dikaryotic mycelium infects the corn plant, and symptoms develop.

Management

Some corn hybrids have a slight amount of resistance to corn smut. In home gardens, corn smut can be managed by simply removing galls when they are first observed and burying them. Crop rotation also may be helpful in small plantings of corn. Both of these management methods are of limited use in large corn-growing areas.

Covered Smut (Stinking Smut or Bunt)

Symptoms

Plants infected with covered smut may be a few to several centimeters shorter than healthy plants. Roots on infected plants usually are poorly developed. When the heads of infected plants emerge, the color of the head is usually bluish green instead of the normal yellowish green. At the time when the grain normally would be mature, the entire content of the kernel is a black, oil, dusty mass of spores. The fungus spores give off a distinctive odor of decaying fish.

Causal Agent and Disease Cycle

Covered smut is caused by three species of the genus *Tilletia,* which all have similar disease cycles. The fungi produce the same two types of spores as the corn-smut fungus. Teliospores are black, sooty spores that fill the grain kernels; when the kernels rupture, the teliospores fall to the ground or onto other healthy kernels. Spores that fall to the ground may survive for several years; spores that land on healthy grain stick to the outer surface of the seeds and overwinter there.

In the spring when infected seed of barley, oats, or wheat germinates, the teliospores also germinate to form basidiospores. After the fusion of the mycelium from two basidiospores, the resulting mycelium penetrate the seedling and grows between the plant cells. The mycelium reaches the growing point of the plant and continues to grow in pace with the plant. The mycelium grows up into the spike and into the developing kernels, where it penetrates the plant cells and is transformed into teliospores. The smutted kernel becomes filled with teliospores, and the cycle is ready to be repeated.

Management

Covered smut can be managed by using smut-free seed obtained from resistant cultivars or by using seed that has been treated with a fungicide. Care should be taken to avoid contamination of healthy kernels with the spores from smutted kernels. Also, because the teliospores can survive in the soil for several years and infect germinating seedlings, the use of a systemic fungicide is helpful.

Loose Smut

Loose smut is caused by *Ustilago tritici* (Pers.) Rostr., which produces teliospores and basidiospores. The fungus overwinters as mycelium in the infected seed. When infected kernels germinate, the fungus also begins to grow between the plant cells until

FIGURE 13.5. Loose smut replaces healthy seed with black masses of fungal teliospores (left and right). Compare with healthy, flowering head (center). *Courtesy NCSU Plant Pathology Photo Collection.*

it reaches the growing point of the plant. The mycelium infects the spikelets and invades most of the cells in the spike except the rachis. As a result, when the head emerges, masses of teliospores are present instead of flowers.

Loose smut is difficult to observe until a plant has headed. One clue to the presence of loose smut is that infected plants grow faster and are taller than neighboring healthy plants, probably as a result of stimulation by the fungus. When the spike emerges, the masses of spores are covered by a thin, silverwhite membrane, which quickly ruptures (Fig. 13.5). The spores are released at a time that coincides with the flowering of surrounding healthy plants. The wind carries spore to flowers of nearby plants. Here the teliospores germinate to form basidiospores. After fusion of the mycelium from two basidiospores, the fungus penetrates the flower and becomes established in the various tissues of the newly forming seed. The mycelium then becomes dormant and awaits the germination of the seed.

Management of loose smut is usually achieved through the treatment of seed with a systemic fungicide. As with all smut diseases, the best management procedure is the use of disease-free seed.

DOWNY MILDEWS

There are many downy-mildew diseases that affect a wide variety of vegetables, ornamentals, cereals, forages, other crops, and weeds, particularly in cool, humid environments. The diseases get their name from the downy growth, consisting of conidia and conidiophores, that cover the underside of infected leaves. Most downy-mildew fungi are obligate parasites. Most of them produce oospores within the host tissue. The mycelium invades all the leaf tissues. In asexual reproduction the mycelium produces branched conidiophores that emerge through the stomates of the host epidermis. The conidia are borne singly at the tips of the branches and, when mature, are readily dislodged.

The downy-mildew fungi are important plant pathogens. In some areas, it is almost impossible to grow a profitable crop unless they are adequately managed.

Late Blight of Potato

One of the most tragic events in human history was the Irish Famine in the middle of the nineteenth century. Two circumstances were responsible for this disaster. First, the impoverished population of Ireland, about four million people, had become almost wholly dependent on the potato for food. Second, the potato crops for 2 years in a row—1845 and 1846—were almost totally destroyed by late blight. The misery and suffering were extensive and terrible.

The Irish people never knew what hit them. They did not know that a microorganism—a fungus—had destroyed the crop. Some blamed it on lightning; others, on bad weather. Many believed God had put a curse on the nation. The English government was so alarmed that the Royal Horticultural Society offered a prize of 50 pounds to the person who first discovered the true cause of the disease. Reverend M. J. Berkeley won the prize when he published his belief that a fungus caused this terrible disease.

A million Irish people died from starvation or from disease following malnutrition in the wake of the late-blight epidemic. One and one-half million more emigrated, many going to America.

Late blight also contributed to the defeat of Germany in World War I. In 1917 it destroyed about a third of the potato crop, which made up a large part of wartime diet of the Germans. Reduction of the already scanty food supply contributed to the breakdown in morale and physical endurance that forced the Germans to surrender.

Distribution and Importance

Late blight of potato occurs worldwide wherever potatoes are grown. Irish potatoes thrive in cool, moist climates, which have just the conditions that favor growth of the fungus. In the northeastern United States and northwestern Europe, late blight is prevalent and destructive because cool, wet summers periodically occur in these areas and provide ideal conditions for late-blight epidemics.

Late blight kills the foliage and stems of potato plants and rots the tubers. Late blight may totally destroy a potato crop if no disease management measures are used. Usually though, losses vary and depend on the weather.

Late blight is also destructive on tomatoes.

Symptoms

An early symptom of late blight is brown, dead areas at the tips or margins of infected leaves (Fig. 13.6). If moist weather prevails, the entire leaf may be killed in less than 4

FIGURE 13.6. Late-blight symptoms on potato leaflets. During cool, moist weather the entire leaf may be killed in less than 4 days. *Courtesy C. W. Averre.*

days. During dry weather the infection advances more slowly. Blighted leaves curl and shrivel when it is dry, whereas in wet weather they remain limp and soon decay with an offensive odor. Generally, the lower leaves are affected first; but in cool and humid weather, when the entire plant can become black and wilted, the entire crop may be destroyed.

Often the late-blight fungus can be seen as a dirty white or gray mold on the underside of the leaves. The white growth consists of the hyphae and sporangiophores of the fungus, on which huge numbers of spores are produced. In dry weather the external fungus growth is seldom seen.

Late blight also affects potato tubers (Fig. 13.7). Death of the plant tops reduces the size and number of the tubers and tubers also may become infected from sporangia washed into the soil or form hyphae in the mother plant. During weather favorable to

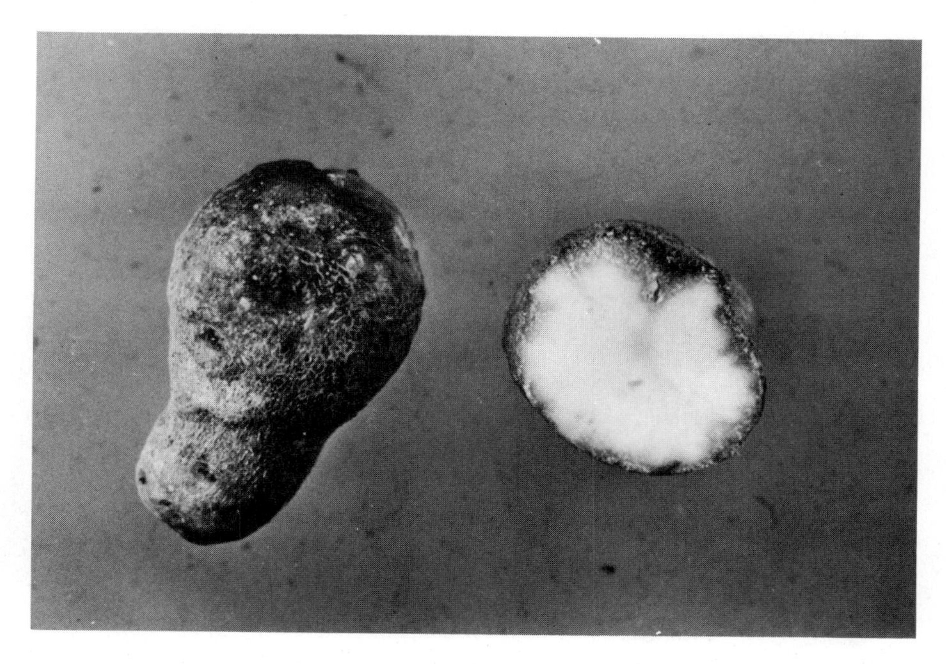

FIGURE 13.7. Potato tubers infected by *Phytophthora infestans.* Only a slight darkening of tissue is seen on the outside of the tuber, but when the tuber is cut, a distinctive blackening of tissue around the outside of the tuber is obvious. If diseased tubers are used for seed pieces, they will serve as inoculum for disease in the next crop. *Courtesy NCSU Plant Pathology Photo Collection.*

the fungus, tubers may rot before harvest; during less favorable conditions, only a portion of a tuber may be infected. Diseased tubers may rot in storage or, if used as seed pieces, furnish inoculum for the succeeding crop.

Causal Agent and Disease Cycle

Late blight of potato is caused by *Phytophthora infestans* (Mont.) d By. It has large thin-walled hyphae, without crosswalls, which grow in the intercellular spaces of the host. Club-shaped *haustoria* penetrate the cells and absorb food from them. Slender aerial branched sporangiophores grow out through the stomates and produce lemon-shaped hyaline sporangia (27-32 μm × 16-24 μm) (Fig. 13.8). The sporangia germinate between 15°C (59°F) and 24°C (75°F). Below 15°C (59°F) the sporangia germinate by forming zoospores, and above 15°C (59°F) by forming germ tubes. There are several pathogenic strains of the causal fungus.

The late-blight pathogen overwinters as mycelia infected tubers. Volunteer sprouts that develop from infected tubers left in the field or thrown into cull piles become infected. The tubers sprout, and soon all aerial parts of the plant are invaded. In a few days sporangia are produced, which are splashed to leaves of adjoining plants or are blown to nearby fields where they germinate to start new centers of infection. Tubers near the ground surface are attacked during wet weather by emerging zoospores,

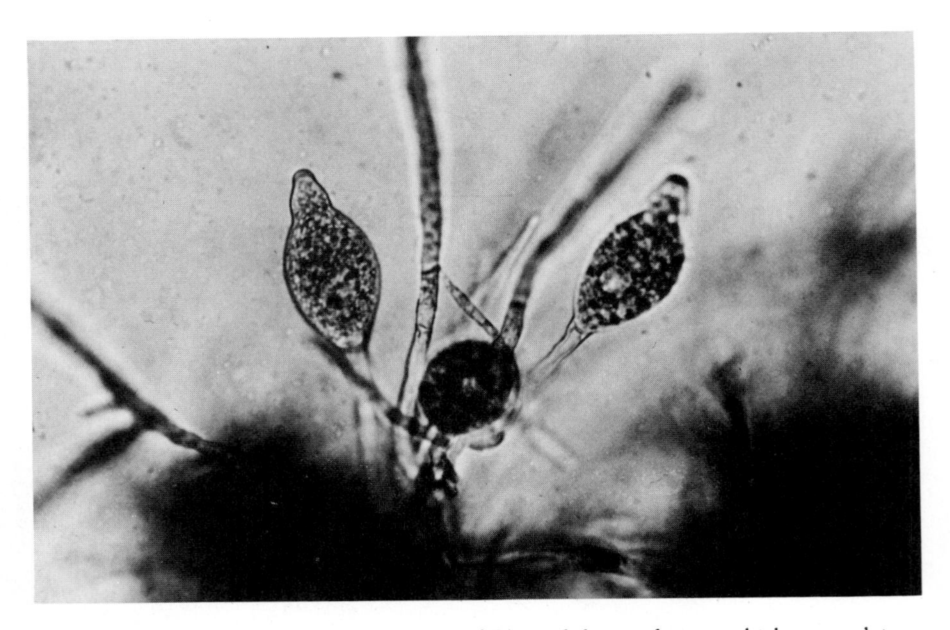

FIGURE 13.8. Lemon-shaped sporangia of *Phytophthora infestans,* which causes late blight of potato and tomato. *Courtesy NCSU Plant Pathology Photo Collection.*

which penetrate through wounds or lenticels. Harvested tubers also may become contaminated with sporangia from infected leaves or soil. Occasionally tubers are infected by hyphae growing from the mother plant.

Late-blight development depends on the weather. The fungus sporulates abundantly at or near 100% RH between 16°C (61°F) and 22°C (72°F). Sporangia die in 3 to 6 hr below 80% RH. The sporangia germinate in 30 min to an hour in free water at 10 to 15°C (50-59°F), and penetration of host tissue may occur within 2 hr.

Temperatures above 30°C (86°F) check growth of the fungus but do not kill it. When the temperature goes down, the fungus will begin to grow again.

Management

Late blight is such a destructive disease that potato growers must use a combination of practices if the disease is to be successfully managed. These practices include the use of sanitary measures, resistant cultivars, effective fungicides, and disease forecasting systems. Moreover, late-blight management practices must be coordinated with an overall IPM program in order to produce the highest yield of a quality crop.

Growers should use only certified, disease-free potatoes for seed. All dump or cull piles should be burned or sprayed with herbicides before planting time. Likewise, volunteer potato plants should be destroyed in or near fields where a crop is to be planted. Use of the most resistant cultivars available for a given area is important in curtailing late blight.

Generally, spraying with fungicides is necessary to manage late blight. Sprays should be applied, starting when plants are 15 to 20 cm (6-8 in.) high, every 5 to 7 days when the weather is damp or rainy. Proper timing and thorough coverage of foliage is necessary. Late-blight warning or forecasting systems are available that predict when late blight will appear based upon rainfall, RH, temperature, and sunshine data. Farmers should use such forecasting systems to help them determine when to apply fungicides or other pest control measures.

Downy Mildew of Grape

Downy mildew of grape is one of the first diseases that was controlled by a fungicide. And the discovery was accidental. In 1880, Dr. Pierre Millardet, a French plant pathologist, was studying downy mildew, which was destroying the grape vineyards in southern France. Millardet was a keen observer. One day as he walked in a vineyard in Bordeaux, he noticed that many vines along the roadside had not shed their leaves. They looked green and healthy. On the other hand, vines away from the road had lost all their leaves. Millardet asked the vineyard owner what had happened. The owner said that passers-by would pick the grapes alongside the road. To stop this pilfering, the owner placed some lime, bluestone (copper sulfate), and water in a bucket, stirred the mixture well, then took a broom and spattered the ugly blue-white liquid on the vines alongside the road. The farmer hoped the vines would look so poisonous and repulsive that thieves would not eat them. Sure enough, this psychological warfare was successful. The grape stealing stopped, and oddly enough the vines remained healthy.

Not only was Millardet observant, but he was also a man of action. He started mixing various proportions of lime and copper salts until he found an efficient combination that would protect grapes against mildew. It took 3 years of testing before he hit on the right combination for Bordeaux mixture, as it became known. For the next 50 years or more this remarkable fungicide was used all over the world to control not only downy mildew but also several diseases on other crops, including late blight of potato. It would be difficult to calculate the worth of Bordeaux mixture. For decades it was the standard fungicide on potatoes, and it is still used to some extent.

Symptoms

Downy mildew affects all parts of the grape plant. It causes losses by killing leaf tissues and defoliation; it rots the ripening grapes and weakens, dwarfs, and kills the young shoots. The first symptoms are pale yellow, irregular spots on the upper leaf surface (Fig. 13.9). On the underside of the leaf surface the downy growth of the fungus appears. The infected leaf areas are killed and turn brown. Often the leaves die and fall off. During favorable weather entire clusters of grapes may be rotted. The fungus thrives best during cool, moist weather.

Disease Cycle

Downy mildew of grape is caused by *Plasmopara viticola* (Berk. & Curt.) Berk & de T. Oospores (30 μm in diameter) overwinter embedded in dead leaves, which disintegrate and release the dormant oospores in the spring. Only oospores that have been frozen can germinate. They form sporangia on short stalks. Each sporangium (25 μm $\times$ 40 μm) produces about 20 zoospores (12 μm in diameter); these are blown or

FIGURE 13.9. Symptoms of downy mildew on a grape leaf. Pale yellow, irregular lesions are visible on the upper leaf surface. The fungus usually sporulates on the lower side of the leaf. *Courtesy NCSU Plant Pathology Photo Collection.*

splashed to nearby leaves, which become infected within 1 to 2 hr if conditions are favorable.

Penetration usually occurs through the stomates on the underside of the leaves. The hyphae quickly spread through the spaces between the leaf cells. The fungus obtains its food from the numerous globose haustoria that grow into the leaf cells and rob them of their contents. The cells begin to lose their green color as the chlorophyll is destroyed, the cell contents disintegrate, the cell walls collapse, and the entire cell turns brown.

Soon the mycelium begins to form sporangiophores, which emerge from the stomates. Great numbers of spores are produced on the treelike branches of the sporangiophores. They are blown or splashed in all directions. If they land on a wet grape leaf or shoot, they germinate quickly and form many zoospores. These zoospores then infect the plant through the stomates and lenticels, and the disease spreads rapidly. When conditions are favorable, a new crop of sporangia may be produced every 5 to 18 days, depending on the temperature, RH, and host susceptibility.

At the end of the growing season the fungus forms oospores inside leaves, and sometimes inside shoots and berries. The next spring the oospores start new infections.

Management

To grow grapes profitably, farmers must manage several diseases and insects through a season-long program of spraying or dusting. Downy mildew control with proper fungicides should be fitted into the overall IPM program.

The most effective way to manage downy mildew of grape is to spray with fungicides at 7- to 10-day intervals beginning before bloom and continuing up to harvest. The time and the number of applications depend upon the weather.

Proper pruning of the canes to permit free circulation of air and rapid drying of the leaves also help growers to manage downy mildew, as does plowing under or removing dead leaves. But the principal management measure is careful application of fungicides at the correct time in the correct way throughout the growing season.

Downy Mildew of Cucurbits

Downy mildew is a destructive disease of cucumber, muskmelon, and watermelon. Occasionally it also causes damage to gourds, pumpkin, and squash. All downy mildews thrive in moist weather, particularly when wet conditions prevail during the growing season. The downy-mildew fungus on cucurbits causes the most damage during warm, wet weather. It is particularly troublesome in the eastern and southern regions of the United States.

Symptoms

Infected leaves have yellow spots on the upper side; on the lower sides, the purplish mildew growth appears when the humidity is high. On cucumbers the affected leaves may die, and the whole plant may be severely stunted or killed. Maturing fruits fail to color properly, have no taste, and usually are sunburned.

Disease Cycle

No oospores of *Pseudoperonospora cubensis* (Berk. and Curt.) Rostow which causes downy mildew of cucurbits have been found. Therefore, because the fungus is an obligate parasite, it must grow on cucurbits at all times of the year in frost-free areas. Conidia are blown northward in the spring and start new outbreaks wherever a susceptible host and favorable climate coincide. The conidia germinate by forming zoospores (10-13 μm in diameter), which penetrate through stomates and quickly form intercellular hyphae with globose haustoria that grow into the host cells and use the contents for food.

Different races of the fungus occur on the various cucurbit crops.

Management

Regular applications of fungicides will generally control downy mildew on cucurbits, although other measures also help manage the disease. Wild cucurbits growing in the vicinity where cucumbers or muskmelons are grown should be eradicated. Production fields should be isolated from one another, and resistant cultivars used, especially in small home-garden plantings where fungicides are not applied regularly. Harvested and abandoned fields should be disked promptly to avoid carryover of inoculum. Rotations should be used. Late plantings should be isolated from areas where early crops were grown. Workers and equipment should stay out of fields when plants are wet from dew or rain. Resistant cultivars have been developed and should be used whenever possible.

POWDERY MILDEWS

Plants with powdery mildew are so conspicuous and striking that almost everybody notices them. Powdery mildew is one of the oldest plant diseases known. Ancient writers regarded it with fear and superstition. It is mentioned in several books of the Old Testament; the ancient Greeks also wrote about it. Powdery-mildew fungi occur all over the world with the exception of the Arctic region, Antarctica, and high mountain regions.

On some crops the disease occurs almost every year and causes great damage. These fungi attack all important crops, including apples, cantaloupes, cereals, flowers, grapes, grasses, legumes, lettuce, muskmelon, ornamentals, peach, shade trees, and many weeds. All broad-leafed plant species are host to some kind of powdery-mildew fungus. Only the pine trees and their relatives do not have powdery mildew.

FIGURE 13.10. Wheat leaf severely infected with powdery mildew. *Courtesy NCSU Plant Pathology Photo Collection.*

Symptoms

Plants infected with powdery mildew have a striking appearance. Infected leaves and stems look as if they have been dusted with flour (Fig. 13.10). All aboveground portions of plants can be attacked, especially young leaves and shoots. Growing leaves become twisted and misshapen. Often infected buds do not open; flower buds become discolored and dwarfed and finally die. On susceptible grasses and red clover, infected leaves turn yellow and brown with resulting unthrifty, stunted growth. Cattle do not like to graze on such forage. On cereal grains, powdery mildew may lead to severe yield losses, sometimes 20% or more, and the quality of the grain is reduced.

Powdery mildew can destroy an entire crop of grapes when temperature and RH are favorable for fungus growth. The grayish-white growth appears on the shoots, tendrils, cluster stems, petioles, and upper surfaces of the leaves. In unusually favorable seasons, even the berries are attacked.

On roses the disease appears year after year. It reduces flower production and weakens the plants by attacking the buds, young leaves, and growing tips. Often infected buds do not open; flower buds become discolored, dwarfed, and ugly and finally die. Growing leaves become curled and distorted. On older leaves, large white patches of fungus growth appear.

Causal Agents and Disease Cycle

Three principal genera of fungi cause powdery mildew: *Erysiphe, Sphaerotheca,* and *Podosphaera.* Contrary to most fungi, which invade the host and grow inside, powdery-mildew fungi grow mostly on the outside surface of the plant. The white powdery coating on the leaves, twigs, and buds consists of white mycelium with slender, many-branched, septate hyphae. Many upright branches bearing chains of egg-shaped, colorless conidia arise from the network of hyphae (Fig. 13.11). At many points the hyphae are attached to the host surface by specialized branches called *haustoria.* These rootlike organs grow into the epidermal cells and absorb liquids from the host sap, which the fungus uses for food. It is a very specialized and unique arrangement.

The powdery, fragile conidia are produced in great abundance. They are easily detached and blown by the wind; thus, extremely rapid spread of the disease can occur. The conidia germinate in less than 2 hr under favorable conditions, penetrate the leaves, and form haustoria. A new crop of conidia is produced every 5 to 10 days.

Perithecia or *cleistothecia* are formed on leaves and twigs in late summer or as the crop matures. They are globose and have a hard, black outer coat. Within each perithecium may be found a small sac or spore case, the *ascus.* Within each ascus may be found two to eight *ascospores,* formed as a result of a sexual union, which may give rise to new pathogenic strains. These sexual spores, safely protected inside the perithecial coat, also act as overwintering spores.

The fungus can overwinter as mycelium in dormant buds. As soon as these buds begin to grow in warm weather, so does the fungus. It forms conidia, which are blown to nearby branches. Perithecia also endure through the winter. Warm temperatures

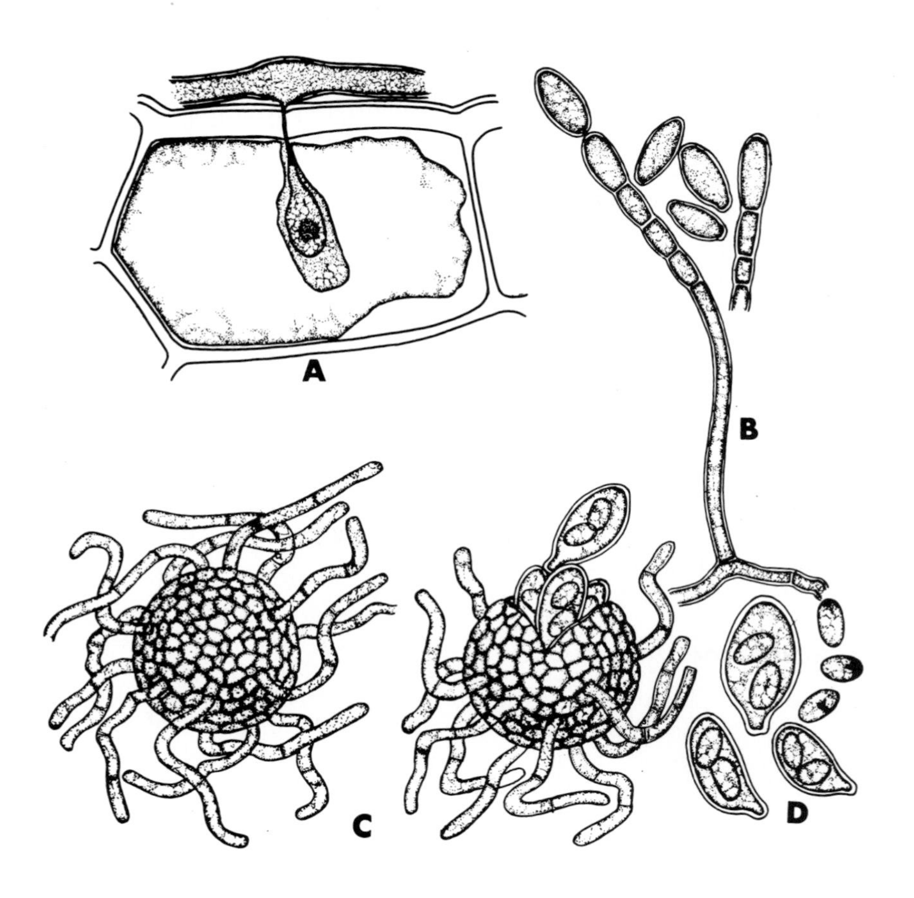

FIGURE 13.11. *Erysiphe cichoracearum,* a powdery-mildew fungus. (A) Haustorium in epidermal cell; (B) conidiophore and conidia; (C) perithecia; (D) asci and ascospores. Magnified about 100 times. *Drawing by A. Husain after Smith and E. S. Salmon.*

and rain make the perithecia swell and burst. The winter spores (ascospores) are splashed to nearby leaves or buds where they germinate quickly to start new infections.

Environmental Effects

Both temperature and RH influence the development of powdery mildew, but RH is more important than temperature. Most fungal spores need free water to germinate; however, free water is unnecessary for and will inhibit germination of powdery-mildew conidia. Maximum germination occurs at 97 to 99% RH and between 14°C (57°F) and 24°C (75°F). Some germination occurs when RH is as low as 25 to 30%;

however, conidia usually die quickly at very low humidities and at very high or very low temperatures.

Periods of low rainfall and moderate temperature favor the development of powdery mildew, as do periods of high RH followed by dry conditions and/or warm days followed by cool nights. High nitrogen levels that produce succulent, rapidly growing tissue, easily attacked by the fungus, also increase mildew severity. Frequently, plants growing in shade also have especially severe mildew. Reduced light apparently favors infection and fungus growth.

Management

Spraying and dusting with fungicides have been used for many years to manage powdery mildews. Under most conditions, weekly applications give good protection; during periods of frequent rains, active growth, or wide temperature changes, more frequent applications may be necessary.

Some fungicides, when used frequently, exert selective pressure on powdery-mildew fungi. As a result, strains resistant to the currently used fungicide appear, and poor management results. When such resistance develops, a different fungicide must be substituted for the one presently used. Some growers use two different fungicides to help prevent the occurrence of resistant strains.

Cultivars resistant to powdery mildew have been developed for some crops (apples, lettuce, small grains, and others). Growers should find out if resistant cultivars are available in their area and use them wherever possible.

ERGOT

Ergot occurs primarily on rye but can occur on wheat, barley, oats, and a large number of wild grasses. The actual amount of damage caused to cereal grains is low, usually no more than 5 to 10%. The disease is important because the dark-purple to black, elongated sclerotia (Fig. 13.12) that replace the seeds of the cereals (grains) or grasses contain chemicals that are poisonous to humans and livestock.

Among all plant diseases, ergot may have had the most evident impact on humans during severe epidemics. When people eat bread that is contaminated with the sclerotia or ergots produced by the fungus that causes the disease, the poisons within the ergot affect their mind and nervous system to produce a range of symptoms including nervous tremors, excessive bleeding, the feeling of burning skin or insects crawling under the skin, miscarriages in pregnant women, gangrene of hands and feet, and violent hallucinations and mental disorders.

Ergotism (the disease caused in humans or animals by eating the sclerotia) was widespread in Europe in the Middle ages and was called holy fire or St. Anthony's fire. Because of laws that prohibit toxic levels of ergot sclerotia in bread, outbreaks of ergotism are rare in modern times. However, when drought struck Ethiopia in the late 1970s and grain crops failed, the people were forced to collect wild grass seeds for food, and ergot sclerotia were plentiful in these wild grasses. As a result, many people suffered from ergotism, and quite a few died.

FIGURE 13.12. Dark, elongated sclerotia or ergots of *Claviceps purpurea* and lighter-colored grains of rye. *Courtesy NCSU Plant Pathology Photo Collection.*

Disease Cycle

The fungus *Claviceps purpurea* (Fr.) Tul. overwinters as sclerotia on or in soil or mixed with seed (Fig. 13.12). In the spring small mushroom-like structures grow from sclerotia on or near the soil surface. The tip of each stalk contains many perithecia, from which ascospores are ejected and carried by the wind to plants. Ascospores also can be exuded in a sticky mass and be spread to cereal or grass flowers by insects. The ascospores that land on the stigma of a flower germinate and quickly infect the ovary, and within a week the fungus is producing the sticky, sweet honey dew that contains conidia and attracts insects to spread the conidia to other flowers. Then an ergot or sclerotium is produced in place of a normal seed. The sclerotium is a hardened mass of fungal tissue that completely replaces individual cereal grains. Sclerotia may be 0.2 to 5.0 cm long, and few to many sclerotia occur on a spike or panicle. The sclerotia mature at about the same time as healthy seeds, and they fall to the ground or are harvested along with the healthy grain.

Management

Cultural and sanitary procedures are vital to the management of this disease. Only clean seed that is free of ergot should be planted. If seed is contaminated, it can be cleaned mechanically or by soaking contaminated grain in water for 3 hours and then placing grain in an 18% salt solution. The sclerotia will float and can be easily removed. In the field, deep plowing will place sclerotia where they cannot germinate and will decompose within about one year. Grassy weeds should be removed from edges of fields because they may be infected by the fungus and serve as a source of inoculum. Rye also should be rotated with nonhost crops such as legumes or corn.

BLAST OF RICE

Blast, one of the oldest known diseases of rice, has been recorded at least since 1637. At that time the trouble was thought to be due to heat absorbed into the grain during drying in hot sunshine. The heat or fever considered the cause of the disease is perhaps the reason why rice blast was called rice fever.

Rice blast is widely distributed and has been reported in some 70 countries throughout the world. Rice grown under irrigation in areas of low RH usually suffers much less damage than crops grown in humid regions.

Blast is considered the principal disease of rice because of its wide distribution and destructiveness. Seedlings often are killed, and panicle infection and leaf infection also cause serious losses. Current annual losses worldwide probably average about 3%; however, in some areas during epidemic years losses may average 50%.

Strains of the fungus that causes blast also are pathogenic on numerous other grasses.

Symptoms

The rice-blast fungus produces spots on leaves, nodes, and different parts of the panicles and grains, but seldom on the leaf sheath. The typical elliptical leaf spots have pointed ends. The centers are gray, and the margin is brown to red-brown. The spots, which usually begin as small water-soaked white to blue dots, enlarge quickly under moist conditions on susceptible cultivars and may be 1.5 cm long by 0.5 cm wide.

Leaf spots caused by the blast fungus often resemble and may be confused with those caused by other pathogens. Severely infected leaves become dry and brown. One of the most conspicuous symptoms is "rotten neck"—lesions on the neck of the culm (stalk) and on the panicle branches near the panicle base. Rotten neck prevents the kernels from filling; often the weakened branches break over where the lesions occur. On highly resistant cultivars, only minute brown specks are seen. The fungus secretes at least two toxic substances—picolinic acid and piricularin—that kill host tissues.

Causal Agent and Disease Cycle

Rice blast is caused by *Magnaporthe grisea* (Hebert) Barr (anamorph-*Pyricularia grisea* Cavara). The gray, slender conidiophores are rarely branched (Fig. 13.13); they are produced on hyphae growing on or in the host tissue. Frequently the conidiophores emerge in clusters through the stomates of infected tissue. The gray conidia are variable in size and shape, but usually are club-shaped with two septae. From one to 20 conidia may be produced on a single conidiophore. Conidial size

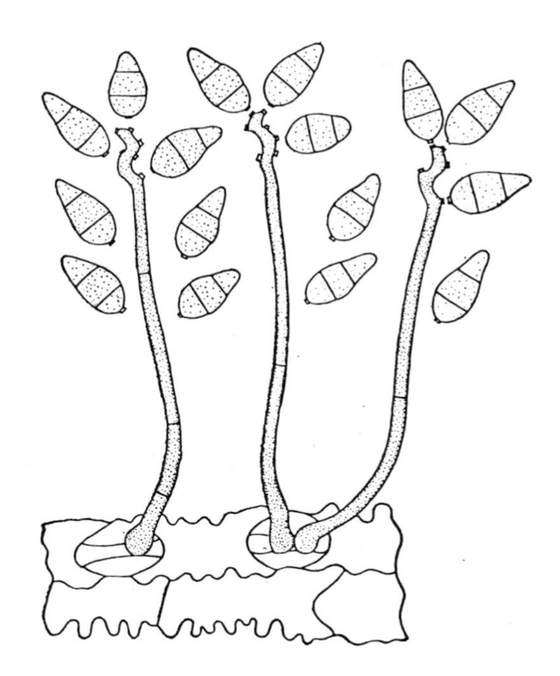

FIGURE 13.13. Conidiophores and conidia of the asexual stage of *Magnaporthe grisea,* which causes rice blast. Magnified about 500 times. *Adapted from D. E. Ellis.*

varies with the fungus isolate and environmental conditions. The fungus also produces ascospores in perithecia, but they are rarely observed in the field.

The optimum temperature for mycelial growth, conidial production, and germination is about 28°C (82°F). Alternate light and dark periods are necessary for conidial production. The conidia germinate freely in water.

In temperate regions, mycelia and conidia on straw and seed are the principal overwintering agents, but the fungus also survives on winter cereals and weed hosts. Under dry conditions conidia will survive for more than a year, but under moist conditions they do not live long enough to infect the next crop.

Wind blows blast conidia for kilometers. On wet leaves they germinate quickly and penetrate the host within a few hours. The mycelium grows rapidly in host tissue, and under ideal conditions produces a new crop of conidia in about 6 days. A typical leaf lesion may produce up to 6000 conidia a day for 2 weeks. Windblown conidia start new infections.

Environmental Effects

When rice seedlings are grown at 20°C (68°F), blast infection is more severe than at 28°C (82°F). Seedlings grown several days at 18 to 20°C (64-68°F) before inoculation also are more susceptible than those kept at 28°C (82°F). Adult plants are more resistant at 28°C (82°F) than at 18°C (64°F). In general, rice cultivars are more susceptible in dry soils than in wet soils. High RH favors disease development. In areas where there are frequent and long periods of rain, blast will cause severe damage. High nitrogen levels increase disease incidence. Many workers have shown that rice plants with a high silica content, especially in epidermal cells, show less than expected damage from blast. Evidently, epidermal cells with high silica absorb less nitrogen and are more resistant to physical penetration by the fungus.

Management

Resistant cultivars have been used for many years as one of the principal methods of managing rice blast. There are many different pathogenic races of the fungus, and new races are continually occurring. Therefore, any given resistant cultivar is effective for only a limited period of time. New cultivars must be bred as old cultivars lose their resistance. Rice is such an important crop, however, that rice scientists continually monitor the prevalence of new races of the pathogen. So far, in many countries rice blast has been held in check by the continual breeding of new cultivars as the old ones break down.

In Japan, where the rice crop is the principal food and is vital to the nation's welfare, fungicides have enabled practical management for a combination of reasons. First, rice is a high-yielding crop, and the government guarantees a high price for the product. The Japanese also have an elaborate disease-forecasting service and a well-developed chemical industry capable of producing relatively cheap fungicides. Finally, the farmers in each neighborhood are well organized to carry out community projects. All of these factors combined permit timely, large-scale application of economical fungicides by aircraft even though individual farm holdings are small. Usually the cost of chemical management is less than 5% of the crop value, and the cost/benefit ratio makes chemical management economically feasible.

In addition to the use of resistant cultivars and fungicides, certain cultural practices also help manage rice blast. Planting can be timed to avoid periods ideal for blast development. Proper use of fertilizers and applying small increments of nitrogen at any one time, proper irrigation, and proper spacing of seedlings also help to decrease disease severity.

GRAY MOLD OR BOTRYTIS BLIGHT

Gray mold is one of the most common diseases of field crops, fruits, ornamentals, and vegetables throughout the world. It can be devastating on greenhouse crops. Host plants include African violet, amaryllis, apple, artichoke, azalea, banana, bean, begonia, beet, cabbage, carrot, chrysanthemum, cucumber, cyclamen, dahlia, geranium, gladiolus, grape hyacinth, lettuce, lily, onion, peony, pepper, rose, snapdragon, squash, strawberry, tobacco, tomato, tulip, and others. Losses from gray mold are difficult to estimate but are in the hundreds of millions of dollars annually in the United States.

FIGURE 13.14. Botrytis blight on grapes. Individual grapes that were infected have become dark and shriveled. *Courtesy NCSU Plant Pathology Photo Collection.*

Symptoms

The disease derives its name from the obvious *gray-mold* layer of fruiting bodies produced by the causal fungus on infected tissues during humid conditions. The fungus can attack many plant tissues. It can blight flower petals and form a cobwebby growth of mycelium that can be seen with the unaided eye. If blossoms are infected, the unsightly blighted tissue may cling to the plant. Flower infections can eventually lead to fruit rots. After fruit or stem tissues are infected, they become soft and watery and appear light brown. Later, infected fruit may shrivel (Fig. 13.14). In addition to petal blights and fruit rots, this fungus can cause leaf spots on certain hosts such as gladiolus, onion, and tulip. Sometimes the fungus also causes seedling damping-off. Disease symptoms may be present and most severe on plants or plant products in storage or, as with strawberries, on the grocery store shelf.

Causal Agent and Disease Cycle

Several species of *Botrytis* cause gray mold; these fungi occur widely as saprophytes, and many pathogenic strains are known. The fungus lives through dry conditions as small black sclerotia, which often develop in old decayed tissue. During wet weather the sclerotia germinate to form a gray mycelium that quickly produces fuzzy, gray masses of conidiophores bearing grapelike clusters of colorless, egg-shaped conidia averaging $10\ \mu m \times 5\ \mu m$ (Fig. 13.15). These masses develop on the surface of rotted

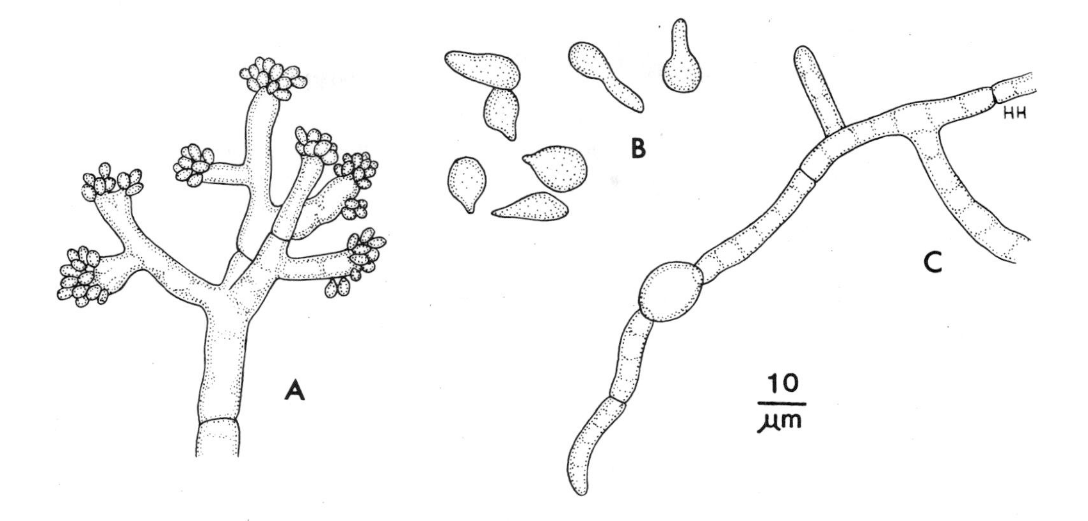

FIGURE 13.15. The gray-mold fungi, *Botrytis* spp. (A) Conidiophore with conidia; (B) conidia; (C) hyphal formation. *Drawing by H. Hirschmann after F. A. Wolf.*

tissue during wet, cool weather or when RH is high. Conidiophores often are seen emerging from cracks in the epidermis of infected plant parts. The slightest air movement blows conidia to nearby plants to start new infections on wet, tender, weakened, injured, or dying tissues. The fungus manufactures enzymes that soften or "tenderize" plant cells before they are actually invaded by the mycelium. This advance attack permits the fungus to grow more rapidly and increases the amount and spread of decay.

Management

Management of gray mold is difficult, but it can be aided by the removal of infested debris from greenhouses, storage areas, and fields. Cultural and storage conditions should allow for good aeration and the quick drying of surfaces of plants and stored plant products. Greenhouse RH should be reduced when possible by ventilation and heating. Repeated application of fungicides gives limited management of gray-mold diseases in the field and rots of fruits and vegetables in storage. Winter application of fungicidal dusts or drenches to soil underneath azalea or camellia plants, severely infected the preceding spring, will help kill overwintering sclerotia.

CERCOSPORA LEAF SPOTS

Cercospora leaf spots are found on a wide variety of crops, particularly in warm, humid regions. On crops such as asparagus, beet (both red and sugar), celery, forage legumes, peanut, and soybean, yields can be much reduced by the disease.

Cercospora leaf spot of peanuts is one of the most important diseases of this crop worldwide, with annual yield losses of 15 to 50%. In the United State where chemical control measures generally are used, losses still may be as high as 15%. The disease reduces the yield of nuts and lowers the quality of peanut hay. In addition, fallen leaves from infected plants provide organic matter to serve as a food source for other fungi, particularly *Sclerotium rolfsii*. As a result there is buildup of inoculum, and the severity of stem rot is increased (see Chapter 13).

Many weeds and ornamentals act as hosts for these ever-present fungi.

Symptoms

Cercospora leaf spot of peanut is easily recognized by the brown or black circular spots on the leaves (Fig. 13.16). Often a bright yellow halo surrounds each spot. As the disease progresses, the spots may enlarge until the entire leaf is affected. So many of these leaves fall off that often the ground beneath the plant is covered.

On the leaves of other crops (beet, celery, soybean, tobacco), the causal agent typically causes numerous small (2-15 mm in diameter) circular brown, tan, or dingy gray spots. Scattered through the centers of the lesions are little black dots composed of masses of conidia borne on short conidiophores. The spots may resemble a "frog-eye" (Fig. 13.17), and Cercospora diseases often are called by this unusual name. Although *Cercospora* spp. primarily cause a foliage disease, infection can occur on any aboveground part of the crop.

During humid weather the spots become so numerous that they merge to form large irregular spots. At this degree of damage the leaves wither, die prematurely, and fall.

Under certain conditions it is difficult to identify the causal agent of a leaf spot simply by looking at it. During hot, dry weather frogeye lesions may be only pinpoint in size and difficult to recognize. On the other hand, during warm, humid weather the spots may enlarge rapidly. Frogeye lesions may be confused with spots caused by other fungi or bacteria. In addition, more than one kind of spot often occurs on the same leaf.

Causal Agent and Disease Cycle

Conidia of *Cercospora* spp. vary greatly in size. They range from 25-300 $\mu m \times$ 3-7 μm, with an average of 50 $\mu m \times 4\ \mu m$. Both temperature and food source affect conidial length. The slender, colorless conidia are straight to curved and have several cross walls. A typical conidium is shaped like a baseball bat. The conidia are found on brown, septate, knobby, sparingly branched conidiophores borne in clusters. The fungus sporulates and grows rapidly at 27°C (81°F).

There are many pathogenic strains of *Cercospora*. Some strains can attack more than one host. Some strains produce substances called toxins that are poisonous to plants and aid the fungus in attacking the plant. The fungus is so variable that it is often difficult to identify precisely.

FIGURE 13.16. Cercospora leaf-spot of peanuts. Note the halo on the leaf lesions on the leaflet to the right.

FIGURE 13.17. Cercospora or frogeye leaf spots on a soybean leaf. Notice the light-colored centers of the lesions with the dark margins. *Courtesy NCSU Plant Pathology Photo Collection.*

These fungi survive from one season to the next on diseased leaves, stems, and seeds. When contaminated seeds are planted, they may produce weak, stunted seedlings with lesions on the cotyledons; in a few days the growing fungus begins producing conidia to start secondary infections (Fig. 13.18). The conidia are blown or splashed by rain or carried by insects and machinery to nearby plants. They germinate within a few hours and quickly penetrate leaf tissue. If the weather remains warm and humid, new leaves become infected successively as they develop. Often it is difficult to find a leaf that has no lesions.

Cercospora leaf spot is more severe when susceptible crops are grown repeatedly in the same field without crop rotation. Early infection starts from conidia produced

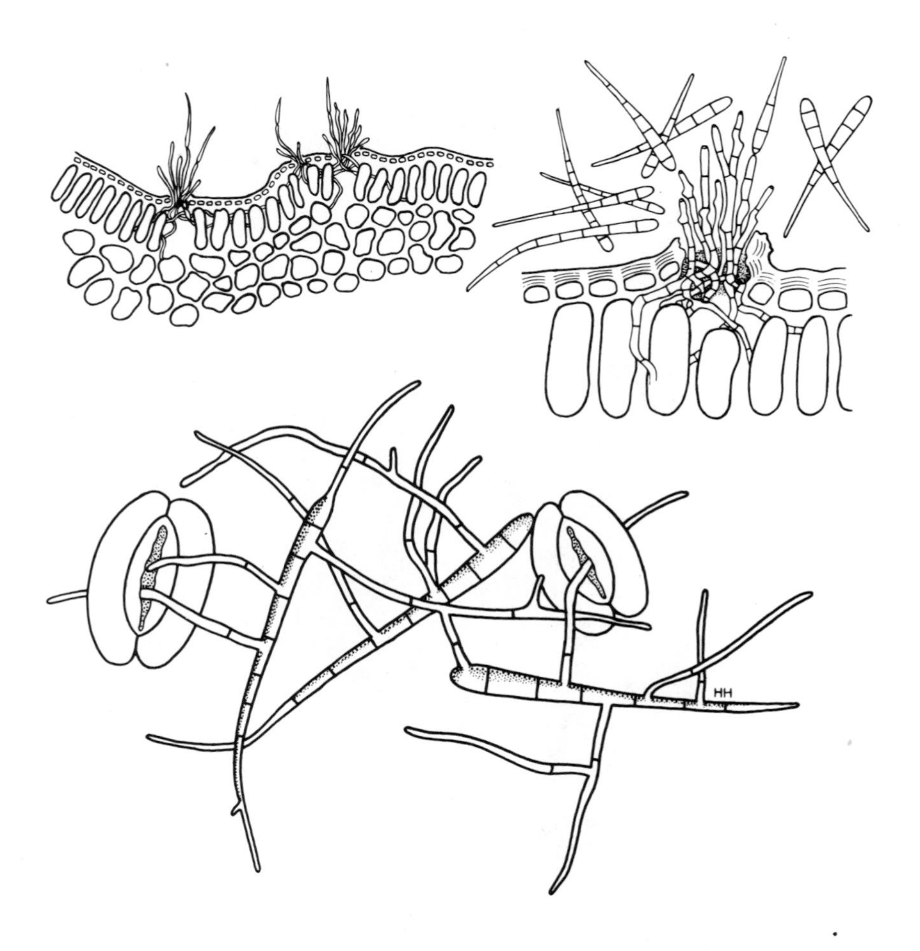

FIGURE 13.18. *Cercospora* spp., the causal agent of frogeye leaf spots. *Top:* Cross section of upper surface of leaf showing the fungal mycelia bursting through the upper surface of leaf and producing spores on conidiophores. *Bottom:* Germinating conidia entering the leaf through the stomata. Magnified about 300 times. *Drawing by H. Hirschmann after W. C. Sturgis.*

on the trash from last year's crop. If the disease gets off to an early start, it becomes difficult to control.

Management

The use of good-quality, disease-free seed helps in managing Cercospora leaf spots. Seed treatment with a fungicide aids in preventing seedling loss, but does not prevent development of the disease in the field. For transplant crops such as celery, fungicides should be applied regularly in the seedbed as part of an IPM program. To delay or prevent the buildup of fungicide-resistant races of the fungus, it is a good practice to alternate or rotate the use of different fungicides.

The fungus is carried over from one season to the next on debris from the previous crop; therefore all trash should be plowed under (or burned) as soon as possible. Crop rotation also should be used to prevent inoculum buildup. On peanuts, the successful control of Cercospora leaf spots at present depends on regular applications of fungicides in the field. Disease forecasts for Cercospora leaf spot are available in some peanut-growing areas to aid growers in timing fungicide applications.

Whenever possible, disease-resistant cultivars (e.g., of soybeans and beets) should be used.

HELMINTHOSPORIUM LEAF SPOTS

The three most important food crops in the world are wheat, rice, and corn. Species of *Helminthosporium* cause a number of serious leaf spots, seedling blights, root rots, culm rots, and seed infection on all three of these crops. Other crops, including barley, forage grasses, oats, sorghum, and turfgrasses, also suffer losses from these versatile fungi.

Originally many diverse fungi were placed in the genus *Helminthosporium* because the conidia of the asexual stage or anamorph of all these fungi were similar. However, as more has been learned about these fungi and the sexual stage or teleomorph has been found for many, new names have been given to the fungi. Although these new names may seem to be confusing initially, they are intended to help us better understand the fungi. Also, we can better understand the life and disease cycles and how to manage the diseases caused by these fungi.

Southern Corn Leaf Blight (SCLB)

In the United States both the land area planted to corn and its total yield of grain are greater than for any other cultivated crop. About 80% of the corn crop is fed to livestock; the remainder is used for human food. Essentially all diseases of corn have been or potentially can be managed by genetic resistance.

The history of SCLB in the United States presents a dramatic illustration of the dangers of genetic uniformity to modern agriculture. (See Chapter 6, Disease-Resistant Cultivars section, for additional discussion of this phenomenon.) The story began in the 1950s with the discovery and introduction of male sterility to produce high-yielding corn hybrids at low cost. The source of male sterility, designated as Texas cytoplasm male sterility (T_{cms}), was transferred into many inbred lines of corn by backcrossing. Inbred lines containing T_{cms} then were crossed with lines possessing a genetic factor (a restorer gene) that restored pollen for kernel development. By 1970, the use of inbred lines and hybrids containing T_{cms} was so widespread that about 85% of the hybrid seed corn sold in the United States was of this type.

In 1969, a disease of corn leaves and ears was observed in localized areas of Iowa, Illinois, and Indiana. The T_{cms} plants were severely diseased, whereas plants having normal cytoplasm showed no ear infection and only a few leaf lesions. Greenhouse pathogenicity tests demonstrated that the pathogen was a distinctive race of the fungus *Bipolaris maydis* (Nisikado & Miyake) Shoemaker, which causes southern corn leaf blight. The old common strain of this fungus on corn was designated race O and the new strain race T.

In January 1970, the disease was reported as causing severe damage on T_{cms} corn in Florida. The spring of 1970 was unusually wet in much of the southern United States, and as a result the disease progressed rapidly northward from the South; by mid-July it was well established in the Corn Belt and a severe epidemic was occurring. In the South many fields of corn were a total loss. This epidemic of southern corn leaf blight (SCLB) received more publicity than any previous plant disease outbreak in the United States in recent times. Losses due to the epidemic were officially estimated at near one billion dollars in the nation as a whole. Average yield losses in the Corn Belt were 20 to 30%; southern states sustained much greater losses.

In 1971 the demand for the disease-resistant, normal-cytoplasm seed was far greater than supply. Race T of *B. maydis* survived the winter of 1970-71 in most of the South, but weather conditions were dry and unfavorable to the fungus in the early spring and summer of 1971. Fortunately, the unfavorable conditions and the reduced usage of T_{cms} corn hybrids halted the SCLB epidemic, and a near-record corn crop was produced that year. Sorghum, a major crop closely related to corn, is now in almost exactly the same vulnerable condition of genetic uniformity as corn was during the winter of 1969-70.

Symptoms

Lesions on *Bipolaris*-infected leaves are tan, spindle-shaped or elliptical, with yellow-green or chlorotic halos (Fig 13.19). Later lesions often develop dark, reddish-brown borders and may occur on leaves, stalks, and husks. A black, felty mold may cover kernels in infected ears (Fig. 13.20). Race O usually attacks only leaves.

Causal Agent and Disease Cycle

The mycelium of *B. maydis,* which causes southern corn leaf blight, is dark gray. The light green to brown conidiophores (60-280 μm $\times$ 6-10 μm), which arise in groups of two or three on the dead leaf spots, produce cigar-shaped, smokey gray conidia (40-150 μm $\times$ 11-27 μm), which have on the average six to eight cross walls (Fig. 13.21). The spores germinate by sending out one germ tube each from the cells at either end of the spore; that is, germination is bipolar. Numerous conidia are produced on the leaf lesions. The perithecia, or sexual stage (the sexual stage is known as *Cochliobolus heterostrophus* Drechs.), containing ascospores appear as tiny specks on or in disintegrating leaf tissue, but are rarely seen in nature.

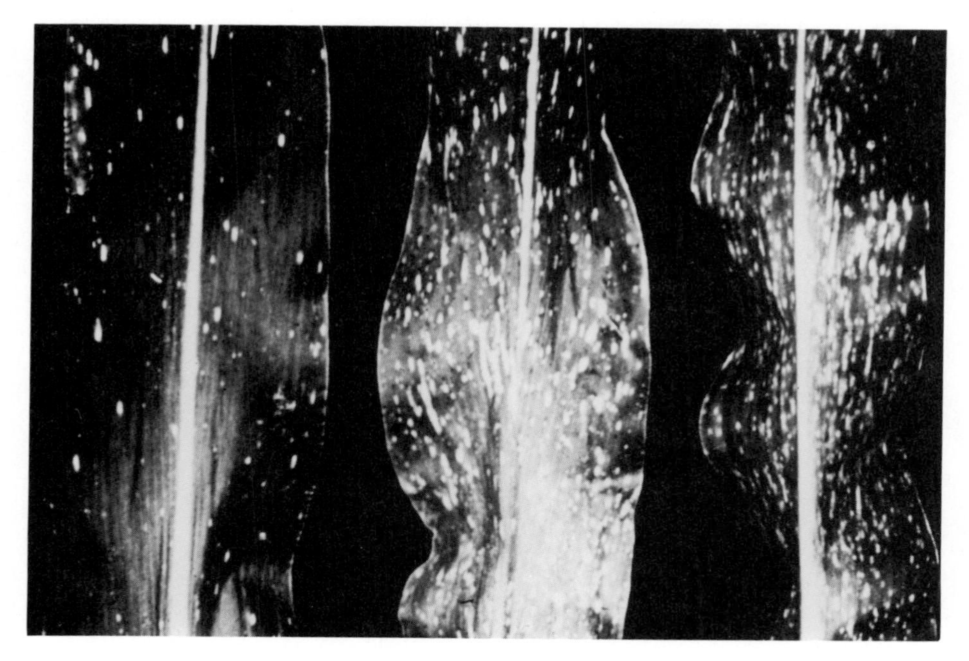

FIGURE 13.19. Lesions of southern maize (corn) leaf blight caused by the fungus *Bipolaris maydis.* *Courtesy R. R. Nelson.*

FIGURE 13.20. Three ears of maize (corn) with darkened, diseased kernels and velvety, black growth caused by the pathogen *Bipolaris maydis* Race T along with a healthy ear of maize. *Courtesy NCSU Plant Pathology Photo Collection.*

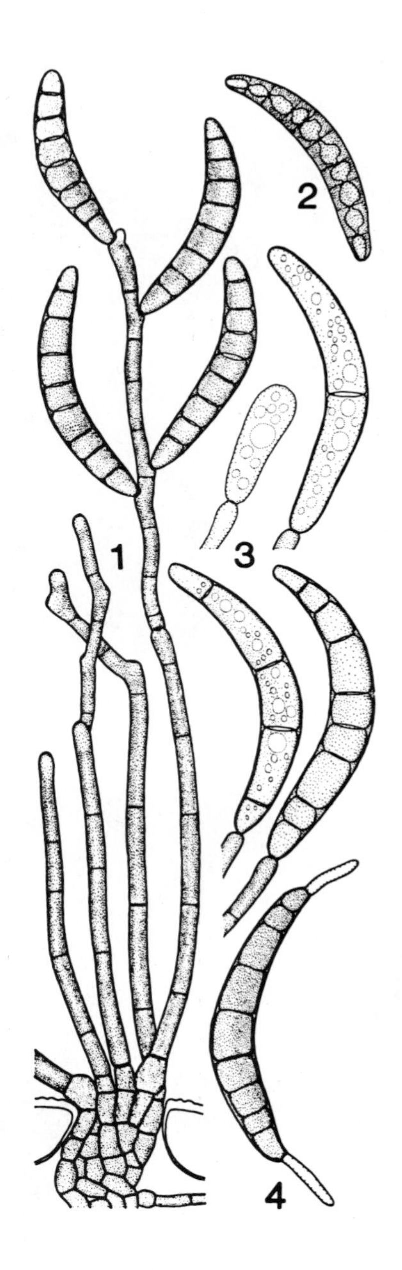

FIGURE 13.21. *Bipolaris maydis*, which causes southern corn maize (corn) blight. 1—conidiophores; 2—mature conidia; 3—formation of conidia; 4—condium with bipolar germination. Magnified about 380 times. *Courtesy E. S. Luttrell.*

The fungus overwinters as mycelium and spores in maize debris in the field and on kernels in cribs, bins, and elevators. Primary infections result from conidia carried by wind or splashing water to growing plants. Sporulation on lesions produces additional primary and secondary inoculum. Under ideal conditions the disease cycles in about 60 to 70 hr. The disease is most severe and spreads rapidly in warm [20-32°C (68-90°F)] damp regions. Race T is virulent on T_{cms} corn but causes only small lesions on corn having normal cytoplasm. This race produces a toxin that is host-specific; that is, it affects only T_{cms} corn. Race O does not differentiate between T_{cms} corn and normal cytoplasm corn.

Management

Management of SCLB is currently achieved by the use of normal cytoplasm corn hybrids and hybrids having disease-resistant sources of male sterility. Plowing under crop residues can reduce early infection by limiting the survival of the fungus. Fungicides have been shown to reduce disease spread and yield losses, but genetic control with resistant cultivars offers the most desirable and least expensive management strategy. The use of certified seed that has been treated with a fungicide is always a good practice.

Brown Spot of Rice

Brown spot of rice is also called sesame leaf blight because the lesions are about the size of seasame seeds. The disease, which has been reported in all rice-growing countries, causes a seedling blight, weakens older plants, and lowers grain quality and weight. Brown spot was considered to be a major factor contributing to the Bengal (India) famine in 1942.

Brown spot is associated only with soil deficient in certain nutrients. The disease serves as an index of such conditions.

Symptoms

The most conspicuous symptoms of rice brown spot are the spots that appear on leaves and glumes; they may also appear on the coleoptile, leaf sheaths, panicle branches, and stems, but rarely appear on roots of infected plants. Typical spots on the leaves are brown, oval, and evenly distributed over the leaf. On susceptible cultivars the spots may reach 1 cm in length. Sometimes the spots are so numerous that the leaves dry out before the plant is mature, and as a result yields are drastically reduced. Under warm, humid conditions dark brown conidiophores and conidia make the spots look like velvet.

Causal Agent and Disease Cycle

Brown spot of rice is caused by *Bipolaris oryzae* (B. de Haan) Shoemaker (synonym *Helminthosporium oryzae* Breda de Haan). The sexual stage is known as *Cochliobolus miyabeanus* (Ito. & Kuribayashi) Drechs. ex Dastur. The dark brown, abundant, many-branched hyphae grow profusely on diseased tissue. Gray conidiophores arise from the hyphae and give rise to slightly curved, cigar-shaped gray to brown conidia (35-170 μm × 11-17 μm) with an average of eight cells. Mature conidia germinate with two polar germ tubes. Many strains exists, which differ in morphology, physiology, sporulation, and pathogenicity.

The brown-spot fungus overwinters in plant trash and infected grain. The conidia can survive 2 to 4 years in infected seed. Primary infection through diseased seed can occur, but infection from trash and soil occurs just as frequently. Conidia germinate on wet leaves to form *appressoria* and infection pegs that penetrate the leaf cells. The pathogen produces a toxin that kills the host cells.

Environmental Effects

Brown spot is more severe on soils deficient in nutritional elements or on those in which toxic substances have accumulated. Deficiencies in potassium, manganese, magnesium, and especially nitrogen late in the growing season all increase brown-spot damage. Plants in soils with high levels of hydrogen sulfide also are prone to brown spot. Often brown spot and deficiency symptoms are difficult to distinguish in the field.

The optimum temperature for infection is about 25°C (77°F) at RH above 90%. Mycelial growth is greatest at 27 to 30°C (81-86°F) although the fungus grows over a wide temperature range.

Management

In areas where infection of seed with brown spot is common, seed treatment with fungicides reduces seedling damage. Where possible, rice should be grown only on soils with a proper nutritional balance; those with mineral deficiencies should be avoided. Proper fertilization is important for management of the disease. Field sanitation, crop rotation, adjustment of planting dates, and good water management help keep brown-spot losses to a minimum. Resistant cultivars should be used where needed.

ALTERNARIA LEAF SPOTS

Fungi belonging to the genus *Alternaria* are commonly found on leaf spots and dead tissue. They are easily isolated from plant trash and seeds. Their identification is not easy because of the large number of strains and their ability to attack different hosts. *Alternaria* spp. cause leaf spots of apple, carnation, carrot, crucifer, cucumber, sweet pepper, potato, pumpkin, radish, tobacco, and tomato. Many weeds also suffer from leaf spots caused by *Alternaria* spp. Not only does the fungus attack crops in the field, but it also causes blemishes and transit and storage rots on cabbages, carrots, onion bulbs, peppers, potato tubers, and tomatoes.

On potatoes and tomatoes, Alternaria leaf spot is often called *early blight* to distinguish it from *late blight* caused by *Phytophthora infestans.* This name is not always accurate, because early blight often appears late in the season.

Symptoms

Leaf lesions caused by *Alternaria* spp. first appear as small water-soaked spots that quickly enlarge to form circular lesions of dead tissue up to 1 cm or more in diameter (Fig. 13.22). In humid weather, the black conidia on the surface of lesions become visible. Often they are arranged in concentric rings to give a "target spot" effect. Linear

FIGURE 13.22. Early blight on Irish potato leaflet, caused by *Alternaria* spp. Note the circular lesions with concentric rings and the darkened, dead tissue. *Courtesy C. W. Averre.*

brown to black spots appear on petioles and stems. As is the case with other leaf-spot diseases, Alternaria leaf spot often increases in severity as the season progresses and leaves become mature. Often leaves die and drop off prematurely, and yields are decreased as a result of the plant's decreased ability to carry on photosynthesis. Typically the lower, older leaves show the most pronounced symptoms.

During transit and storage, crops such as carrot, cabbage, and potato are also injured by *Alternaria* spp. The outer leaves of cabbage heads often are overrun by the sporulating fungus and turn black. A black mold, consisting chiefly of conidiophores and conidia, develops on the surface.

Causal Organism and Disease Cycle

The much-branched mycelium of *Alternaria* species has numerous cross walls and becomes dark with age. The short conidiophores produce tennis-racket-shaped, brown to black conidia that vary greatly in length, averaging 30 μm $\times$ 12 μm (Fig.

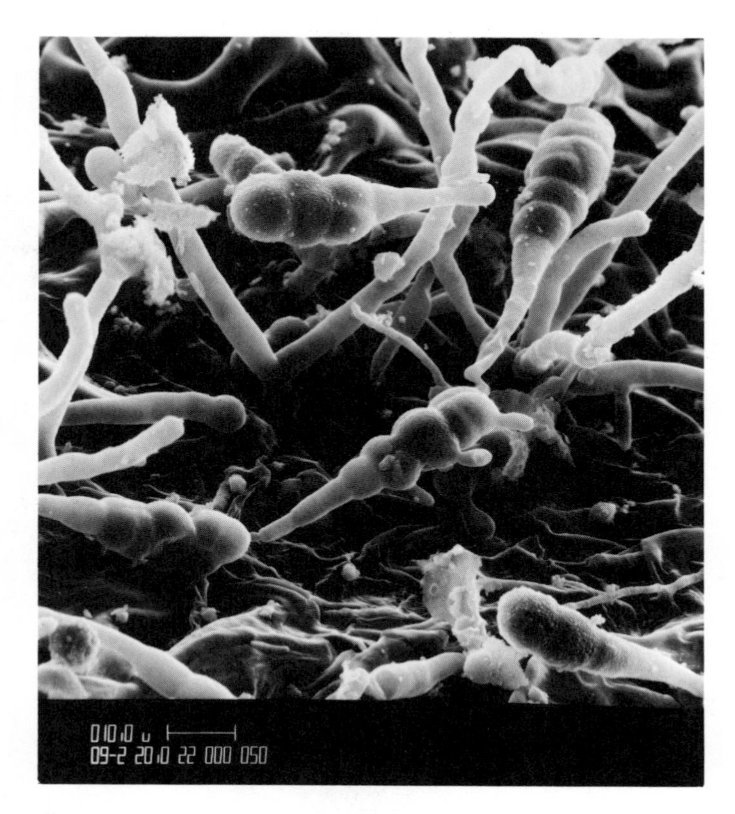

FIGURE 13.23. Scanning electron micrograph of conidia of *Alternaria* spp. on a leaf surface. Magnified about 730 times. *Courtesy G. Van Dyke and H. W. Spurr.*

13.23). The conidia may or may not have beaks (some call them tails). Conidial production depends upon the isolate, light, moisture, and temperature. Several toxins produced by *Alternaria* species have been identified, and their role in premature yellowing and death of host tissue has been determined.

Alternaria species are carried on seed and plant refuse. They also overwinter as mycelium in woody debris such as stems and stalks. Conidial production starts with warm spring weather, and the spores are scattered by wind and water. Infected transplants also spread the fungus. The spores germinate readily and grow over a wide temperature range [4-35°C (39-95°F)]. During warm, wet weather with prolonged periods of high RH or heavy dews, numerous lesions appear on leaves, petioles, and stems. Conidia produced on these lesions build up to high numbers as the season progresses. After the host plant dies, the fungus persists as mycelium in dead host tissue.

Management

Because *Alternaria* species often are carried on the seed, seed production in dry areas helps to minimize seed infection; however, seed treatment also is often desirable, especially on crucifers. Hot-water treatment at 50°C (122°F) for 30 min is satisfactory for treating seed of members of the cabbage tribe.

Transplant crops should be sprayed with a fungicide in the seedbed to prevent transportation of the fungus on diseased transplants. Periodic spraying or dusting with fungicides is the principal method of managing Alternaria leaf spot diseases. Often repeated applications are necessary; the use of fungicides must be incorporated into an overall IPM program as recommended by the local extension service.

Crop debris should be destroyed or plowed under as soon as possible and rotations used where possible. Maintaining proper fertility will help prevent severe damage on some crops.

ANTHRACNOSE

The name *anthracnose,* derived from a Greek word meaning ulcer, apparently refers to the ulcer-like lesions caused on many hosts. Anthracnose fungi are worldwide in distribution. Their many hosts include alfalfa, bean, coffee, clovers, corn, cotton, cucurbits, grasses, lupines, onion, ornamentals, peach, small grains, soybean, strawberry, tobacco, and watermelon.

Anthracnose is a common and widespread disease that infects the foliage and fruits of many crops, particularly in warm, humid regions. On alfalfa, clover, and lupine, anthracnose is one of the most destructive diseases in warmer regions. It reduces yields of hay and seed and can destroy stands of clover and other pasture crops.

Anthracnose fungi thrive where wet, warm growing seasons prevail. It is almost impossible to produce anthracnose-free seed in such regions. Therefore, certified seed must be produced in semiarid, irrigated areas such as the western United States, where rainfall is very low during the growing season.

Symptoms

Anthracnose may occur on plants at any stage of development. The appearance of the disease varies somewhat on different hosts. It develops most commonly on the young, succulent parts of stems, petioles, and fruits, but is not limited to them.

On beans the disease appears on all aboveground parts of the plant, but rarely on the roots. Usually the lesions are dark brown; in moist weather pinkish spore masses often can be seen. Elongate spots appear on the veins on the lower side of the leaf and eventually spread to the top side. Petioles, stems, and hypocotyls often are affected, causing the plant to collapse or damp off. The most striking symptoms occur on the immature bean pod (Fig. 13.24). Small brown spots appear, which enlarge rapidly up to a diameter of 1 cm. They become dark brown to black and are depressed in the center. Sometimes the fungus penetrates through the pod, and the developing bean becomes infected.

On alfalfa and clover the first noticeable symptoms usually are dark brown or black spots on the petioles. The spots soon enlarge and cut off translocation to the upper parts of the leaf, causing the leaflets to wilt and die. Dark tufts of black setae (hairlike structures) (Fig. 13.25) often are conspicuous in the older lesions and help confirm a diagnosis of anthracnose.

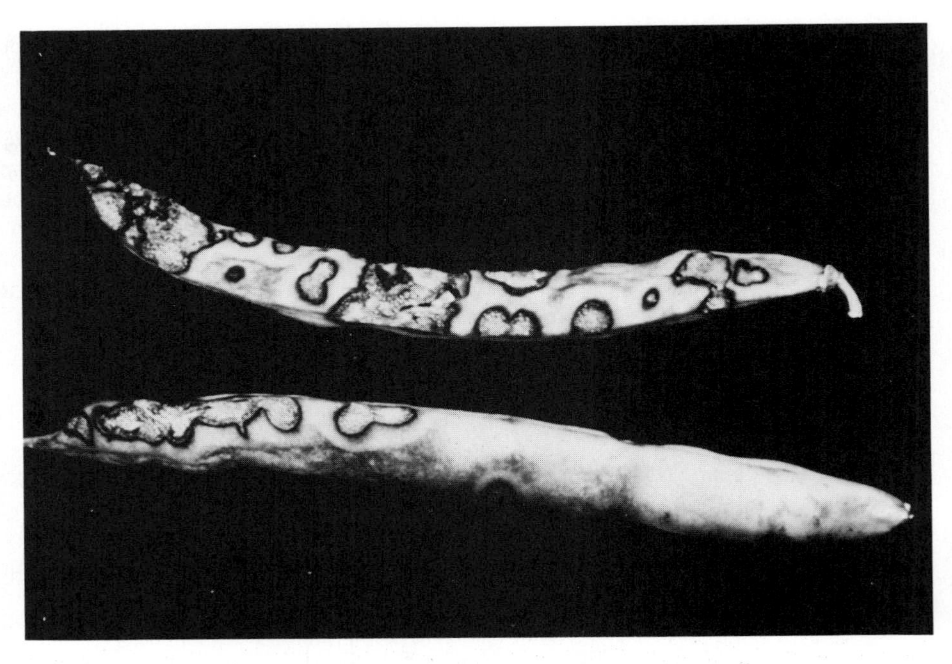

FIGURE 13.24. Symptoms of anthracnose on bean pods. *Courtesy NCSU Plant Pathology Photo Collection.*

On cucumber leaves the spots commonly start on a vein and expand into brown spots reaching 1 cm or more in diameter. Growing leaves may become distorted, and the fusing of many spots may blight entire leaves. Petioles and stem lesions are shallow, elongate, and tan. Lesions appear on ripening fruits as roughly circular, sunken, water-soaked areas, which turn black. In the center, pink spore masses interspersed with black setae can be seen with the use of a hand-held magnifying lens.

On strawberries, the foliage and stems are infected, and leaves turn brown and shrivel. Plants can be killed by the disease in a short time during severe epidemics. The fruit also can be rotted by the anthracnose fungus. In some cases, the strawberry plants may become infected in nurseries by inoculum that lives on other fruit plants such as apples, blackberries, and wild grapes. When transplanted to growers' fields, the infected plants then serve as a source of inoculum that can be spread easily to uninfected plants.

Causal Agent and Disease Cycle

Anthracnose diseases are caused by species of *Colletotrichum*. Typically they have septate mycelium, colorless when young, dark green or gray when old. Often the mycelium forms masses of thick-walled cells called *stromata*. From these the fruiting structures, called *acervuli*, arise. Typically the acervuli form in the host tissue just below the epidermal cells, causing them to rupture. Short, oblong, one-celled, colorless

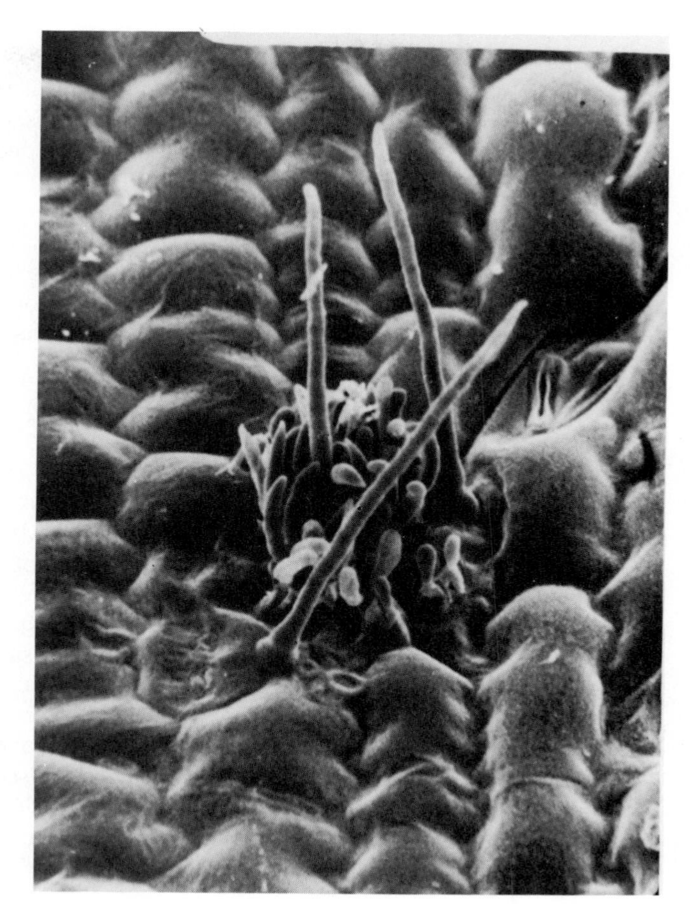

FIGURE 13.25. Conidia and setae (longer, pointed structures) of *Colletotrichum* spp. erupting from the surface of an infected leaf. *Courtesy NCSU Plant Pathology Photo Collection.*

conidia (5 μm × 20 μm) form on short conidiophores, and among them black hairlike *setae* are produced. Often the spores are so numerous that they form pink, shiny masses.

Anthracnose fungi live over from one growing season to the next on infected plant debris and on seed. When infected seed germinate, spores produced on infected cotyledons serve as secondary inoculum sources to infect other parts of the growing plant.

Spattering and windblown rain spread the spores to nearby plants. If a conidium lands on a wet leaf or petiole, it germinates and penetrates the surface by a short hypha that forms an infection peg to force its way through the cell wall. Optimum conditions for infection occur at about 20°C (68°F) when RH approaches 100%. There are many strains of *Colletotrichum* species, differing in their pathogenicity; thus the breeding of resistant cultivars is a continuing task.

Management

Important ways to manage anthracnose include the use of anthracnose-free seed and disease-free transplants (especially with strawberries). If necessary, seed should be treated with a suitable fungicide. Often certified seed dealers sell seed that already has been treated.

Rotations of 2 to 3 years and field sanitation help eliminate the fungus on overwintering trash.

Row crops should be cultivated only when dry if the disease is present because cultivation when foliage is wet spreads spores from one plant to another. Fungicides used in the overall complete disease management program also are of help. Resistant cultivars should be used if they are available.

APPLE SCAB

Apple scab—the most important disease of apples—occurs wherever apples are grown except in very dry or warm climates. In years when scab is severe, crops may be a total loss unless management measures are applied. Even the most efficient management programs may fail to give satisfactory results when the disease is severe or weather conditions hinder the application of fungicides.

The primary effect of scab is a reduction in fruit size and quality. But the disease also causes premature fruit drop, leaf fall, and poor fruit bud development for the next year's crop. Scab also increases the cost of production because efficient scab management requires timely and repeated fungicide application. Scab is such an important disease that several states now have established scab-forecasting services to tell growers when conditions are favorable for scab development. Fungicides then can be applied at the proper time for the most efficient management.

Symptoms

Scab spots usually appear as small, olive areas on the sepals or young leaves (Fig. 13.26) of flower buds. Soon the lesions become olive green and assume a velvety appearance caused by the abundant production of spores on the ends of short, erect, threadlike branches of the fungus. Later the circular lesions turn black; some may reach 1 cm or more in diameter. Often the lesions are so numerous that they grow together (coalesce).

On the fruits, black circular spots (scabs) with deep cracks frequently develop, causing malformations that make the fruit worthless for the fresh market. Late-season infections result in small lesions, which enlarge into dark scab spots during storage.

Causal Agent

The apple scab fungus, *Venturia inaequalis* (Cke.) Wint. emend. Aderh., has both asexual and sexual spores. Either kind can germinate in water on a susceptible part of the apple tree and penetrate the cuticle of the plant by a microscopic peglike growth. Hyphae then grow between the cuticle and the cellulose wall of the outer layer of cells

FIGURE 13.26. Numerous scab lesions on the underside of apple leaves. *Courtesy C. N. Clayton.*

of the apple plant until a solid layer of fungal growth, which may be several cells deep, is formed over the lesion surface.

The asexual spores (conidia) are produced on short, erect, brown conidiophores, which successively form several one- to two-celled, red-brown, ovate to lanceolate conidia, averaging 16 μm $\times$ 8 μm. The sexual spores (ascospores) are produced in the spring in fruiting bodies (pseudothecia) of the fungus that develop in dead apple leaves on the ground. The tiny, black pseudothecia are 90 to 150 μm in diameter. They resemble perithecia in their outward appearance, but are formed differently. They are spherical and have a small opening (*ostiole*) at the top through which spores are discharged (Fig. 13.27). Inside each pseudothecium 50 to 100 asci are formed. Each hyaline ascus is shaped like a bag or sack and contains eight ascospores (13 μm $\times$ 6 μm), each consisting of two cells of unequal size. The ascospores are colorless at first but turn brown when mature.

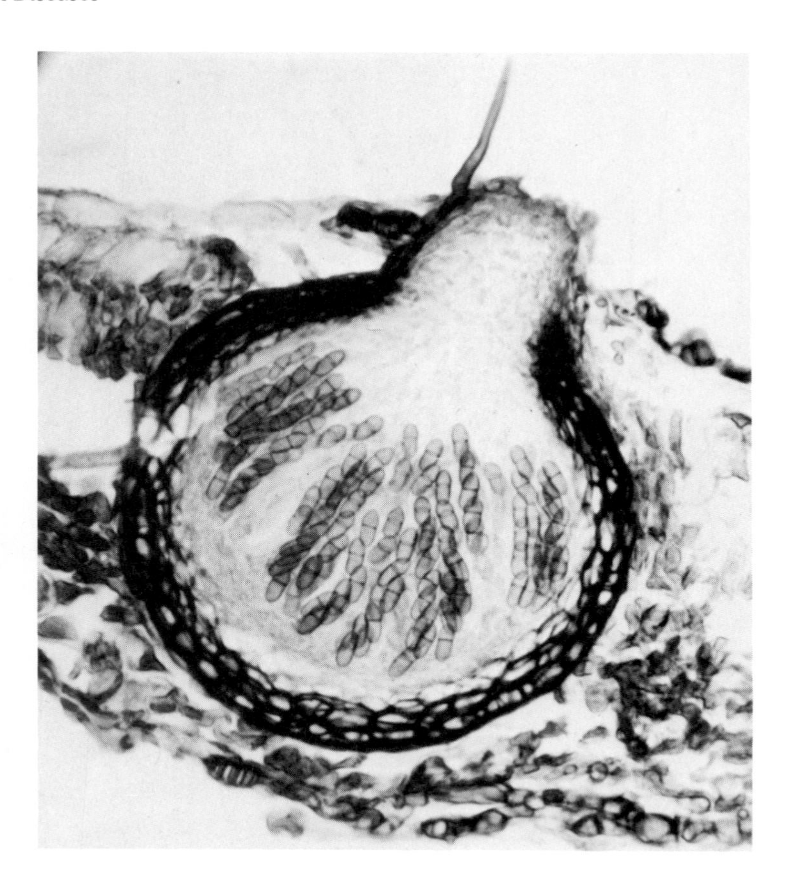

FIGURE 13.27. Pseudothecium of the apple-scab fungus containing mature ascospores, magnified about 750 times. The ascospores are two-celled and mature in early spring when they are discharged into the air. If an ascospore falls on a wet apple leaf or flower, a new infection is initiated. *Courtesy J. R. James.*

Disease Cycle

Pseudothecia of the apple scab fungus overwinter in dead leaves on the ground. Cool fall weather, snow cover, and a cool, moist spring favor the production and discharge of ascospores. In areas favorable for scab development, the ascospores ordinarily are mature and ready for discharge during rainy periods by the time the first susceptible apple buds start to open in the spring.

When dead leaves with pseudothecia become thoroughly soaked in the spring, the asci elongate, push through the ostiole, and forcibly discharge the ascospores into the air. The ascospores are carried by air currents to nearby leaves and flowers. For infection to occur at 5°C (41°F), the ascospores must be continuously wet for 18 hr; at 18 to 24°C (64-75°F), they must be wet only for about 6 hr. Not all pseudothecia ripen at the same time; so ascospore discharge may occur continuously over a period

FIGURE 13.28. Lesions of apple scab on nearly mature apples. *Courtesy T. B. Sutton.*

of 4 weeks or so. This discharge often happens just at the time when apple leaves and sepals are expanding rapidly.

The ascospores penetrate host tissue by means of peglike hyphae much like the conidia; the developing mycelium absorbs nutrients from the leaf cells and produces enormous numbers of conidia within 8 to 15 days. Conidia are easily detached when wet with rain and are washed or blown to other leaves and fruits to establish secondary infections, especially during cool, wet weather.

After infected leaves fall to the ground, they die; and the hyphae form perithecia, which survive during the winter to start the disease cycle again the following spring.

Management

Although the breeding of scab-resistant apple cultivars holds promise for the future and some cultivars in use today exhibit some resistance (Fig. 13.28), the most important method of scab management is timely spraying with the proper fungicides. You cannot grow pretty, unblemished apples unless you spray. But fungicides are expensive and must fit in with other aspects of orchard management such as removal or destruction of fallen leaves in the winter to reduce initial inoculum.

The spray programs most widely used depend chiefly on protection of the susceptible parts of the apple tree from infection by the fungus. As new leaves expand and fruits enlarge, they should be protected with fungicides. This protection should begin before the first infection occurs in the spring and should continue through repeated application of fungicides as long as there is danger of infection. Thus a scab management program may require 6 to 10 fungicide applications per season, beginning in the spring when weather conditions become favorable for infection and continuing almost until harvest. Scab-forecasting services, based on computer analysis of weather conditions, are becoming more accurate each year, allowing many growers to apply fungicides at more precise intervals or when conditions ideal for scab outbreaks develop. Several fungicides give excellent scab management. All can act as protectants that prevent infections, and some have the ability to stop infections that have already started; other newer fungicides can eradicate young scab infections.

RHIZOPUS SOFT ROT

A common bread mold is caused by species of *Rhizopus*, which also affect fleshy organs of flower crops, fruits, and vegetables, usually during storage, transit, and marketing. Rhizopus rot is the most destructive disease of sweet potatoes in storage (Fig. 13.29) and also causes losses on cherries, corn, cucurbits, gladioli, peaches,

FIGURE 13.29. Rhizopus rot on sweet potato (right) and disease-free sweet potato (left). Note the fuzzy mycelium of the fungus on the diseased sweet potato. *Courtesy J. W. Moyer.*

peanut, squash, and tulips. As a storage rot of tobacco, *Rhizopus* species must be guarded against every year.

Soft rot losses vary with the crop, storage temperature, RH, amount of wounding of the fleshy organs or fruits, and amount of inoculum present. When conditions are favorable, losses can be great in just a few days.

Symptoms

The first symptoms of infection by *Rhizopus* species is a soft, watery rot, which progresses rapidly. A squash fruit may become completely rotted in 2 to 3 days; a sweet potato root, in 4 to 5 days. If the skin of infected tissues is unbroken, the softened fleshy organ gradually loses moisture and shrivels into a mummy. More frequently, the softened skin ruptures, fluid leaks out, and tufts of whisker-like, gray sporangiophores develop rapidly. The bushy growth extends to healthy parts of the affected fruit, to adjacent fruits, and sometimes to the wet surface of containers. At first the rotting fruits have a mildly pleasant smell, but soon a sour odor develops. Affected fruits sometimes dry out rapidly, and dry rots develop.

Causal Agent and Disease Cycle

Rhizopus spp. are in the Zygomycotina. They are found almost everywhere in nature and can use many kinds of crop debris as food. The mycelium has no cross walls. When the conspicuous aerial mycelium grows on a surface, it produces arching aerial branches (*stolons*). Wherever they touch the surface, the stolons anchor themselves with rootlike hyphae (*rhizoids*). Opposite the rhizoids, erect hyphae grow in tufts up into the air. Each aerial hypha becomes a sporangiophre, which produces a black, spherical sporangium at the tip containing thousands of tiny black sporangiospores (5 μm $\times$ 5 μm). When the sporangium breaks open, the sporangiospores float about in the air (Fig. 13.30). If they fall on a moist surface or wound of a susceptible plant organ, they germinate and produce mycelium again.

When unfavorable environmental conditions (e.g., low or high temperatures, diminished food supply) occur, the fungus produces black, warty sexual spores called *zygospores,* formed by the fusion of two hyphae. The zygospores are resistant to adverse environmental conditions and serve as the overwintering stage. Each zygospore germinates by forming a short hypha bearing a sporangium.

Both zygospores and sporangiophores may survive for many months. The fungus is such an efficient saprophyte that it undoubtedly subsists on crop debris between crop seasons. The optimum temperature for fungus growth and fruit decay is about 25°C (77°F) with 75 to 85% RH.

Rhizopus spp. are strictly wound parasites. The hyphae produce enzymes that dissolve the cementing substances (*pectins*) between walls of plant cells. The cells become unglued from each other, and a "soft rot" develops. The fungus also produces other enzymes that break down the cellulose of cells, which then disintegrate. The fungus does not invade cells until after they have been killed by the enzymes; that is, the fungus kills cells first, then feeds on the dead remains like a saprophyte.

Soft rot develops quickest on mature fruit. Unripe fruits do not rot rapidly. Infection and invasion of tissues also are slowed by cool temperatures (below 20°C) and low RH.

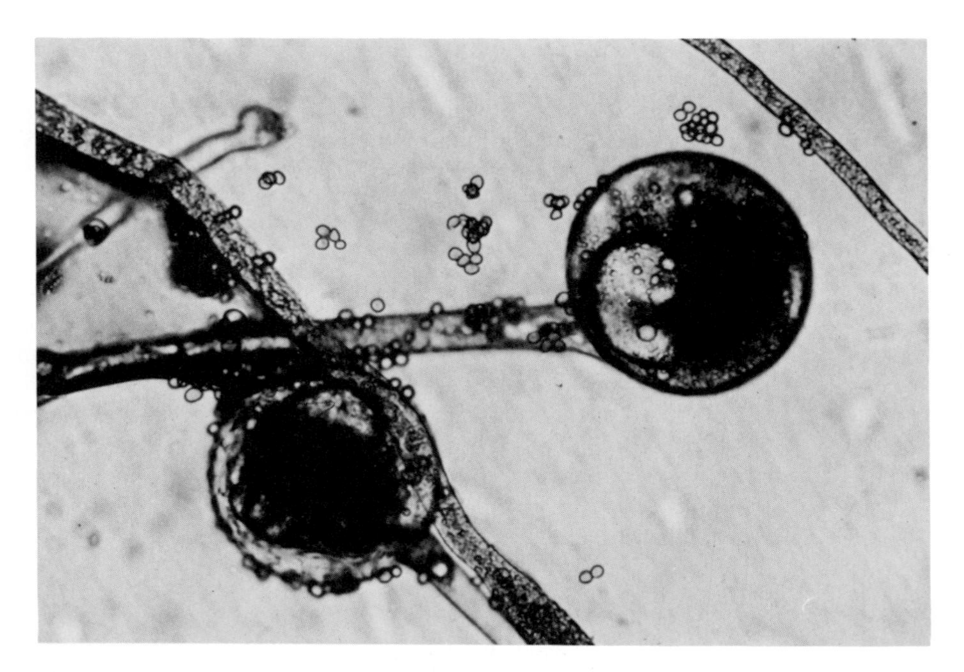

FIGURE 13.30. Sporangia and sporangiospores of *Rhizopus* spp., the bread mold fungus that causes storage rots of many crops. *Courtesy NCSU Plant Pathology Photo Collection.*

Management

One way to reduce Rhizopus soft rot is to avoid wounding fleshy fruits, roots, and tubers as much as possible during harvest, handling, and transportation. Wounded fruit should be discarded or packed and stored separately. Sanitation is important. Storage containers and warehouses should be cleaned of crop debris before use and disinfected.

Very succulent fruits, such as strawberries, should be stored at about 12°C (54°F). Sweet potatoes may be protected by holding them at 25 to 30°C (77-86°F) and 90% RH for 6 to 8 days; during this period the cut surfaces cork over so the fungus cannot penetrate. After the "curing" period, the storage temperature is lowered to 12°C.

Good management also is obtained by wrapping individual fruit in paper impregnated with fungicides.

14

Fungal Diseases of Shade and Forest Trees and Decay in Wood

The shade and forest trees of North America are a precious natural resource that humans must protect and treasure. These silent, majestic inhabitants cover glens, glades, and mountain slopes, roof over cathedral groves in hidden valleys, anchor hillsides against the wash of sudden downpours, shelter violets and wild columbine, purify freshets as they rush toward lazy rivers, provide thickets filled with birds' song and hidden animals, supply lumber and paper, splash vistas with mosaics of orange, yellow, gold, and red in autumn, and etch bare trunks and limbs against the deep mantle of winter snow. Trees restore the air for human and animal use, and restore human souls with their beauty, diversity, and nobility.

Like cultivated crops, trees can be attacked by pathogenic fungi, which stunt their growth and eventually may kill them. Management of tree diseases is difficult, in part because the host plants are widely dispersed. Three major diseases—chestnut blight, Dutch elm disease, and oak wilt—have severely devastated three of the most important hardwood species during the past century. The fungi causing these diseases were introduced to the United States from other countries, where they were of relatively minor importance. However, in this country they spread rapidly, causing widespread destruction.

Softwood trees, such as firs, pines, and spruces, are not immune to fungal diseases. Infection of pine trees, in particular, can cause serious economic losses because of the importance of pine species in the pulp and lumber industries. Pine trees often are grown in plantations (especially in the South), where management of pine diseases is more feasible than it is in natural forests.

Even after trees have been cut and processed into lumber, the fungi do not give up. Wood decay induced by fungi may reduce the appearance and lifespan of wood structures. Fortunately, wood decay can be prevented or reduced by various measures.

A summary of the diseases discussed in this chapter is presented in Table 14.1.

TABLE 14.1. Some important fungal diseases of shade and forest trees.

Disease	Pathogen	Host(s)
Chestnut blight	*Cryphonectria parasitica*	American chestnut
Dutch elm disease	*Ceratocystis ulmi*	American elm
Oak wilt	*Ceratocystis fagacearum*	Many oaks
Annosus root and butt rot	*Heterobasidion annosum*	Coniferous trees
Fusiform rust	*Cronartium quercuum* f. sp. *fusiforme*	Oak/loblolly, slash, longleaf and short-leaf pines
Anthracnose	*Discula* spp.	Dogwood
Little leaf	*Phytophthora cinnamomi*	Pine
Decay	Many fungal species	Wood in buildings

CHESTNUT BLIGHT

The American chestnut, a valuable hardwood species, has been virtually eliminated as a forest tree by chestnut blight throughout its natural range on the upper slopes of the Appalachian Mountains. In 1900, this handsome, useful tree grew in almost pure stands for miles on end. In Appalachian forests nearly one in every four trees was a chestnut. Trees killed by chestnut blight comprised 50% of the total timber value in the eastern hardwood forest. By 1940, nothing remained but the dried trunks, the bare racks of a few dead monarchs high up on some inaccessible hillsides. Small chestnut sprouts continue even today to grow from old stumps for several years, but then fall victim to the blight. Although many stumps refuse to succumb to the disease and continue to send up new sprouts, the sprouts always become infected and eventually die.

The fungus *Cryphonectria parasitica* Murrill and Barr that causes chestnut blight originated in the Orient and evidently was introduced unknowingly into New York on nursery stock about 1904. It spread to the highly susceptible American chestnut, and within 30 years thousands of square kilometers of chestnut trees were decimated.

Symptoms

The first conspicuous evidence of chestnut blight is brown shriveled leaves on twigs that have been girdled by bark lesions. Swollen or sunken cankers forming on twigs, branches, and the main trunk quickly kill all tissue beyond the cankers. Young cankers are yellow-brown in contrast to the normal olive-green of the bark. As the cankers become older, the bark splits and falls away. In older cankers, red-brown spore pustules, up to 2 mm in diameter, form.

Causal Agent and Disease Cycle

During the summer the chestnut-blight fungus produces spores in pycnidia; the colorless, one-celled oblong spores (averaging 8.6 μm $\times$ 1.6 μm) are squeezed out like toothpaste from a tube, following warm summer rains. These spores stick to the beaks and feet of bark-feeding birds. Bark beetles also become contaminated. Both birds and bark beetles can carry the spores considerable distances to infect other trees. Ascospores are produced in perithecia embedded in the chestnut bark. The ascospores are shot out in tiny clouds and are blown to nearby trees in the spring. Both kinds of spores infect through feeding wounds made by birds and bark beetles. Hyphae from germinating spores penetrate into the inner bark, killing vital cells.

Management

Adequate control measures are not known for chestnut blight. Large sums of money have been spent on research to find ways to combat this disease, and extensive attempts at eradication have been tried with little success. Sanitation measures involving the removal and destruction of infected trees are expensive and unsatisfactory. Managing the bark beetles by spraying with insecticides delays spread of the fungus, but spraying must be continued year after year. Resistant cultivars are now available for home plantings, but the trees are too expensive for planting in forests. Systemic chemicals injected into the sap stream give some control, but it is too early to determine the long-term effectiveness of chemical control, which is economically feasible only on trees with a high dollar value.

One hope in managing chestnut blight is the use of isolates of the causal fungus, *Cryphonectria parasitica,* that have a low degree of virulence; that is, they cause very little disease on a very susceptible host even under the most suitable environmental conditions. These *hypovirulent* (hypo = low) isolates can be used to infect existing chestnut sprouts and will compete with the highly virulent isolates that have been responsible for destroying trees. Not only would there be a competition for infection sites, but if the hypovirulent isolates fuse with the highly virulent isolates, then genetic material is transferred among the isolates, and a highly virulent isolate may be transformed into a hypovirulent isolate. It is hoped that introduction of hypovirulent fungal isolates into forests will permit the mighty chestnut tree to thrive once again.

DUTCH ELM DISEASE

Dutch elm disease, first reported in the United States in 1930, owes its name to the fact that it was first described in Holland. (There is not a tree known as Dutch elm.) The disease is caused by the fungus *Ceratocystis ulmi* (Buism.) C. Mor., which attacks and kills the American elm. This stately tree, which once lined the residential streets of most towns in the eastern and midwestern United States and inhabited the meadows and fence rows, has now largely disappeared from many areas. Aside from the aesthetic loss of the natural beauty of this shade tree, the cost of cutting down dead elm trees amounted to millions of dollars per year.

Symptoms

Trees suffering from Dutch elm disease show a variety of symptoms. Leaves become yellow, wilt, and turn brown. The leaves fall prematurely, and the branches die, causing the crowns to appear thin and sparse. Internally the outer sapwood turns brown (Fig. 14.1). Many dead branches may appear on a tree. The trees may take several years to die, or they may die in the same season in which they are infected.

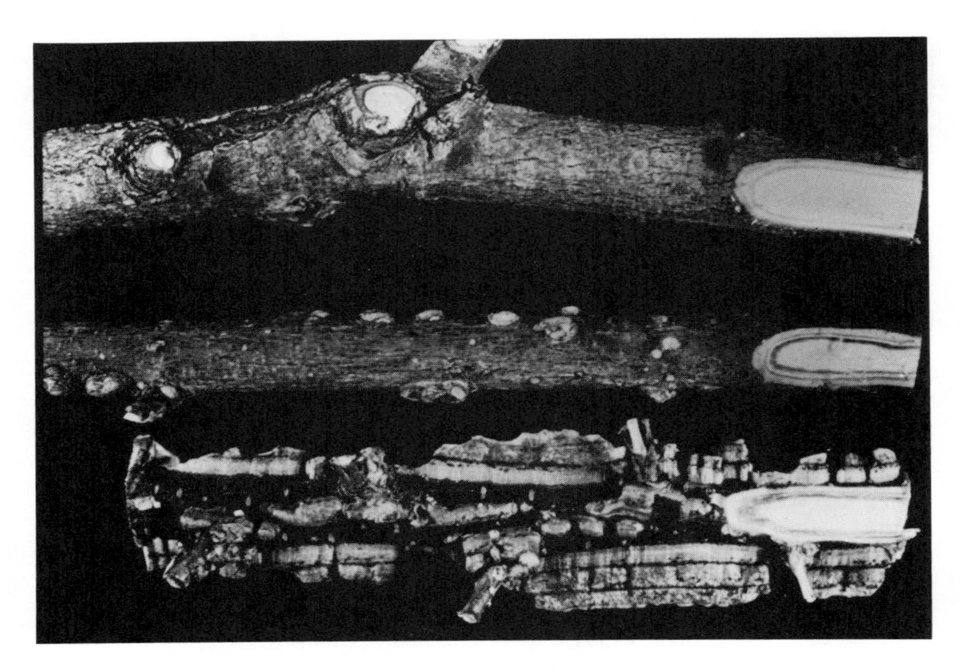

FIGURE 14.1. Twigs of winged elm infected by *Ceratocystis ulmi*. Note the distinct darkening of the outer sapwood, which is visible where the outer layers have been cut away. *Courtesy NCSU Plant Pathology Photo Collection.*

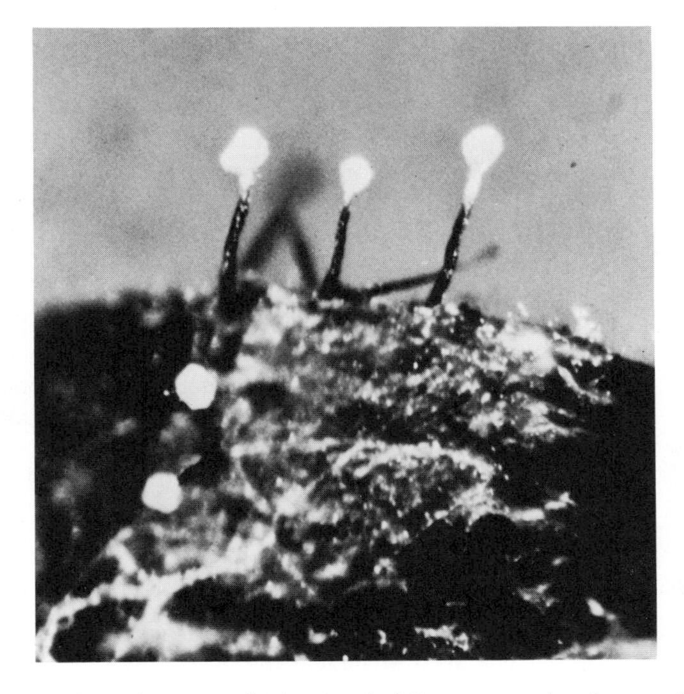

FIGURE 14.2. Spore-bearing stalks (coremia) of *Ceratocystis ulmi*, the causal agent of Dutch elm disease. Magnified about 25 times. *Courtesy NCSU Plant Pathology Photo Collection.*

Causal Agent and Disease Cycle

Although *Ceratocystis ulmi,* the causal agent of Dutch elm disease, produces three kinds of spores, the ones most commonly seen are produced on short black stalks, called *coremia* (Fig. 14.2), which are scarcely visible to the unaided eye. Each is about 1 mm high with an enlarged head bearing great numbers of egg-shaped, colorless spores (3 μm $\times$ 2 μm) embedded in a clear sticky fluid. The coremia grow in cracks, underneath loosened bark, and in the tunnels made by the European elm bark beetle and the native elm bark beetles (Fig. 14.3). If a beetle touches a coremium, the spores stick to the insect. Contaminated beetles emerge from the feeding tunnels, fly to healthy trees, feed on small tender twigs, and unknowingly deposit the spores in the feeding wounds in the sapwood. Here the spores germinate and start new infections.

Usually there are two broods of bark beetles each year. The beetles reproduce in great numbers. As many as 25,000 beetles have emerged from 0.09 m^2 (1 ft^2) of bark. The overwintering brood emerges in May or June and is chiefly responsible for spreading the fungus. The second brood emerges in August. After feeding, these beetles may fly more than 3 km to find a breeding place on dead or dying elms.

Once the fungal mycelium is growing inside the water-conducting vessels of the tree, it produces small, oval, colorless conidia; these are carried in the sap stream to other parts of the tree where they germinate. As a result of these multiple infections, the tree usually dies.

FIGURE 14.3. Brood galleries, each about 5 cm long, made by a female bark beetle between the bark and the wood of an elm tree. The eggs are laid in the central tunnel. The eggs hatch, and the feeding larvae open tunnels at right angles to the central gallery. The fungus that causes Dutch elm disease grows in the tunnels and produces spores that stick to the adult beetles wherever they go, thus spreading the fungus. *Courtesy NCSU Plant Pathology Photo Collection.*

Birds also carry the fungus long distances. To a limited extent water and wind also spread the spores, but bark beetles are the principal vectors. The fungus also spreads from diseased elms to nearby healthy elms by root grafts. Spread by root grafts is particularly important because elms were frequently planted in rows along streets, and in parks.

Management

As with chestnut blight, there are not yet satisfactory methods for managing Dutch elm disease. See the discussion in the Chestnut Blight section.

OAK WILT

A third fungus that is quietly devastating the forests is *Ceratocystis fagacearum* (Bretz) Hunt, a close relative of the fungus that causes Dutch elm disease. At least 50 species and cultivars of oaks are susceptible. Oak wilt has caused the most damage around the Great Lakes, but it now occurs widely in the eastern United States and continues to spread.

Symptoms

Oak-wilt symptoms are most noticeable during the late spring or early summer. Red oaks may be killed in 3 weeks. In white oaks, symptoms usually are confined to a few branches, and infected trees may live several years. Leaves turn yellow or brown, dry up, and fall prematurely. A definite characteristic of oak wilt is the raising and cracking of the bark due to the pressure exerted by fungal mats growing between the bark and the wood. Each mat consists of a central, black, furrowed, gelatinous sclerotium-like pad or cushion surrounded by mycelial growth. As the pad develops, it puts pressure on the bark, causing a bulge and forming a moist chamber, which is excellent for spore production. The continued growth of the pad eventually ruptures the bark. Because the mat produces an odor similar to that of fermenting apples, insects are attracted to the mats on which they feed. The spores stick to the insects' bodies and may be carried to other trees to start new infections.

Causal Agent and Disease Cycle

The oak-wilt fungus, *Ceratocystis fagacearum,* overwinters as mycelium in infected trees and as fungal pads on dead trees. In the spring the pads produce great numbers of one-celled, hyaline, cylindrical conidia (average $13\ \mu m \times 3\ \mu m$) in the bark-beetle galleries (Fig. 14.3). The beetles feed on the pads, become contaminated, and may carry the spores long distances, introducing the fungus into healthy trees through feeding wounds. The fungus spreads rapidly within the water-conducting tissues of the host, causing the leaves to wilt. After the tree is killed, the fungus grows throughout the outer wood and into the roots. Adjoining oaks can become infected through root grafts. Pressure from mycelial pads formed under the bark of dead trees forces it to crack. Insects enter the cracks and feed on the pads.

Management

See the discussion on management in the Chestnut Blight section.

ANNOSUS ROOT AND BUTT ROT

Annosus root and butt rot is common throughout coniferous stands of the North Temperate Zone. Hardwoods may be attacked, but damage is usually of little importance. In the southern United States, the disease is most serious in pine plantations on sandy soils low in organic matter. All species of southern pine are susceptible. Slash and loblolly pine plantations often are severely affected.

Symptoms

The disease gains entry into plantations when spores infect freshly cut stumps exposed during thinning operations. The fungus then spreads from infected stumps to nearby trees by growth along the roots to points of root contact. Residual trees usually begin to die within a few years after thinning.

The *sporophores* or conks (fruiting bodies) generally are found at ground line or in the root crotch, partially hidden by ground litter (Fig. 14.4). Pines in initial stages of the

FIGURE 14.4. Conks or sporophores (about one-half natural size) growing at the base of a pine tree infected with *Heterobasidion annosum*. *Courtesy USDA.*

disease usually have fewer leaves than normal; however, white pines with full crowns may have extensive butt and root decay. Occasionally trees may die rapidly with a sudden red discoloration of a nearly full crown. Diseased trees often are found in groups or circular areas in the forest. Symptoms of the disease may include the pink to violet stain of new decay, narrow white pockets and scattered black flecks of decay in the wood, and the yellow stringy rot of the late stages.

Causal Agent and Disease Development

Annosus root rot is caused by *Heterobasidion annosum* (Fr.) Bref. The fungus forms thin mycelial mats on infected trees between the bark and the wood and small dirty-white outgrowths of mycelium on the undersides of the roots. The gray-brown perennial conks or sporophores are flat to bracket-like and measure 1-15 cm × 2-25 cm × 0.5-7 cm. The colorless microscopic basidiospores (3.5-5 μm × 3.4 μm) look like tiny teardrops. They are produced in great numbers in the conks and are blown by the wind to infect recently cut stumps.

Decay progresses from the infected stump into the roots. Roots of adjacent trees are infected by contact with decayed roots and through root grafts, which allow the fungus to grow from a diseased to a healthy root. After stump invasion, it usually takes from 4 to 10 years for symptoms to appear in nearby trees. Although most infection occurs through wounds, the fungus can penetrate nonwounded bark. The fungus survives in a stump for many years but does not exist long in the soil. Once established in a stand, the fungus remains indefinitely.

Management

To minimize the chances of infection by the causal fungus, growers should avoid planting pine trees on soils low in organic matter and, if possible, eliminate thinning on disease-free sites. Immediately after trees are cut, technical-grade granular borax should be applied to stump surfaces with a large "salt shaker." This fungicide serves as a chemical barrier to any fungal spores that happen to fall on the stump. Borax is effective, safe, easy to apply, inexpensive, and not harmful to the environment.

In infected stands, tree stumps can be sprayed immediately after cutting with a spore suspension of a competing fungus, *Peniophora gigantea* (Fr.) Mass., which prevents infection. This use of a competing, nonpathogenic fungus to prevent infection by a pathogenic one is another example of biological control.

FUSIFORM RUST

Fusiform rust is one of the most destructive diseases of forest trees in the United States. The disease occurs from Maryland to Florida and westward to Texas and Arkansas. Serious economic losses occur in nurseries, seed orchards, young plantations, and natural stands. During the past 50 years the disease has increased in intensity and frequency. Loblolly and and slash pines are very susceptible, longleaf pine is moderately resistant, and shortleaf pine is highly resistant. Damage to oaks, the alternate host for the fungus, is negligible.

Symptoms

The most characteristic symptom of fusiform rust is a tapered. spindle- or fusiform-shaped gall that forms on branches and stems of young pines (Fig. 14.5). These galls often encircle the stem or branch causing it to die. In early spring these swellings appear yellow to orange, as the fungus produces many spores. The fungus stimulates excessive branches in young trees, which may develop so many branches that they resemble brooms. As host tissue is killed, older stem cankers—sometimes reaching 30 cm in length—become flat, sunken, and soaked with pitch. Mortality is heaviest on trees less than 10 years old, but the galls and resulting cankers deform older trees, reducing growth and weakening the stems so that they are easily broken during wind or ice storms.

FIGURE 14.5. Galls of fusiform rust on pine. Each of the three major galls has encircled the stem on which it occurs and is covered by bright orange spores. *Courtesy R. I. Bruck.*

Causal Agent and Disease Cycle

Fusiform rust is caused by *Cronartium quercuum* (Berk.) Miyabe ex Shirai f. sp. fusiforme (Hedge & N. Hunt) Bursdall and G. Snow. It has two hosts—oak and pine—and five kinds of spores. Although the fungus takes 2 years or more to complete its life cycle, most of the active growth occurs in the early spring. From late February to early April, the fusiform galls produce enormous numbers of orange aeciospores. The aeciospores are blown to the young oak leaves that are emerging about this time, where they germinate and penetrate through the stomates. Only very young oak leaves, less than 2 weeks old, can be infected.

Within 7 to 10 days after infection, the fungus produces small pustules containing red urediniospores on the lower surface of oak leaves. The urediniospores can also infect oak leaves and are called repeating spores. After about a week, brown, hair-like structures (telia) appear on the underside of infected oak leaves. The telia contain the teliospores. During favorable conditions [15-27°C (59-81°F) and 97-100% RH], the teliospores germinate, each producing four delicate, short-lived, basidiospores, which are blown to susceptible pine hosts where infection occurs on cotyledons, needles, or succulent tissue.

Most infection occurs at night and in the early morning following a day or an evening of rain. Once established in host tissue, the fungus stimulates excessive growth in infected tissues, resulting in gall formation. If the disease does not kill the pine host during the first few months after infection, the newly formed galls exude yellowish drops of fluid full of pycniospores. Aeciospores are produced on the galls the next spring, completing the life cycle. The fusiform galls produce aeciospores successively for many years. Old galls or cankers often are invaded by insects and wood-rotting fungi that reduce a tree's value. Such weakened trees break easily or snap off during windstorms, often falling or lodging against other trees, thus making cutting operations more difficult and expensive.

In oaks, only the leaves are infected by the fusiform-rust fungus, and the fungus is killed when the temperature remains over 30°C (86°F) for several days. The oaks are free of disease in winter when they cast their leaves. In pines the fungal mycelium lives in infected tissue throughout the year.

Environmental Effects

Favorable conditions for infection by *C. quercuum* f. sp. fusiforme infection on pine occur when the RH remains at or near 100% and temperature ranges between 15°C (59°F) and 27°C (81°F) for as few as 12 hr; such conditions are common on many spring nights in the southern United States. When rain, fog, or overcast skies extend the time these conditions occur, during a period that coincides with flushes of new growth on pine, infection rates can reach epidemic proportions.

Management

Fusiform rust can be managed in nurseries by spraying with recommended fungicides. If adequate weather forecasts are available, the nursery can be sprayed prior to periods favorable for rust infection. Usually three to four applications are necessary.

Infected seedlings should not be used. Some diseased seedlings may not have visible galls. If a given bed contains more than 1% of visibly galled seedlings, it may be desirable to discard the entire bed.

Neither slash nor loblolly pines should be planted on sites with histories of severe rust infection or where there is an abundance of oaks nearby. Growers should plant longleaf or shortleaf pine instead.

The most promising method for managing fusiform rust in future years is the use of seed from resistant trees or hybrid seed produced by crossing shortleaf pine with either loblolly or slash pine. Seed orchards of rust-resistant clones of loblolly and slash pines are now established. As seed from these orchards become available, they should be used in areas of high rust hazard.

DOGWOOD ANTHRACNOSE

The dogwood tree is one of our most beautiful natural resources. Its fragile white blossoms decorate woods, roadsides, and homesteads in most southern and eastern states and in the northwestern United States and Canada. Each year dogwood blossoms herald the advance of spring.

Since the late 1970s, however, a new disease has been killing flowering dogwoods (*Cornus florida*). By 1991 this new disease had been reported on these beautiful trees all the way from New York to Florida. The disease now also occurs on the Pacific dogwood (*C. nuttallii*) in the western United States and Canada.

Symptoms

Initial symptoms are small, purple-rimmed leaf spots or larger tan blotches that may enlarge to kill the entire leaf. Infected leaves often cling to the stems after normal leaf fall. In the winter the most noticeable symptom is dead leaves clinging to the dying trees. Twigs also are infected and may be killed back several inches. On some trees, twigs can be killed back all the way to the main stem or trunk. The dead portions of the twigs are tan, and there may be a purple border between dead and healthy twig tissues. Often numerous epicormic shoots form up and down the main stem and on major branches of infected plants. These shoots become infected and die. The disease results in brown, elliptical, annual cankers on the main trunk that may girdle the tree and kill it.

Trees often are killed 2 to 3 years after the first symptoms are observed. If the weather is unfavorable for plant growth, trees will be killed more quickly. Consecutive years of severe infection have resulted in extensive mortality of both woodland and ornamental dogwoods in the vicinity of the mountainous regions in Georgia, North Carolina, South Carolina, and Tennessee. The disease is more severe in forested areas near streams than in landscape or homeowner plantings.

Causal Agent and Disease Cycle

Dogwood anthracnose is caused by a newly identified fungus, *Discula* spp. The fungus produces masses of orange conidia (5.5-10 μm $\times$ 1.5-3 μm) on infected leaves or bark on twigs and branches. Abundant conidia for spring infection are produced on blighted leaves or twigs that were killed the previous season. Infection by the *Discula* spp. is favored by cool, wet spring and fall weather. The fungus overwinters in infected leaf, twig, and stem tissues.

Management

Management is centered around cultural practices and fungicidal sprays in landscape plantings. No adequate management practices are currently available for stopping the epidemic of anthracnose in natural dogwood stands in forests.

The maintenance of vigorous dogwood trees by mulching, watering during periods of drought, and avoidance of mechanical injury helps to minimize the impact of the disease. The pruning and removal of diseased twigs and branches, the removal of epicormic branches, and the raking and disposal of leaves may be of some value. Pruning low branches of taller trees and thinning other understory plants to improve air movement and decrease the duration of leaf wetness also may help. Fungicides should be used only as a supplement to a cultural control program. Several applications of registered fungicides may be needed at 10- to 14-day intervals during leaf expansion in the spring. Additional fungicide applications may be needed later in the growing season if weather conditions are favorable for infection.

LITTLE LEAF OF PINE

Symptoms

Little-leaf disease is most evident in older pine trees; rarely is it seen in stands less than 20 years old. The disease is caused by a malfunctioning of the root system resulting from a combination of biological and physical factors. The fungus *Phytophthora cinnamomi* Rands attacks and kills the root tips. When conditions of moisture, fertility, and drainage are not good, such trees are not able to rapidly replace the destroyed root tips. Trees on good sites with adequate moisture, fertility, and drainage reportedly are also attacked by the fungus, but their vigor is such that they quickly overcome the disease by producing new root tips.

The disease usually progresses rather slowly. Some trees may live 15 or more years after the appearance of initial symptoms. In general, trees live only 5 to 6 years after attack, but they may die in as little as 1 year. Symptoms are those typical of trees stressed by a malfunctioning root system. In the early stage of the disease, the foliage may turn yellow-green, and the current year's needles are shorter than normal. In later stages, diseased trees have sparsely foliated crowns with shortened needles and dead branches. Abundant foliage sprouting on the boles of infected shortleaf trees is common.

Causal Agent and Disease Cycle

Phytophthora cinnamomi, which causes little leaf of pine, is similar to other species of this fungus that cause various root and stem rots. For a discussion of these *Phytophthora* species, see Chapter 12, sections on Phytophthora root rot and stem rot of soybeans.

Management

At present, effective management of little-leaf disease is not available. Losses can be minimized by salvage cuttings and by using loblolly pine and hardwoods in reforestation plantings.

However, research since 1970 on some fungi that grow in association with the roots of many trees offers hope for biological management of little leaf of pine in the future. Some of these fungi—called *mycorrhizae*—grow inside roots; these are called *endomycorrhizae*. Others are attached to the outside of roots and form a cover or mantle around the roots; these are called *ectomycorrhizae*. Oddly, pine seedlings supporting ectomycorrhizae on their roots are resistant to infection by the fungus that causes little leaf of pine, whereas seedlings with nonmycorrhizal roots are heavily infected by *Phytophthora cinnamomi* under similar conditions. Perhaps someday soon roots of pine seedlings can be inoculated with beneficial mycorrhizal fungi and thus be protected against little leaf. If this practice proved feasible and effective, it would be an excellent example of biological management of a forest tree disease and might be applicable to the management of root diseases in other crops.

WOOD DECAY IN BUILDINGS

Wood is one of the most versatile and durable building materials. When properly used, it will give long and satisfactory service, but if a few simple rules are ignored, wood-decay fungi can attack it. Management then can be very costly. Termites and some other insects also cause damage to buildings, but management and prevention of insect attack will not be discussed here.

Causal Agents and Symptoms

Two groups of fungi cause wood decay. Sap-stain and mold fungi discolor wood but do not seriously weaken it. Sap-stain, surface molds, and mildews can be black, blue, gray, green, orange, or pink. They all grow under the same conditions and are sometimes found together, but not always. The second group, the decay fungi, attack wood that remains moist because of rain or groundwater seepage, leaky plumbing, or condensation. Such decay, commonly known as wet-rot, dry-rot, or brown cubical rot, does not extend beyond the moist areas.

Below 5°C (41°F) decay of wood is slow. When its moisture content is below 20%, wood will not decay. Wood that is either properly treated with a good preservative or comes from the heartwood of durable tree species will not decay even when its moisture content is above 20%.

Occasionally, wood is attacked by a fungus that first decays moist wood and then conducts water to normally dry wood. Consequently, the fungus advances more rapidly than ordinary decay fungi. This fungus is *Poria incrassata* (B & C) Burt., the water-conducting or dry-rot fungus. It can destroy the wood in a building within 2 years. Although this does not occur very often, it is possible for this fungus to grow where ordinary decay fungi will not grow.

Poria species can conduct or pump water several meters from soil or other sources, sometimes through brick or concrete, by means of thick rootlike strands (rhizomorphs) up to 1 cm or more wide. These strands, which are often hidden behind wooden structures, distinguish the water-conducting fungus from ordinary wood-decaying fungi. If wood decay is caused by a *Poria* species, greater care is required to control attack than for ordinary decay fungi. If the cause of decay is not clear, one should seek advice from the local county extension office, pest control company, or building inspector.

Darkening of wood, along with cracking and splitting across the grain, are the most common symptoms of more advanced decay in structural timbers. Finally, the wood will appear quite dry and crumbly.

Management

Arresting Wood Decay

The first step in halting the spread of wood decay in buildings is to find and correct all sources of excessive moisture. Some common sources include the following:

- Poor drainage or poor foundation waterproofing around the building so that water runs into or through the foundation.
- Inadequate ventilation in crawl spaces or other enclosed areas of construction.
- Leaky plumbing.
- Untreated wood in contact with soil, particularly in earth-filled porches.

If possible, eliminate excessive moisture, remove all obviously decayed wood, and replace it with dry, sound lumber. It is not necessary to remove sound wood around the decayed area. If excessive moisture cannot be fully eliminated, pressure-treated wood should be used to replace the decayed wood and surrounding wood for a distance of at least 0.7 m beyond any evidence of decay.

Localized decay in wood structures such as joists, porch floors, or bases of columns can be arrested by treating the wood surface with a water-repellent preservative, provided that the wood can be dried before treatment.

Helpful Hints to Prevent Wood Decay

To prevent wood decay, it is necessary to keep untreated wood dry. The following practices are recommended (Fig. 14.6):

- Provide adequate drainage and foundation waterproofing to prevent water accumulations under or near the building.
- Cover at least 70% of the crawl space floor in basementless houses with polyethylene (4 or 6 mil), or place polyethylene directly over or preferably directly under concrete slabs.
- Provide adequate ventilation in crawl spaces and attics.
- Install a moisture barrier between sills and foundations.
- Avoid all forms of construction that will trap moisture.
- Insulate any pipes likely to cause condensation of water onto wood.
- Keep gutters clear of leaves and other plant debris.
- Avoid the discharge of water or water vapor (e.g., from dryers or air conditioners) into crawl spaces.
- Keep untreated wood well above the soil.

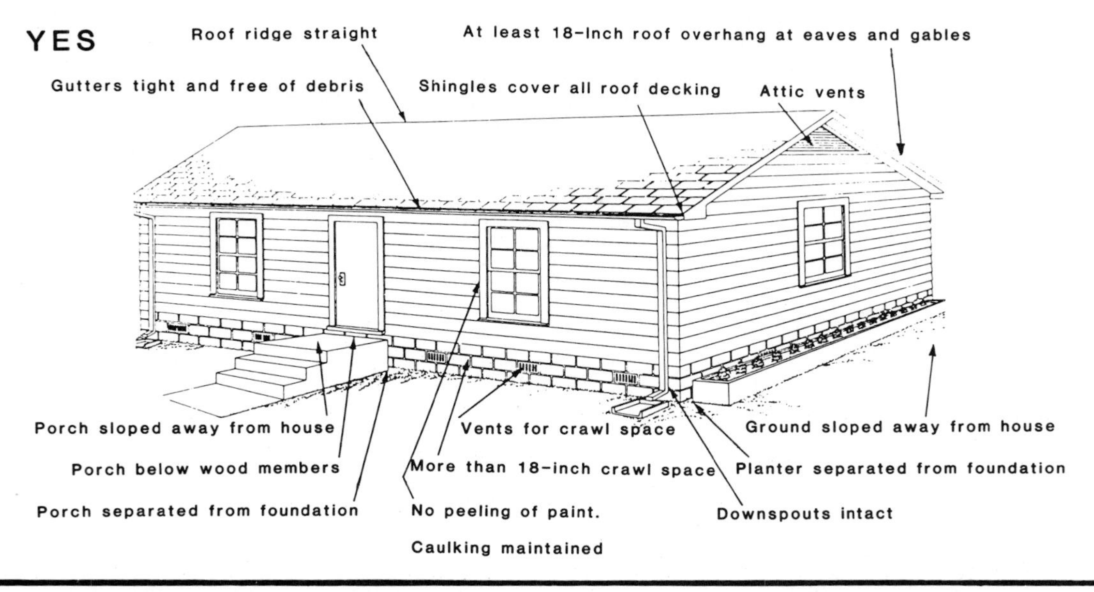

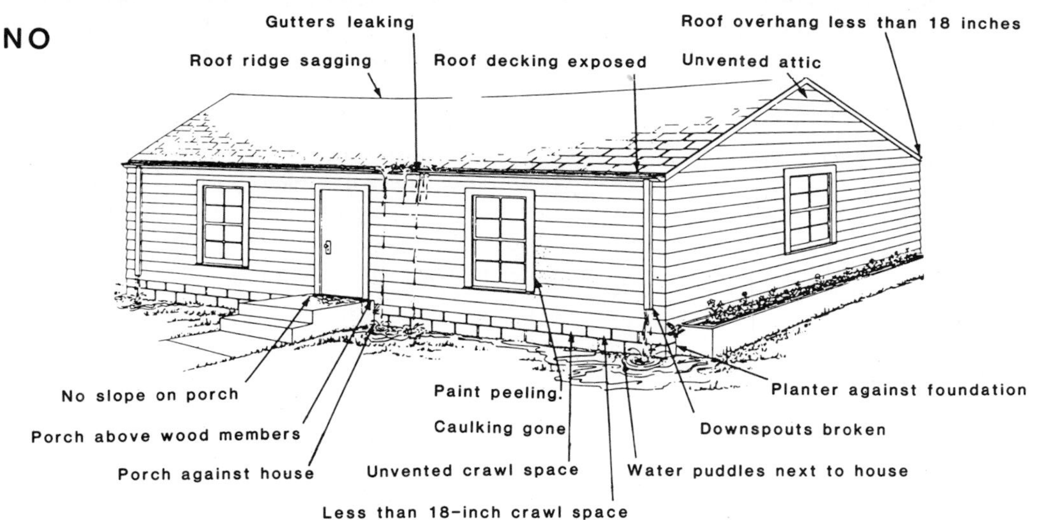

FIGURE 14.6. Proper and improper features of house construction for prevention of wood rot. The labeled features in the bottom drawing invite decay. *Courtesy USDA.*

- Remove all forms used in pouring concrete and wood debris from under a building.
- Keep caulking and paint around joints well maintained.

Crawl spaces should be cross-ventilated and contain at least 0.1 m^2 of clear vent space per 15 m^2 of crawl space. Vents should be kept open and should not be blocked by leaves, shrubs, wood piles, or trash. Attics should contain 0.1 m^2 of vent space per

15 m^2 of the ceiling below; however, when more than 50% of the vent space is near the top of the gable, 0.09 m^2 space per 27 m^2 of the ceiling below is sufficient.

Foundation walls supporting wood frame construction should be at least 20 cm above the outside finish grade. In buildings without a basement, a crawl space should have at least 45-cm clearance under wood sills and joists. Wood siding should be at least 15 cm above the outside finished level of soil. Earth-filled porches should be avoided unless the earth is at least 20 cm away from wood construction.

Buildings should be inspected regularly for the warning signs of decay and for the conditions that may lead to decay. Ideally, inspections should be made twice yearly, in midsummer and midwinter.

Wood Preservation

The most efficient way to prevent or control decay when wood is used where it will get wet is to treat it with a chemical. Fence posts, pilings, railroad ties, and utility poles usually are pressure-treated with *creosote.* In the southeastern United States, untreated fence posts will often rot in 3 to 4 years, whereas properly treated posts will last 20 years. Creosote-treated posts usually are available at most building material companies or farm supply outlets.

The odor and the appearance of wood treated with creosote make it undesirable for some uses; therefore, other oil-soluble preservatives are used for many products. *Pentachlorophenol* and *copper naphthenate* are commonly used to treat wood used for benches, doors, interior framing, fence posts, floor joists, sundecks, windows, and other wood products exposed to the weather. Some uses of creosote and pentachlorophenol may be eliminated, but other chemicals are available or are being developed.

For wood used above ground and where the decay hazard is relatively small, dip and brush-on treatments provide adequate protection. These dip or brush-on treatments should be used on wood with a moisture content of less than 12%. Wood that has been treated with certain chemical preservatives is difficult to paint properly.

15

Bacteria

Bacteria—humankind's most useful servants and most destructive masters—are essential to human existence. Bacteria produce useful food and chemicals, increase the fertility of soil, fix nitrogen from the air, aid in digestion, decompose the complex bodies of dead plants and animals, and salvage the vital chemical elements and compounds for reuse. However, as causes of disease, bacteria have caused more deaths than all wars. Bacteria have defeated armies and destroyed nations. Yet most bacteria are beneficial, not harmful.

Discoveries about the nature and activities of bacteria are among the most important scientific accomplishments of modern times. Anton von Leeuwenhoek, a Dutch lens grinder, first saw bacteria about 1683 in a crude microscope he built himself. In 1876 Robert Koch, a German country doctor, showed that a specific bacterium was the cause of a specific disease—anthrax—in animals. We now know that bacteria cause numerous diseases in human beings, including typhoid fever, tuberculosis, diphtheria, lockjaw, and syphilis. A few years later, in 1882, Louis Pasteur identified bacteria as the cause of fermentation. Many plant parts, when deprived of free oxygen, still continue to give off carbon dioxide and often produce alcohol and other compounds. This process is called *anaerobic respiration* or *fermentation*. The theory that living agents cause fermentation became the basis of modern bacteriology and laid the foundation for aseptic surgery, the control of epidemics, preservation of foods, purification of water and milk (heating milk to destroy bacteria is called *pasteurization*), and sewage disposal.

As a result of the work by Koch, Pasteur, and others, the *germ theory of disease* became widely accepted. This theory states that diseases are caused by living agents (*germs*). Thomas Burrill, a professor of botany at the University of Illinois, first discovered that bacteria caused plant diseases in 1881, when he showed that these microscopic germs caused fire blight of pear and apple.

SIZE AND STRUCTURE

Bacteria are microscopic, unicellular organisms, often known as germs or microbes. Bacteria do not have roots, stems, or leaves and do not contain chlorophyll. They are placed in the kingdom Prokaryotae, which means that their genetic material (DNA) is not organized into a nucleus with a membrane. Some are so small that 25,000 laid end for end would be only 2.5 cm (1 in.) long. A bacterium consists of only one cell, which varies in size from 0.3 μm to 6 μm in length (see Fig. 2.1). Thousands of bacteria might be contained in a single drop of water and not be crowded at all for space. A milliliter (20-30 drops) of sour milk contains millions of bacterial cells.

Bacteria are classified into three general types, according to shape (Fig. 15.1): (1) spherical or globular (*coccus*), (2) rod or cylindrical (*bacillus*), and (3) spiral (*spirillum*). Only non-spore-forming bacilli (about 1 μm in length) are plant pathogens. The cells of the different types may be held together as pairs, long or short chains, long spirals, or clusters.

Each bacterium consists of a cell wall, cytoplasm, and nuclear material (Fig. 15.2). Many bacteria secrete a sticky or gummy material called *slime,* which accumulates around the cells and helps to hold masses of cells together.

Some bacteria have long slender, hairlike, coiled appendages called *flagella.* By a lashing movement, the flagella propel a bacterial cell through water or nutrient solutions. Bacterial movement may be straight, wiggling, or tumbling. Spiral forms may whirl like a ship's propeller. Some bacteria have only one flagellum attached to one end of the cell; others have many flagella distributed over the cell surface.

Mycoplasmas are similar to bacteria but do not have a rigid cell wall. These micro-organisms are discussed in Chapter 16 in the Mycoplasma Disease section.

IDENTIFYING BACTERIA

Many plant-pathogenic nematodes and fungi are big enough and different enough in size, shape, and/or color to be identified by just looking at them. But not so with bacteria. Most plant-pathogenic bacteria are short rods that one cannot tell apart even with the most efficient light microscope. However, it is essential to identify different bacterial species to study them and take proper measures to manage them. Because bacteria are so similar and cannot be identified by shape, size, and/or color only, scientists use other characteristics to differentiate bacteria; these include chemical composition, enzymatic action, serological specificity, pathogenicity to plants, and susceptibility to certain bacterial viruses called *bacteriophages.*

Chemical composition can be detected with specific staining techniques. Gram's method, developed in 1884 by Christian Gram, involves the use of crystal violet and iodine to determine if bacteria will absorb these chemicals and turn dark (gram-

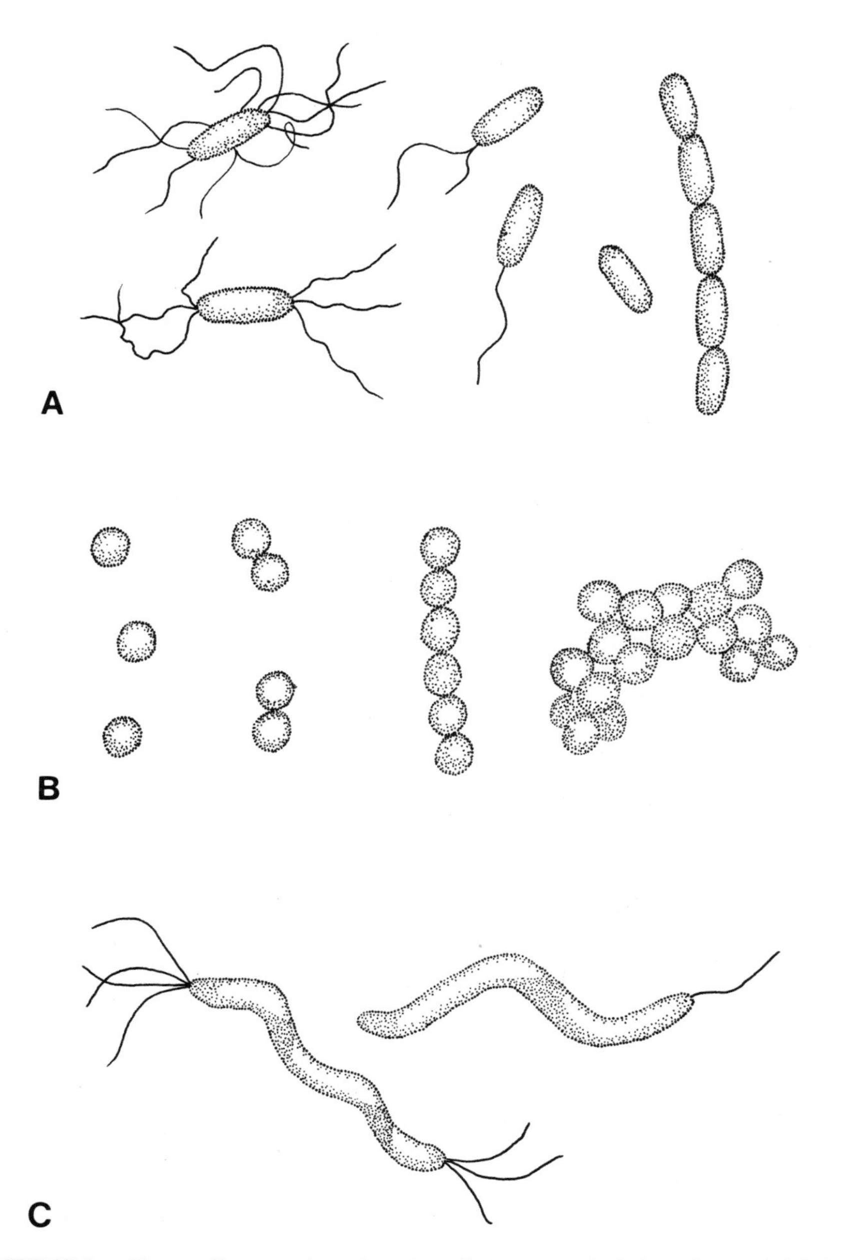

FIGURE 15.1. Types of bacteria, based on their shape, magnified about 900 times. (A) Rod. (B) Spherical. (C) Spiral. *Drawing by Pi-Yu Huang.*

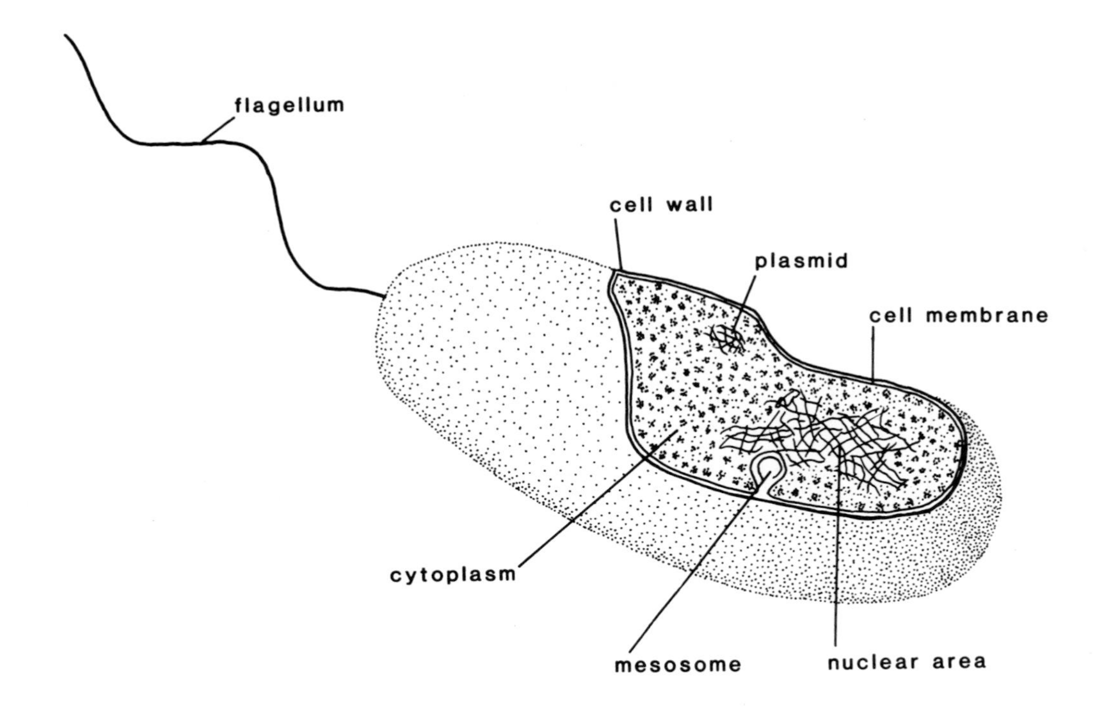

FIGURE 15.2. Diagram of a bacterial cell with wall cut away to show anatomy. *Drawing by Pi-Yu Huang.*

positive). Most plant-pathogenic bacteria do not retain the crystal-violet stain and are called gram-negative.

Enzymatic action is determined by recording the substances that bacteria can use for food and also the substances produced when bacteria grow on certain substrates. For example, some bacteria may change nitrates to nitrites; others produce hydrogen sulfide, ammonia, or many other substances. Determinations of the minimum, optimum, and maximum temperature and pH for growth, and the temperature at which bacteria are killed also are used to identify bacterial species.

When an unknown bacterium has been put through all these tests (which may total 50 or more), it can be identified by comparing its reactions to those listed in tables available for the known species of bacteria. The unknown is considered to be the same as the listed species that gives the same reactions, especially when it can be shown to attack various species and cultivars of host plants.

Recently, new techniques using DNA and serology have been developed to identify bacteria. DNA can be isolated and compared to known species for rapid identification. Substances have been identified that are specific for certain bacteria. Some of these substances can be used to produce monoclonal antibodies in animals that can be used to identify bacteria in relatively simple and quick tests.

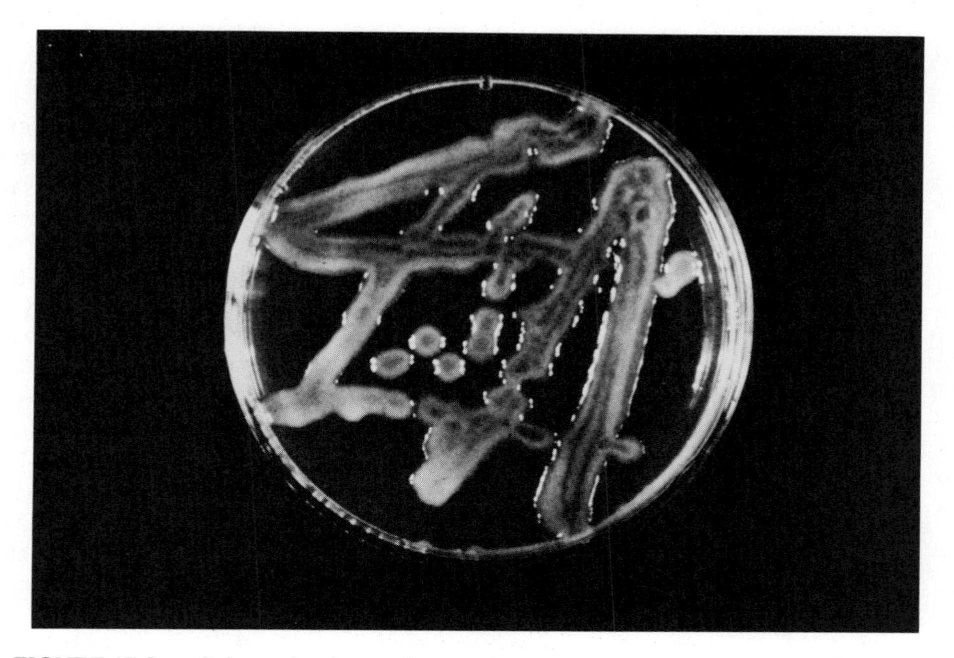

FIGURE 15.3. Culture of a plant-pathogenic bacterium growing on a nutrient medium in a petri dish. As long as the bacteria can obtain nutrients from the growth medium, they will continue to divide and multiply. *Courtesy NCSU Plant Pathology Photo Collection.*

GROWTH AND REPRODUCTION

Bacteria multiply by dividing! Each bacterium grows until it reaches a certain size; then the cell becomes pinched in the middle and separates into two equal parts or daughter cells. This process is called *binary fission*. Each daughter cell quickly enlarges and then divides again. Some bacteria divide every 20 to 30 min. Such a rate of division can be maintained as long as sufficient nutrients are present (Fig. 15.3). Crowding, depletion of food and water, accumulation of poisonous waste products, and changes in the pH of the environment can limit bacterial populations.

The ability of bacteria to produce tremendous numbers of individual cells in a short time and the production of toxins and enzymes that harm the host explain why bacteria are so destructive to animals and plants. A single bacterial cell is harmless, but severe damage occurs when a great many develop in wounds or other natural openings.

Bacteria do not contain chlorophyll and must have a source of organic matter for growth.

Many bacteria grow on the dead remains or the products of plant and animal life and do not cause disease. These bacteria are *saprophytes*. Some species are parasitic and attack the living cells of plants or animals as a source of food. Some parasitic bacteria such as the root nodule bacteria are useful, and some cause diseases. Many of the saprophytic bacteria are beneficial. Bacteria that can grow on a living or nonliving host are called *facultative parasites*. Many of the bacteria that cause plant diseases are in this group.

Bacteria produce many enzymes that digest or change complex food materials into simpler compounds and also help build simple compounds into larger compounds. Certain bacteria are able to synthesize all compounds necessary for growth from sugars and essential elements. Others must obtain certain growth factors, vitamins, and/or amino acids from the environment.

Some bacteria require free oxygen for growth; others can grow in the absence of oxygen. Some can do both. In addition, bacteria require water and suitable temperature, osmotic concentration, and pH for growth.

Many saprophytic bacteria are capable of producing profound chemical changes in the substances used for growth. Decay of plant and animal materials and processes known as fermentation are changes of this type. Decay includes the decomposition of organic material into its constituents or to such simple compounds as water, carbon dioxide, ammonia, and hydrogen sulfide. Many bacterial species may be involved in the decay of a dead plant or animal, each breaking down some special compounds into simpler substances. Other species carry the destruction a step further, until it is completed. One kind of bacteria could not completely decay the carcass of an animal or a whole plant.

Decay processes are important to humans from at least two standpoints. Decay prevents the accumulation of organic matter—both plant and animal—on the earth, and it results in the formation of simple compounds or free elements that are returned to the soil, water, and air to be used again by plants.

It must be remembered that the cell walls and membranes of bacteria are not hard and stiff like steel, but are soft and pliable. Water and some compounds dissolved in water easily pass through the cell walls into the cytoplasm and are transformed by the bacterial enzymes to supply food and energy. Materials also can move outward from the bacterial cell. For example, toxins and enzymes formed inside bacterial cells may leak through the cell walls and react with plant juices to kill the tissues.

DISTRIBUTION AND DISPERSAL

Bacteria occur everywhere—in soil, water, air, food, dust, the oceans—and are found on and in plants and animals. Bacteria enter plants through natural openings (stomates, lenticels, hydathodes) and wounds. Once inside, the bacteria are carried in the sap stream or swim in the water between the cells.

Bacteria cannot spread long distances on their own but are easily spread in many ways. Soil, refuse from a previous crop, compost, diseased transplants, contaminated seed, splashing rain, insects, nematodes, birds, humans, water, and wind—all may carry bacteria to healthy plants, from field to field, or over long distances from country to country.

MANAGING PLANT-PATHOGENIC BACTERIA

Bacterial diseases of plants are managed by use of resistant cultivars, crop rotation, sanitation, disease-free seed and transplants, and chemical sprays or antibiotics. The most effective antibiotics against bacteria are obtained from actinomycetes, bacteria, and fungi. One of the most commonly used antibiotics for the management of bacterial diseases on plants is streptomycin, an antibiotic produced by a strain of the actinomycete *Streptomyces.* (An actinomycete is a bacterium that forms branching filaments.) This antibiotic also can be used to control disease caused by some bacteria in animals and humans. Bacterial diseases also can be managed by reducing wounding or damage to plant parts.

Bacteria can cause rot or decay when excessive moisture is present. In addition, many bacteria require warm temperatures to grow and multiply. Therefore, bacterial rots of harvested grains (corn, oats, rice, wheat) and other crops are controlled by first drying the harvested product and then storing it under very dry conditions. Harvested vegetables and fruits are stored at temperatures below 10°C (50°F).

Bacteria may also be attacked by viruses called bacteriophages. These bacteriophages attack bacteria and eventually kill the cells. In the future bacteriophages may become a useful tool in the management of bacterial diseases.

Biological control methods for bacteria that cause plant diseases have been discovered. Investigators are conducting research in order to understand these processes and to develop practical uses of biological techniques for the control of bacterial diseases. Biotechnology techniques (see Chapter 8) that can be used to transfer genetic information from one organism to another also are being used to develop control methods for bacterial plant diseases. Genetic material has been transferred to some useful bacteria on plants to cause the bacteria to produce substances harmful to pathogenic bacteria. These techniques may be used in the future to transfer genes to plants from microorganisms that produce antibodies. These genes in plants may make the plants produce antibodies and prevent disease-causing bacteria from growing and harming the plants. Genes already have been transferred from other organisms to other bacteria to produce useful products. One example is the transfer of the gene for the production of insulin from animals to bacteria. These bacteria produce large quantities of insulin, which can be used to treat the disease diabetes in humans.

16

Diseases Caused by Bacteria and Mycoplasmas

When bacteria fill the water-conducting cells of a plant and kill plant tissues by means of toxins, the plant suffers a *wilt* or *blight* disease. If the bacteria kill leaf cells, leaf spots may develop. When fruits, tubers, or roots are infected, *soft rots* may develop, which result in a slimy, foul-smelling mass. Other bacteria stimulate tissue growth to form *galls* or tumors that resemble cancerous tissues of animals. In addition to killing plant tissues, bacteria can stunt the growth of plants, change plant color from green to brown or black, or cause distortion of leaves, stems, flowers, or fruits.

Several typical bacterial diseases of plants are discussed in this chapter. These diseases are summarized in Table 16.1

TABLE 16.1. Several important diseases caused by bacteria.

Disease	Pathogen	Hosts
Wilt or Granville wilt	*Pseudomonas solanacearum*	Tobacco, tomato, potato, eggplant, and others
Fire blight	*Erwinia amylovora*	Apple, pear
Soft rot, barn rot, storage rot	*Erwinia carotovora* pv. *carotovora*	Potato, cabbage, lettuce, carrot, tobacco
Wilt	*Erwinia tracheiphila*	Cucumber
Canker	*Corynebacterium michiganense* pv. *michiganense*	Tomato
Spot or canker	*Xanthomonas campestris* pv. *pruni*	Peach
Crown gall	*Agrobacterium tumefaciens*	Fruit trees, woody ornamentals

Mycoplasmas, organisms smaller than most bacteria, cause a group of diseases distinguished by stunting, chlorosis, yellowing of leaves, changing of flowers to leaflike structures, production of numerous leaves, or some combination of these symptoms. A discussion of mycoplasma diseases appears later in this chapter.

BACTERIAL WILT

The wilt disease caused by the soilborne bacterium *Pseudomonas solanacearum* Smith, also called Granville wilt, is one of the most important and widespread bacterial plant diseases. During the past 50 years the total losses from this disease on all crops have been worth hundreds of millions of dollars. Bacterial wilt has been reported from most countries in the temperate and semitropical zones of the world. It is an annual threat on sandy soils to such crops as banana, eggplant, peanut, pepper potato, tobacco, and tomato. Certain ornamental plants—including dahlia, marigold, and nasturtium—and many weeds are hosts to this wilt bacterium.

Symptoms

Typical external symptoms of bacterial wilt include stunting, yellowing, and wilting of the aboveground parts (Fig. 16.1). Leaf *epinasty* (i.e., turning down of leaves) and adventitious roots may appear on certain host plants. The vascular tissue invaded by *P. solanacearum* usually becomes dark brown. Many roots of infected plants decay, becoming dark brown to black in color.

If the stem of a diseased plant is cut in cross section, an exudate of bacterial slime in the form of tiny droplets may become evident on the cut surface. The presence of the wilt pathogen also can be demonstrated by floating a razor-thin slice of a longitudinal section containing vascular tissue from a diseased stem in a container of clear water. Within a few minutes, fine milky-white strands composed of masses of bacteria will stream from the margin of the tissue.

Causal Agent

The wilt bacterium, *Pseudomonas solanacearum,* is a gram-negative, non-spore-forming rod, about 0.5 μm × 1.5 μm, with a single polar flagellum. The bacterium is aerobic, reduces nitrate, and produces ammonia. Colonies on solid media are small, irregularly round, white in reflected light, and tan in transmitted light. In certain simple synthetic media, the bacterium can utilize inorganic sources of nitrogen and most of the common carbon sources. The organism is sensitive to desiccation. A characteristic brown pigment is formed on plant-extract media and on potato plugs. Rapid growth *in vitro* is favored by relatively high temperatures [30-35°C (86-92°F)] and a neutral to slightly acid medium (pH 6.7). Unless special procedures are followed, such as storage under sterile mineral oil or sterile distilled water, pathogenicity and viability may be lost rapidly in culture.

Strains of *P. solanacearum* differing in cultural and biochemical characteristics and in levels of virulence have been reported. Bacteriophages (viruses) that attack and kill the bacterium have been found.

FIGURE 16.1. *Left:* Healthy tomato plant. *Right:* Stunted, wilted tomato plant with yellowing leaves that is infected with *Pseudomonas solanacearum*. *Courtesy S. F. Jenkins.*

Disease Cycle

Infection by *P. solanucearum* usually occurs below ground through wounds in the root system. The root wounds made by the invading larvae of the root-knot nematode and by the emergence of new roots provide excellent places for the wilt bacterium to enter the roots. The bacterium moves upward in the vessels from secondary roots into larger roots and the stem. Bacterial cavities may be formed in local areas in the phloem, cortex, or pith. When the number of vessels blocked by masses of bacterial cells, and probably also by gums and slimes of bacterial and plant origin, becomes great enough, the infected plant suffers irreversible wilting followed by death. Toxic materials such as growth inhibitors and hormones probably play an important role in the development of wilt symptoms and death of host tissues.

There are many known cases in which bacterial wilt has caused heavy losses of potato, tobacco, and bananas in newly cleared land. Apparently the bacterium is a normal inhabitant of many soils Although the bacterium survives longer in some soils than others, for all practical purposes a field will remain contaminated indefinitely once it has become infested with the wilt organism. Most of the bacteria are found in the plow layer (top 30 cm of soil).

The wilt pathogen can be spread in seedlings used as transplants, in potato seed tubers, in banana suckers, by various cultural practices that involve the spread of soil or plant debris, and by the movement of soil by surface water. Dispersal by insects is of minor significance, and spread in true seed has not been conclusively demonstrated. In general, the rate of dispersal in a given area is relatively slow.

Disease development is favored by relatively high temperatures and high soil moisture levels. As a result, seasonal fluctuations in disease severity may be very great. In greenhouse experiments an interrelationship between disease severity, day length, temperature, and host nutrition has been demonstrated.

The bacterium occurs throughout the world in many different soil types, ranging from sandy to heavy clay and over a wide range in pH levels. It causes the greatest damage in the temperate and tropical regions of the world where rainfall is high. In a given field the disease is usually more severe in localized areas, often associated with inadequate drainage. Soils that dry out quickly are unfavorable for bacterial wilt, as are soils high in organic matter that permit the growth of organisms antagonistic to the wilt bacterium.

Management

Successful management of bacterial wilt requires an integrated management program using every sensible measure possible. Root-knot nematode management, discussed in Chapter 9, is necessary. Only healthy transplants and wilt-resistant cultivars, where available, should be used. Crop rotation is recommended. Seedbed and greenhouse soils should be sterilized before susceptible crops are planted.

FIRE BLIGHT OF APPLE AND PEAR

When Koch, Pasteur, and others first proposed the germ theory of disease, they based their conclusions on observations of animal diseases. A few years later, in 1881, Thomas Burrill of the University of Illinois showed that bacteria caused fire blight of apple and pear, the first plant disease attributed to a bacterium. This discovery caused much argument, and it took 20 years or more to convince some scientists that bacteria caused plant diseases. Fire blight also was one of the earliest diseases demonstrated to have an insect vector.

Fire blight is still one of the most important diseases of apples and pears. In certain areas, including the coastal plains of the eastern United States, commercial production of pears, even if other diseases are absent, is impossible because of this disease. Many ornamental trees and shrubs, especially those in the rose family, are susceptible to fire blight. The attractiveness of these plants is decreased by the ravages of this disease. Plants infected by the fire blight bacterium can serve as an inoculum source for commercial orchards nearby.

Symptoms

Fire blight often appears suddenly in the spring, killing the blossoms, shoots, and fruit spurs on apple, crabapple, hawthorn, pear, and pyracantha. The flowers and leaves become watersoaked, then shrivel rapidly, and finally turn brown and then black as though scorched by fire. The symptoms spread rapidly to nearby twigs. Branches also become blighted and turn black; the dead, curled, black leaves that cling to blighted branches throughout the summer contrast starkly with the healthy green of nearby parts of the tree (Fig. 16.2). Watersprouts ("suckers") usually wilt from the tip downward. The bark cracks, turns brown-black, hard and shrunken. Fruit set of blighted trees is reduced. Spurs, branches, and sometimes whole trees are killed in one year.

Under humid conditions, droplets of milky, sticky ooze appear on the surface of blighted twigs or branches. The ooze turns brown as it hardens and dries. Threads of dried ooze can be blown by the wind to other trees. Or the droplets may run together to form large drops, which drip onto and contaminate other parts of the tree.

FIGURE 16.2. Characteristic symptoms of fire blight on an apple tree. The leaves have wilted and become dry and dark brown to black in color (arrows). Flowers also have become darkened and shriveled.

Rot-producing fungi often invade blighted twigs and branches, which makes fruit-rot and leaf-spot management more difficult.

Causal Agent and Disease Cycle

Erwinia amylovora, the fire-blight pathogen, is rod-shaped and about 0.6 μm $\times$ 1.5 μm. It is motile, possessing numerous flagella attached over its entire surface. It grows with or without free oxygen. Cultures often lose their virulence when maintained on artificial media.

The occurrence and the severity of fire blight vary from season to season and orchard to orchard. The disease starts in the spring from bacteria that survive in holdover cankers. The bacteria ooze out of the cankers (old infected areas) when the weather is mild and moist. The development of the ooze corresponds to the initiation of new growth and flowering. Rain spreads the bacteria to the blooms and increases the moisture level in the flowers, twigs, and leaves so that the bacteria can infect, often directly through the stomates, and grow in them. Strong winds during rains blow the bacteria into wounds and cracks on twigs or branches.

Temperatures between 20°C (68°F) and 28°C (82°F) are most favorable for the growth of fire-blight bacteria and disease development. Young, succulent tissues are most susceptible. Aphids can spread the bacteria, especially on watersprouts. Pollinating insects such as honey bees frequently carry the bacteria from infected to healthy blooms, from tree to tree, and from orchard to orchard, often during warm, sunny weather following rainy periods.

Management

A simple, practical, and fully effective management program for fire blight has not been developed. However, several practices can help growers to manage fire blight.

Holdover cankers and blighted shoots should be eliminated; branches with cankers should be cut off about 30 cm (1 ft) below the cankers during the winter. Another way to reduce the number of cankers on large branches and trunks of trees is to rub off all watersprouts during the spring and summer. Such shoots and spurs should not be allowed to develop on main branches. The source of blight for commercial orchards often is maintained in neglected pear and apple trees in yards, gardens, and home orchards. Such uncared-for trees in the vicinity of commercial orchards should be removed.

Blossom blight infection can be greatly reduced by spraying with the antibiotic streptomycin during bloom, at rates to give 60 to 100 ppm streptomycin. Applications should be made at 5-day intervals through bloom. The 100-ppm rate is more effective than the 60-ppm rate. The first spray is applied in the late pink stage before many of the center blooms open. Leaves and blooms should be thoroughly sprayed, but overspraying increases leaf injury and should be avoided. Perfect control of blight should not be expected from bloom spraying because it is impossible to cover all the parts of the blooms and shoots that develop during rainy periods when blight infection takes place.

Postblossom spraying can help control twig and shoot blight. Streptomycin at 50 ppm is now registered in many areas for this use in postblossom sprays, to be applied up to 50 days before harvest.

Trees subject to fire blight should not be overfertilized or stimulated in other ways to make extremely vigorous growth during and just after the bloom period. Blight is often less serious in sod orchards than in clean cultivated ones.

If blight is noticed early, blighted shoots may be broken off 25 cm below any sign of discoloration of the bark. Breaking, rather than cutting, infected shoots results in less chance of spreading the bacteria to other branches. This procedure is practical only with young trees. Apple and pear cultivars that are extremely susceptible to fire blight should not be planted. All cultivars can be attacked by fire blight at times, but the more susceptible ones are Golden Delicious, Grimes, Jonathan, Yellow Transparent, Lodi, and Stayman. Some pear cultivars, such as Bartlett, are so susceptible to fire blight that they usually are killed if grown in the eastern United States. With other cultivars (Kieffer and Orient), the degree of injury from blight usually is less severe. Magness and Moonglow are two relatively resistant pear cultivars.

Fire blight also attacks ornamental pears that are used in landscapes. Cultivars that are resistant to the bacterium should be used to limit this disease.

SOFT ROTS

Bacterial soft rot is one of the most destructive diseases of vegetables in storage and transit. It occurs most commonly on those vegetables that have fleshy storage tissues or fruits such as cabbage, carrot, celery, cucumber, eggplant, lettuce, onion, potato, radish, squash, sweet corn, sweet potato, and tomato. Hyacinth and iris bulbs also are affected. The soft-rot bacterium causes a disease on tobacco in the field called *hollow stalk* and a *barn rot* when the tobacco is being cured.

Soft rots occur all over the world and can cause serious losses during harvest, on the way to market, and especially in storage. Nearly all fresh vegetables may start to rot within a few hours in storage or during marketing. Soft rots cause economic losses by reducing the amount and the quality of produce available and by greatly increasing expenses for preventive measures during storage and shipment.

Symptoms

The name soft rot comes from the characteristic soft decay of fleshy tissues associated with the disease. Soft-rot symptoms are similar on all hosts. Carrot roots, celery stalks, onion bulbs, and potato tubers are dormant, storage organs high in carbohydrates. On such tissues, soft rot first appears as a small watersoaked spot which enlarges rapidly. The affected tissues become mushy and soft, appear cream-colored, and leak water. A whole fruit or tuber may be changed into a soft, watery, rotten mass within 3 to 5 days. As the tissues collapse, a foul odor is released. Rotten cabbage and onions smell particularly bad. Once you smell tissue in an advanced stage of soft rot, you will never forget it. The actual smell is not caused by the soft-rot bacterium, but by other contaminating bacteria.

Black leg, the seedling form of soft rot, occurs on some plants such as tomato and tobacco during wet periods. The rot spreads through the petioles into the stem, which turns black. Frequently the disease appears in areas up to 1 m in diameter in seedbeds. The black leg form of the disease also can be a serious problem on potato plants.

Causal Agent and Disease Cycle

Soft rots are caused by *Erwinia carotovora* pv. *carotovora,* a non-spore-forming, gram-negative, motile, rod-shaped bacterium (2.0 μm $\times$ 0.8 μm). It occurs singly or in chains; above 27°C (81°F) it forms longer rods and filaments. Each rod usually has two to six flagella uniformly distributed over the surface. The bacterium can grow aerobically or anaerobically over a wide range of temperatures, but does not grow below 2°C (36°F). The optimum growth temperature is about 25°C (77°F); the bacterium is killed at 50°C (122°F). *Erwinia carotovora* produces pectolytic enzymes that dissolve the cementing substances that hold plant cell walls together. There are strains of the bacterium that cause rapid rotting of some plants but not others.

Soft-rot bacteria overwinter in infected storage organs in plant trash, on seed, in soil, and in the pupae of several insects. Harvest bruises, freeze injury, and insect wounds increase the chances of decay. Abundant moisture at the surface of wounds is necessary for invasion. High RH favors disease development. Larvae of certain maggot flies carry the bacteria and introduce the bacteria into the tissues when such insects attack healthy plants or storage organs. After soft-rot bacteria invade plants through wounds, the bacteria multiply in the liquids released from wounded cells. The bacteria multiply rapidly, producing increasing amounts of enzymes that cause the cell walls to soften and dissolve; the contents leak out, and the cells collapse and die. Other contaminating bacteria that are always present help in the decay process. In a few days much of the stored crop can be a rotten, stinking, slimy mass. In some instances Irish potatoes harvested, washed, and packaged on Friday and shipped by truck over the weekend were completely rotted when the truck was opened on Monday morning. Moreover, the truck smelled so bad that the driver had trouble getting rid of the mess and cleaning the truck.

Management

Proper handling of crops at harvest is the most important measure for managing soft rot of vegetables. Growers should avoid bruising plant parts and keep plant surfaces dry. Potatoes, if washed at harvest before shipment, should be dried before being sacked. Do not store wet, wounded, or decayed tubers, fruits, or plant parts. Crops should be stored at low RH and at the lowest temperature at which the particular product can be expected to retain its edible quality. Maintain good air flow in storage areas to ensure dry plant surfaces. Leafy vegetables should be cooled to 4 to 6°C (39-43°F) immediately on arrival.

Crops very susceptible to soft rot should be rotated with cereals, grasses, or other nonsusceptible crops. Spraying or dusting to control the insects that spread the pathogen also helps control soft rot.

BACTERIAL WILT OF CUCUMBERS

Bacterial wilt is a common and destructive disease on cucumber and musk-melon; damage to squash and pumpkin is less severe. This disease is unusual in that the pathogen overwinters in the cucumber beetle and could not survive without this insect host.

Symptoms

The first evidence of cucumber wilt usually appears as dull green patches on individual leaves that wilt in sunny weather. As the disease progresses, more leaves wilt; eventually an entire branch becomes permanently wilted with shriveled, dead leaves. Occasionally a sticky ooze can be seen on the fruits. When an infected stem is cut crosswise, the bacterial ooze can be drawn out in strands 2 to 5 cm or more in length. This is a useful technique to help identify the disease.

Causal Agent and Disease Cycle

The causal agent of cucumber wilt, *Erwinia tracheiphila,* is a motile, gram-negative rod with four to eight scattered flagella. It forms capsules but no spores. Colonies in agar medium are small, glistening, circular, smooth, and white.

The bacteria overwinter in the bodies of adult cucumber beetles. The only means of natural infection is by beetles feeding upon young leaves or cotyledons in the spring. After infection, the bacteria invade the water-conducting vessels and multiply to enormous numbers. The sticky slime plugs the vessels, preventing the flow of sap, and wilting results. Secondary spread occurs only in wounds made by feeding beetles.

Management

Because the bacterium that causes wilt is solely dependent on the cucumber beetle, the most efficient to manage the disease is to manage the beetle. Insecticides must be applied early to forestall feeding by overwintering adults, and must be continued as necessary. Some cucumber cultivars are more resistant to wilt than others. These should be used in localities where the disease is a problem.

BACTERIAL CANKER OF TOMATO

Bacterial canker, although listed as a major disease in several countries, is erratic in occurrence. It generally is not so prevalent as early or late blight on tomato, which are caused by fungi (see Chapter 13). However, when canker does become a problem, it is very serious or devastating. Canker is an important disease only on tomato, but it occurs worldwide in the major tomato-producing areas.

Symptoms

Canker symptoms vary according to environmental factors and the age of the plant when infected (Fig. 16.3). Seedlings wilt and die. Young plants not yet bearing fruit show a one-sided wilting of the lower leaves. Leaves on older plants die at the edges

FIGURE 16.3. Symptoms of bacterial canker of tomato. (A) Bird's-eye spotting on ripe tomatoes. (B) Yellow to brown streaks in the water-conducting tissue of stems and petioles. (C) Wilting of leaflets. (D) Cankers along midribs of veins of leaflets. *Courtesy D. L. Strider.*

and dry up, and a general withering or firing of the foliage follows. Yellow to black streaks are found on stems, petioles, and leaf veins. These areas rupture to form brown, pimple-like cankers. Usually the roots are not infected.

During wet weather, secondary spread from cankers is common with infections occurring on the foliage and fruits as water-soaked spots (3 mm in diameter) surrounded by a white halo. This condition on the fruit is called *bird's eye* spot and greatly reduces market value.

Causal Agent and Disease Cycle

Unlike most plant-pathogenic bacteria, *Corynebacterium michiganense* pv. *michiganense*, which causes bacterial canker, is a gram-positive rod. It averages 0.6 μm ×

1.0 μm, forms yellow colonies on nutrient agar, and has no flagella. Nonpathogenic strains have been reported.

The pathogen enters the host through wounds or natural openings (stomates). All parts of the tomato plant are susceptible including the seeds. Rapid succulent growth of the host, warm weather [28°C (82°F)], low light intensity, sandy soils, and adequate moisture increase the severity of the disease. The bacterium survives up to 5 years in soil and 2 years in compost. The primary means of survival and spread is in seeds. The worldwide distribution of *C. michiganense* has resulted from contaminated seed being carried from one country to another.

Local spread is possible from diseased transplants, contaminated soil or debris, splashing rain, and workers handling infected plants.

Management

Bacterial canker can be managed effectively by the use of bacteria-free seed or transplants, crop rotation, and sanitation.

Use only certified seed obtained from areas where canker is not present, or treat the seed in hot water [56°C (133°F)] for 25 min or in 5% hydrochloric acid for 5 hr. Seedbeds should be fumigated or steamed before sowing of the seed. Roguing, the careful removal of diseased plants, also reduces spread. Antibiotic sprays, especially streptomycin, and copper materials help reduce disease incidence. Because the organism persists in the soil 3 to 5 years, long rotations are recommended.

BACTERIAL SPOT OF PEACH AND PLUM

Bacterial spot appears to be native to North America, where it occurs on wild plum. The disease probably spreads from wild plum trees to peach orchards planted nearby. Not only does it weaken the trees, but also it disfigures the fruit. Sometimes in years favorable for the disease, 15 to 30% of the spotted fruit must be discarded.

Bacterial spot does the most damage on peaches and plums, but almonds, apricots, cherries, and nectarines also are hosts.

Symptoms

Bacterial spot affects leaves, twigs, and fruit. It occurs on the leaves as small circular or irregular areas. The spots gradually enlarge from 1 to 5 mm in diameter, become angular, and turn brown to black. Usually this dead tissue falls out, giving the leaves a ragged, shot-hole appearance. Sometimes several spots may merge and involve large areas of the leaf. Young succulent twigs are killed, and the green shoots turn black, particularly on susceptible cultivars. With severe infection trees may be partially defoliated (Fig. 16.4). Small circular brown spots appear on the fruit surface (Fig. 16.5). A few spots or many spots that cover large areas may develop on fruit. Later the fruits "pit" or "crack" and become useless for sale.

FIGURE 16.4. Peach tree (left) with bacterial spot caused by *Xanthomonas campestris* pv. *pruni* is partially defoliated. Peach tree on the right does not have the disease. *Courtesy NCSU Plant Pathology Photo Collection.*

FIGURE 16.5. Small lesions of bacterial spot on peach fruit. *Courtesy NCSU Plant Pathology Photo Collection.*

Causal Agent and Disease Cycle

Bacterial spot of peach is caused by *Xanthomonas campestris* pv. *pruni*, a short gram-negative rod ($0.5\ \mu m \times 1.0\ \mu m$), which has one to several flagella attached to the end of the cell. It is aerobic and forms yellow colonies on agar. Exposure to the sun's rays for about 30 min kills the bacterium.

The pathogen overwinters in twig lesions and in the bud scales near the ends of the twigs. Primary infection of young leaves and twigs occurs early in the spring. The pathogen can enter uninjured peach tissue through natural openings. Warm weather with light, frequent rains accompanied by strong winds and heavy dew is most favorable for severe infection. When the disease is present, premature leaf fall occurs; the trees are weakened and predisposed to winter injury. If wet weather prevails during the few weeks before harvest, when the fruits are enlarging rapidly, damage is extensive on fruits.

Management

Because bacterial spot is more severe on weak trees that are growing poorly, the first step in managing the disease is to keep trees vigorous. This means that trees must be pruned correctly, fertilized properly, and sprayed to manage other diseases and insects. In many areas, especially in light sandy soils, it is hard to keep trees growing vigorously. Therefore, resistant cultivars should be used as recommended by the state experiment station whenever bacterial spot is a problem.

CROWN GALL

Crown galls are diseased growths or tumors that appear on apples, grapes, peaches, roses, and many other woody and herbaceous plants belonging to 140 genera of more than 60 families. The galls appear commonly on the stems of plants (the crown of roses or brambles) where they come out of the ground. That is why it is called crown gall.

The disease occurs all over the world. Infected nursery stock probably carried the causal agent from one country to another. On some crops, crown gall can cause significant losses; on others, it is just a curiosity.

Crown-gall tumors resemble human and animal cancers. In fact, some scientists have called crown gall a "plant cancer." When the cause of crown gall was first discovered about 1900, some researchers thought that by studying crown gall they could unlock the secret of human cancer. Therefore, the causes and mechanisms of plant galls have been studied extensively. However, crown gall tumors and tumors of humans and animals are really quite different. The changes that convert a healthy plant cell into one that produces tumorous cells have only recently been found to involve a *plasmid* carried by the causal agent. The plasmid is called the Ti or "tumor-inducing" plasmid. This Ti plasmid is a relatively short, circular piece of DNA in the cells of the crown gall bacterium. This plasmid DNA is independent of the main chromosomes in the bacterial cell. The plasmid has become a vital link in modern biotechnology and genetic engineering because it can be used to transfer pieces of

DNA between certain plants or between microorganisms and plants. Much has also been learned about the biochemical processes of growth stimulation and inhibition and the chemical compounds that control them.

Symptoms

Crown gall first appears as small overgrowths or swellings on the stems and roots, particularly near the soil line (Fig. 16.6). Ordinarily, the galls are quite soft at first. Their surfaces become wrinkled and furrowed as the galls enlarge and the color turns brown or black with age. Some tumors are spongy; others become hard and knobby and may reach 30 cm in diameter. Sometimes the tumors rot partially, and new growths appear the next growing season. Several galls may occur on the same root or stem, singly or in groups. Secondary tumors also develop.

In general, tumorous plants are stunted, have small chlorotic leaves, and are more susceptible than healthy plants to adverse conditions, particularly winter injury.

Causal Agent and Disease Cycle

The crown-gall bacterium, *Agrobacterium tumefaciens* (Smith & Townsend) Conn, is a soil-inhabiting, gram-negative rod ($0.5\,\mu m \times 2.0\,\mu m$) with two to four flagella at one end. It forms capsules but no spores. The optimum temperature for growth is 25 to 30°C (77-86°F). In some soils the bacterium remains alive for years.

FIGURE 16.6. Crown gall tumor on stem of tomato plant. *Courtesy NCSU Plant Pathology Photo Collection.*

The bacteria survive well in the soil as a saprophyte for many years and enter plants through wounds and occupy the intercellular spaces. Juices from ruptured host cells diffuse several millimeters from the wound site and bathe other cells. Bacterial cells in contact with wounded cells, transfer genetic material from the Ti plasmid to cytoplasmic genetic material of the host. The bacterium, *A. tumefaciens,* thus has been operating as a natural genetic engineer by modifying the genetic material of its host. The bacterium, via the plasmid, adds some of its own genetic material to that of the host cell. The genetic information from the Ti plasmid directs the plant cell to produce substances that cause uncontrolled growth of cells. These host cells begin to produce tumor cells within 4 to 5 days. Growth hormones that control cell enlargement and cell division also are involved in tumor development. Cells in the tumors continue to grow in an abnormal manner until maximum size is reached. Once the cells have been transformed to tumor cells by the bacterial plasmid, the bacterium is not necessary for gall growth.

Management

To effectively manage crown galls, growers should adhere to the following recommendations:

- Use nursery stock free of crown gall.
- Bud instead of graft nursery stock to avoid injuries near the soil.
- Do not wound crowns or roots while cultivating.
- Rotate infested fields with corn or small grains before planting nursery stock.
- Remove and burn infected trees.
- Do not mix diseased and healthy plants.
- Dip or spray plants with solutions containing antibacterial chemicals.
- Treat seedlings with closely related bacteria that prevent infection by the crown-gall bacterium. These bacteria grow on the plants and prevent infection by strains that can cause crown gall. This is an example of biological control of a plant disease.

MYCOPLASMA DISEASES

For a long time a group of diseases distinguished by stunting, chlorosis, yellowing of leaves, changing of flowers to leaflike structures, the production of numerous leaves, or combinations of these symptoms were thought to be caused by viruses. Some common diseases of this type are aster yellows, big bud of tomato and tobacco, corn stunt, and lethal yellowing of coconut palm. However, investigators now know that these diseases are not caused by viruses but by organisms, smaller than most bacteria, called *mycoplasmas.*

Japanese scientists, in 1967, were the first to show the presence of mycoplasmas in the cells of infected plants. Mycoplasmas do not have a rigid cell wall, only a cell membrane (Fig. 16.7). These organisms are soft, fragile, and subject to distortion under slight shifts in osmotic pressure. They assume many different shapes and reproduce by forming chains, budding, or fission.

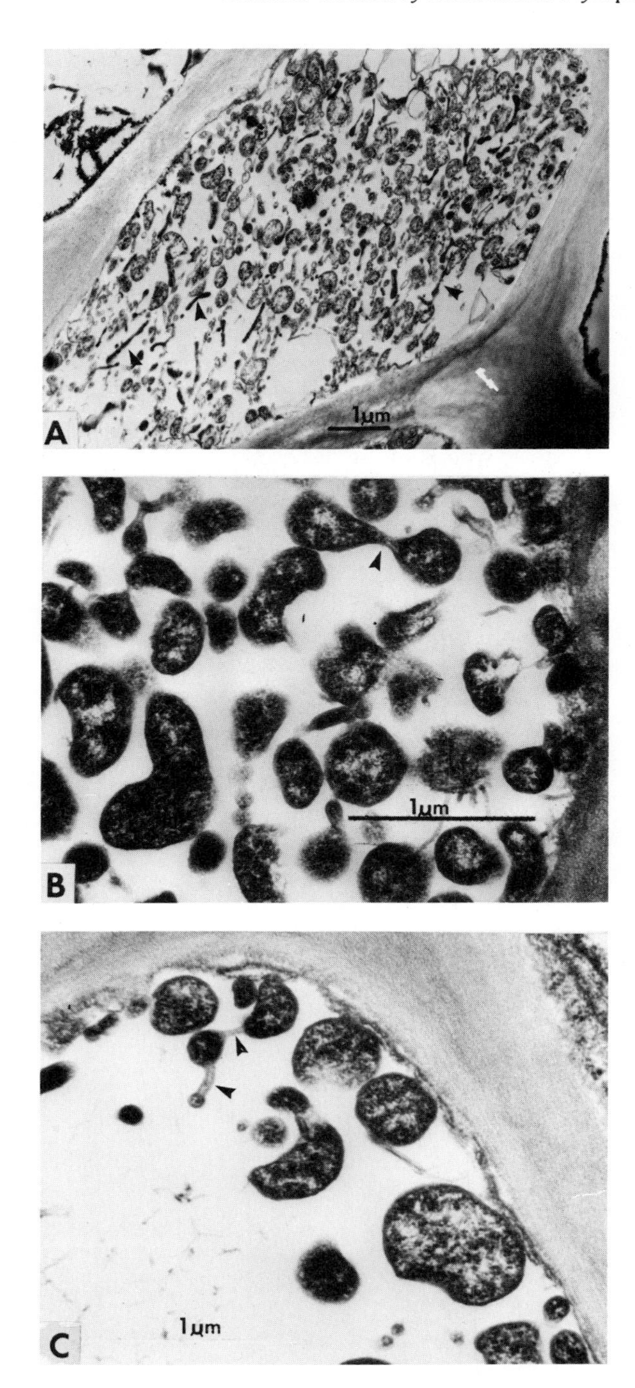

FIGURE 16.7. Mycoplasma (arrows) in food-conducting tissue of a tobacco plant. These microorganisms are soft and fragile and are easily distorted by pressure changes in the host cell. *Courtesy J. F. Worley.*

Mycoplasmas are transmitted from plant to plant by leafhoppers. An unusual property of mycoplasmas is their ability to multiply in leafhoppers, in some cases with harmful effects such as decreased hatchability of insects from infected eggs and decreased life span of infected adults.

Corn stunt is caused by a spiral mycoplasma or *spiroplasma* transmitted by leafhoppers. It is a serious disease on maize in Mexico, but it probably will be a serious problem only in areas with mild winters or where the leafhopper vector can survive because corn-stunt spiroplasma must survive in overwintering leafhoppers. If cold temperatures kill these vectors, the spiroplasmas die, and inoculum is not available to start a new epidemic.

Mycoplasmal diseases of expensive perennial ornamentals and other valuable plants can be managed or delayed by treatment with injections of certain antibiotics such as tetracyclines. In some cases, management of the leafhopper vector helps delay spread of the disease.

17

Viruses

The discovery of viruses as causal agents of disease ranks as a giant step in our understanding of the life sciences and diseases. We now know that all living things—animals, plants, even fungi and bacteria—are attacked by viruses. More than 400 plant viruses have been described since 1898.

The invention of the light microscope permitted scientists to see and study bacteria and helped produce the germ theory of disease. But until about 1935 many human diseases—such as chicken pox, colds, infantile paralysis, influenza, measles, mumps, smallpox, scarlet fever, and yellow fever—for which no causal organism could be observed, were believed to be caused by "ultra-microscopic organisms" (i.e., invisible bacteria) that were beyond the range of the microscope.

In plant pathology, too, despite careful and extensive research the causes of some plant diseases baffled the best-trained and most efficient workers. Some of these diseases were known to early workers by symptom names (e.g., chlorosis, dwarf, etch, leaf curl, leaf roll, mosaic, mottling, vein banding, and yellows). Many of these puzzling diseases caused a partial yellowing of plants, and some have been shown to be caused by mycoplasmas. (See Chapter 16.)

DISCOVERING PLANT VIRUSES

Research workers learned in the 1880s that some diseases were readily transmitted by grafting a diseased plant to a healthy one or by rubbing sap of a diseased plant onto a healthy plant. Other similar diseases were rarely or never transmitted to healthy plants by either method. In the 1890s, scientists determined that tobacco mosaic was caused by a "contagious living fluid" that did not contain bacteria.

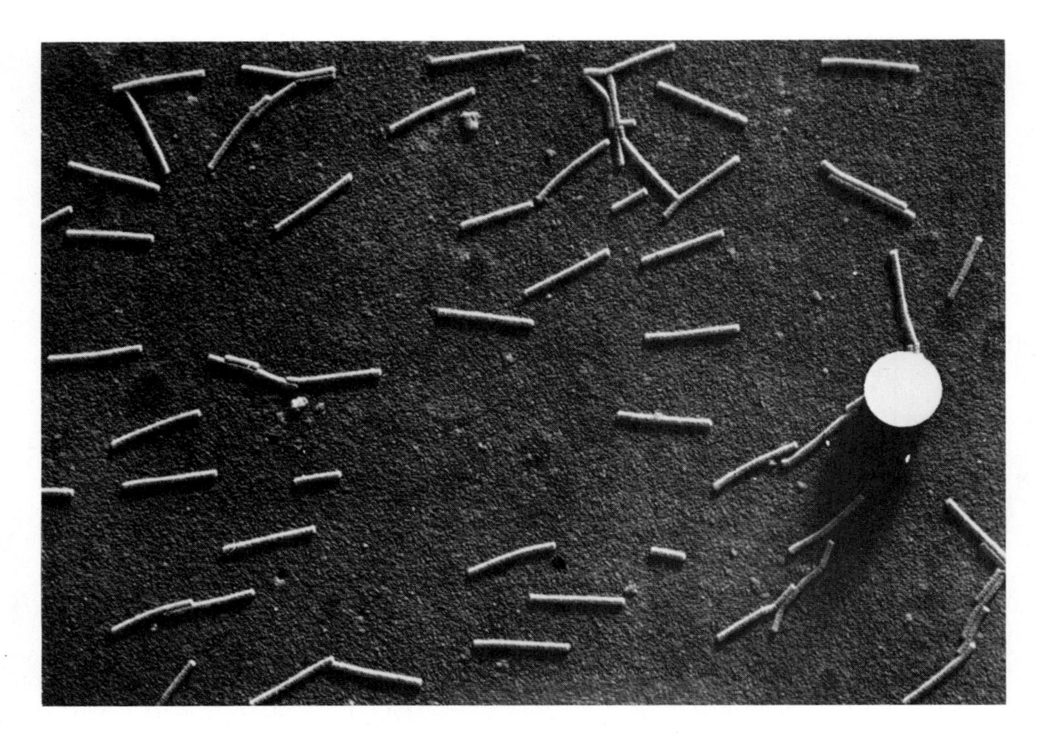

FIGURE 17.1. Electron micrograph of tobacco mosaic virus particles. The white sphere is a polystyrene latex particle that has a diameter of 88 ± 8 nm (nanometers). *Courtesy M. K. Corbett.*

In 1935 W. M. Stanley, an American, isolated and purified some tiny white crystals from the leaves of mosaic-infected tobacco plants. He had crystallized the proteinaceous part of tobacco mosaic virus. He wrote: "Tobacco mosaic virus is regarded as an autocatalytic protein, which, for the present, may be assumed to require the presence of living cells for multiplication." He was correct that a protein was involved and that living cells were essential for virus multiplication. For this important work Stanley received the Nobel Prize for chemistry in 1946. However, Stanley did not have a complete understanding of the nature of plant viruses. F. C. Bawden and N. W. Pirie, of England, demonstrated that a virus is really a nucleoprotein and, therefore, contains both protein and nucleic acid. Bawden also found later that the nucleic acid was required for infection of plants, and that the protein served only to protect the nucleic acid component.

About 1945 the electron microscope was invented. With the aid of this new tool, scientists were able to "see" virus particles, which are so small that they are invisible under the highest power of ordinary microscopes. (See Figs. 17.1 and 17.2).

Since the 1940s, developments in virology have been rapid. In the early 1950s. F. H. Crick and J. D. Watson, working in England, set forth their ideas on the double-helix structure of deoxyribonucleic acid (DNA) and its replication. The work of Crick

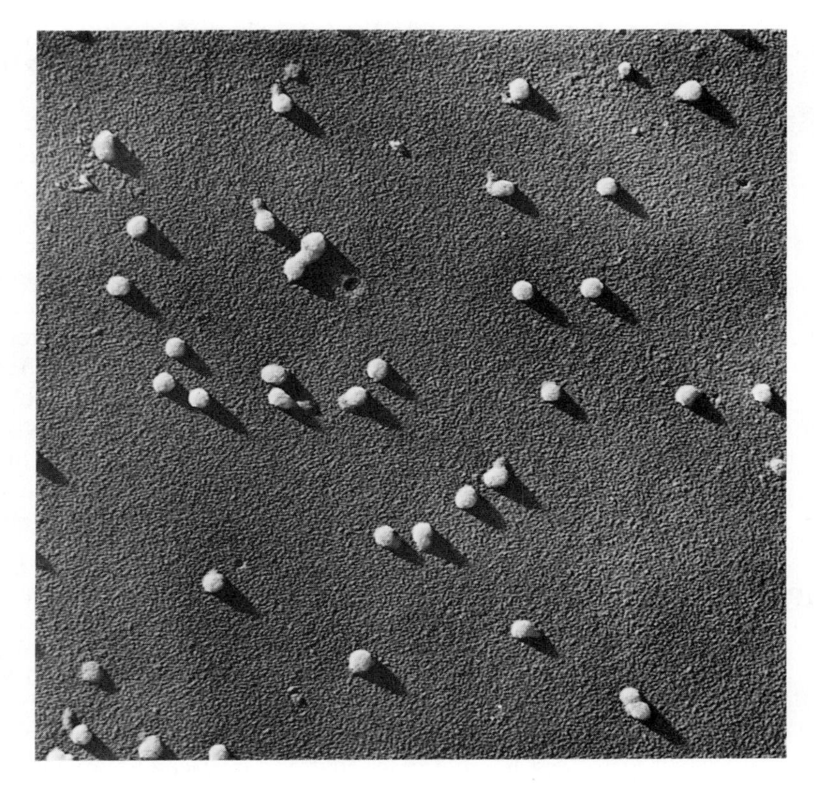

FIGURE 17.2. Electron micrograph of tomato spotted wilt virus magnified about 120,000 times. *Courtesy L. M. Black.*

and Watson permitted scientists to visualize how viruses reproduce and, more than that, how genes and chromosomes replicate.

In 1971, T. O. Diener, an American, described a new class of infectious units called *viroids*, composed only of nucleic acids. These pathogens are discussed in Chapter 18.

NAMING OF VIRUSES

Ever since viruses were discovered, scientists have debated about how to name them. Even today no one system has been accepted by all. The same virus has often received different names from different workers. Sometimes the same name has been given to distinct viruses. This has occurred largely because researchers named viruses on the basis of symptoms, without realizing that the same symptoms are not always produced by the same virus or that similar symptoms may be caused by different viruses.

At first, little attempt was made to distinguish between cause and effect. The virus and the disease usually were referred to as the same thing. Later a distinction was made by adding the word virus to the name of the disease to indicate the cause. For example, tobacco mosaic is caused by the tobacco mosaic virus; however, this system of naming viruses on the basis of the host attacked and the symptoms produced often is inadequate. And sometimes it is confusing when names are translated into different languages.

As more has been discovered about viruses, different systems of naming them have been proposed. Viruses can be classified into 26 families or groups on the basis of several characteristics: properties of the viral nucleic acid, viral protein, structure of the virus particles, physicochemical properties, activities in the plant, and method of transmission. With the approval of the International Committee for Taxonomy of Viruses, viruses are placed in families or groups based upon common characteristics. A type virus is designated for each family or group. Viruses other than the type virus but with similar characteristics belong to the specific family or group but retain an accepted name and acronym. For example, one of the families with rod-shaped particles is the Potyvirus family. The type member is potato virus Y, which has the acronym PVY. Clover yellow vein virus (ClYVV), maize dwarf mosaic virus (MDMV), pepper mosaic virus (PepMoV), and papaya ringspot virus (PRSV) also are potyviruses.

In this book, virus diseases are referred to by the name most commonly used. For example, the disease tobacco mosaic (TM) is caused by tobacco mosaic virus (TMV).

SYMPTOMS OF VIRUS DISEASES

The symptoms of virus disease were very important in disease identification in the early days of virus research. However, many factors, such as weather, plant age, and virus strain, can affect the symptoms that result from the infection of a plant by a single virus. In some cases a virus can infect a plant and not cause any visible symptoms. In other cases, infection may lead to rapid plant death. Thus, symptoms alone may not be enough to diagnose the presence of a specific virus. Other important techniques for virus diagnosis and identification are discussed in the following sections.

The most common symptom that results from virus infection is reduced plant size. Severity of plant stunting is generally correlated with the severity of other symptoms. Stunting usually results from reduced leaf size and shortened internode length.

Individual lesions may be caused on some plants by viruses. Individual or local lesions may be chlorotic or necrotic. In a few cases the lesions may be darker green than the surrounding tissues. Also, in some hosts ring spot lesions may appear. With ring spots concentric rings of chlorotic or necrotic tissue are separated by bands of green tissue. Ring spots may occur on leaves or on fruit. Although such symptoms are not usually of economic importance with regard to yield loss, these symptoms may be important in disease diagnosis.

A very common and obvious effect of virus infection on many plants is the development of a pattern of light and dark green areas that give a mosaic effect in infected leaves (Fig. 17.3). The specific pattern of the mosaic symptom varies with the host-virus combination.

Some viruses cause a general yellowing of the leaves. Soon after infection, veins of the younger leaves become clear or yellow, and a general yellowing of leaves follows. In some leaves there may be sectors of yellowed tissue and of normal, green tissue.

FIGURE 17.3. Distinctive mosaic pattern of symptoms on a virus-infected rose leaf. The light areas were a brilliant yellow. *Courtesy NCSU Plant Pathology Photo Collection.*

PROPERTIES AND MORPHOLOGY

Viruses are noncellular, ultramicroscopic particles that multiply only in living cells. Viruses withstand freezing, and they can remain infective for many months if stored in dry ice or at temperatures below zero (below 32°F). R. E. F. Matthews defines a virus as "a set of one or more nucleic acid template molecules, normally encased in a protective coat or coats of protein or lipoprotein, that is able to organize its own replication only within suitable host cells. Within such cells, virus replication is (i) dependent on the host's protein-synthesizing machinery, (ii) organized from pools of the required materials rather than by binary fission, (iii) located at sites that are not separated from the host cell contents by a lipoprotein bilayer membrane, and (iv) continually giving rise to variants through various kinds of changes in viral nucleic acid."

Most plant viruses consist of protein shells surrounding a core of nucleic acid, which is usually single-stranded ribonucleic acid (ssRNA) but in a few cases is double-stranded ribonucleic acid (dsRNA) or deoxyribonucleic acid (DNA). Some viruses consist of more than one size of proteins and nucleic acids. The amounts of nucleic acid and protein vary with each virus. Nucleic acid can make up 5 to 40% of a virus, with protein making up the remaining 60 to 95%.

Some viruses are shaped like rods or flexible threads (Fig. 17.1); others are spherical (17.2). The rods or flexible threads generally have less nucleic acid and more protein. The spherical viruses often have more nucleic acids and lower percentages of proteins. The elongated viruses are usually 10 to 15 nm wide and range in length from 480 nm (potato virus X—PVX) to 2000 nm (citrus tristeza virus—CTV). Spherical viruses range in diameter from 17 nm (tobacco necrosis satellite virus—TNV) to 60 nm (wound tumor virus—WTV). Flexible, spherical particles of tomato spotted wilt virus (TSWV) range in size from 70 nm to 80 nm.

IDENTIFICATION AND DETECTION

Viruses are identified and detected with a number of different techniques that range from the use of expensive equipment such as the electron microscope to simple bioassays. The identification of plant viruses often combines a bioassay such as virus indexing with serology.

Two approaches to identifying viruses do not involve their isolation from infected tissues. In one method the light microscope is used to examine host plant tissues for inclusion bodies. Many plant viruses induce distinctive intracellular inclusions that contain virus particles and/or other products of the viral genome and, in some cases, modified cell constituents. Most inclusions are induced with great constancy over a range of hosts. Their detection can provide a rapid, reliable, and relatively inexpensive method for determining viral infection. In many cases, inclusions have such a distinctive appearance that they may be used to identify a specific virus (Fig. 18.1).

To examine cell inclusions, the epidermis of a virus-infected leaf is peeled off and immersed in a specific stain. After a few minutes, the leaf segment is examined with a light microscope for suspected viral inclusions. The inclusions differ from the surrounding cytoplasm and organelles and are easily recognized by a trained observer.

In another approach, called *virus indexing,* the juice from an infected plant is used to inoculate one or more kinds of *indicator plants.* Indicator plants also are known as differential hosts. This means that based upon the symptoms caused on a number of different plants called indicators, a virus can be differentiated from other viruses. Indicator plants are sensitive to specific viruses, and when inoculated with these viruses develop characteristic symptoms. If the symptoms caused by an unknown virus on the indicator plants are the same as those caused by a known virus, then the two viruses are considered to be the same. However, it is often desirable to have a more precise identification than this, and additional tests can be used.

Viruses can be obtained in purified form by grinding up the leaves of infected plants and squeezing the juice through a filter made of glass, paper, asbestos, or other materials that remove bacteria and tissue cells. Then the virus may be precipitated with chemicals, or the plant juice can be placed in a test tube and whirled in a centrifuge at very high speed to separate viruses from leaf sap. Viruses then can be identified by visualizing them under electron microscopes.

One of the most important techniques used to identify viruses is *serology,* which involves the use of antibodies or antisera produced in an animal, such as a rabbit, that react with a specific protein from a specific virus.

Serology currently is the favored means for identifying plant viruses, and numerous techniques are available, including precipitation, microprecipitation, latex agglutination, agar-gel double-diffusion, enzyme-linked immunosorbent assay (ELISA), immunoblots, and immunoelectron microscopy. The technique of choice is determined by the availability of antisera, the number of identifications to be made, equipment availability, and personal preference. Virus identification methods developed after 1985 are based on nucleic acid homology and include dot-blot hybridization and squash-blotting.

Immunoassays are specific, quantitative for a certain chemical, and sensitive to several parts per billion (ppb). They take minutes rather than days to complete, and are quicker, simpler, and cheaper than many standard lab tests.

The ELISA technique is widely used. It is extremely sensitive; only a small amount of antisera is required, results are quantitative, many samples can be tested concurrently, the procedure is semiautomated, and antigens (the proteins that react with antisera) of many different pathogens can be tested. Portable immunosassy test kits employ antibodies that, in the presence of the virus (antigen) being tested for, cause a color change or some other measurable reaction. Within minutes the test can be "read" visually or with a colorimeter The presence of color indicates antigen (virus) in the sample, and, with the appropriate control tests, the degree of coloration indicates the amount of antigen present. With appropriate antisera, ELISA also can be used to identify fungi, bacteria, and even toxins.

TRANSMISSION

Plant viruses rarely spread from one plant to another spontaneously; virus particles usually are not transmitted by wind or in water or soil. Also, to be successfully transmitted from one plant to another, virus particles must come into contact with a wounded living cell. The transmission of viruses, then, is quite different from the movement of fungi, bacteria, and nematodes in nature.

Plant viruses are transmitted from plant to plant by vegetative propagation, mechanically through sap, and by other biological agents such as seed, pollen, insects, mites, nematodes, and dodder (a parasitic plant—see Chapter 19) and by fungi. Some mechanisms are more important in virus transmission from one generation of a plant to another, such as vegetative propagation and transmission by seed and pollen. Other methods can be important in the transmission of viruses between plants in the same generation, such as mechanical transmission through sap and by biological agents.

Transmission by vegetative propagation is especially important with many horticultural crops because they are propagated almost exclusively by budding or grafting, by cuttings, or by the use of tubers, bulbs, or rhizomes. Many fruit and ornamental trees are propagated by grafting, and many ornamentals are propagated by cuttings. Potatoes are propagated as tubers. Many flower crops are propagated by bulbs (tulips and daffodils) and rhizomes (iris). Thus, if a mother plant used for propagation stock is infected with a virus, many of the vegetatively propagated offspring also will be virus-infected.

Mechanical transmission of viruses is relatively rare in nature. However, one important virus, tobacco mosaic virus (TMV), can be easily transmitted by mechanical contact from the rubbing together of two leaves or by movement of people or equipment through fields. Mechanical transmission also is very important in the study of plant viruses, especially in the inoculation of indicator plants.

Fungi and nematodes are important as reservoirs of virus and as biological agents (*vectors*) that transmit viruses to plants. Most of these fungi and nematodes are soilborne and transmit the virus(es) to plant roots.

Insects, such as aphids and leafhoppers, are the most important agents (vectors) for transmitting viruses. When an insect feeds on leaves of virus-infected plants and sucks out the juice, the insect's stylet or mouth parts become contaminated with virus particles. Then if the insect should happen to fly or crawl to a healthy plant to feed on a leaf, some virus particles are injected into the leaf cell through tiny punctures made by the probing insect or through damage caused by the insect's chewing.

MANAGEMENT

Virus diseases are hard to manage. Once a plant is infected with virus, it becomes a source of contamination for other plants; therefore, growers should try to produce virus-free plants.

Virus-free seed and propagation stock that has been inspected and is certified to be virus-free should be used.

Seedbeds should be fumigated with methyl bromide or other chemicals that kill weeds (see Chapter 7, Soil Treatment section). Weed management with herbicides or cultivation markedly reduces the possibility of virus spread by eliminating weeds that can serve as hosts or reservoirs for important viruses. A seedbed area should be kept free of virus-susceptible weeds that serve as a virus source.

Because milk inactivates many plant viruses, it is a good idea for workers to dip their hands in milk every 20 to 30 min when handling plants. Virus contamination also can be removed from the hands by thorough washing with soap and water.

Viruses do most of their damage when they infect young plants. Therefore, crops should be kept virus-free as long as possible. This is difficult to do for insect-transmitted viruses, but spraying young plants with insecticides before transplanting can help. Growers should not grow crops near other host crops, and should plant crops during seasons of the year when insect populations are low or before they build up.

Sometimes valuable plants can be freed of viruses by growing the plants at high temperature or by making tiny cuttings or tissue cultures from tips or buds of the growing plants. Often these actively growing bud tissues are virus-free; healthy virus-free plants can be propagated from such tissue cultures. This method requires the skill of trained technicians but has been highly successful in the commercial production of virus-free cuttings of carnations, chrysanthemums, geraniums, and other greenhouse crops and some vegetable and field crops. See Chapter 8 (Vegetative Propagation section) for a detailed discussion of this technique.

Breeding plants for genetic resistance to virus diseases is very important. Many crop cultivars and varieties are available with resistance to specific viruses or virus vectors. Also, rapid progress is being made in the genetic engineering of plants that are resistant to specific viruses.

18

Diseases Caused by Viruses

The symptoms of plant disease caused by different viruses often are quite similar and difficult to distinguish. Symptoms vary widely, depending upon virus strain, host, cultivar, plant age, nutrition, weather, and infection by more than one virus. Accuracy in field diagnosis frequently is improved with knowledge of such factors as the pattern of infected plants in a field and the cultivar being grown. The pattern of symptom distribution, for example, can be of great value in separating mosaic symptoms induced by tobacco mosaic virus (TMV) from similar symptoms caused by some herbicides. Plants in a field with mosaic symptoms induced by a virus seldom display uniform symptoms, whereas plants with mosaic symptoms induced by herbicide injury will appear uniform; that is, all plants in a row or adjacent rows will be the same size, and symptoms will be similar.

Identification of a particular disease may require isolation of the causal virus from infected plants and its identification (Fig. 18.1). Various characteristics of isolated virus particles may be used to pinpoint their identity, including their serological specificity, size and shape as determined by electron microscopy, reactions to different hosts, and chemical composition. (See Chapter 17.)

Several of the most important virus-caused plant diseases are discussed in this chapter, and are summarized in Table 18.1. Readers should note that the effectiveness of measures for managing virus diseases often depends upon accurate diagnosis. Although some plant viruses can be identified by the visible symptoms and general behavior of the diseases they cause, the identity of many viruses can be determined only by trained personnel using the techniques discussed in Chapter 17.

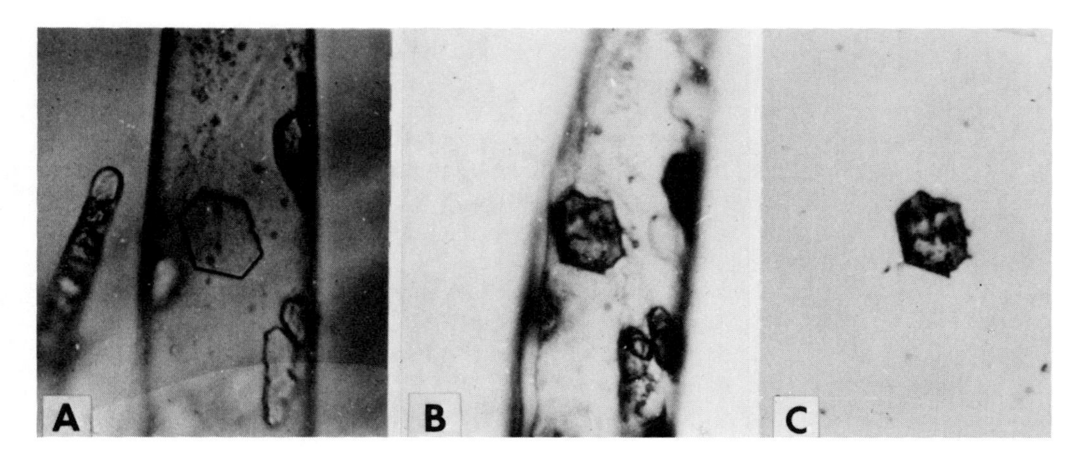

FIGURE 18.1. Isolation of a TMV crystal from a hair cell of a tobacco leaf. (A) Leaf hair before freeze-drying. (B) Leaf hair after freeze-drying. (C) Isolated crystal extracted by micromanipulation with a thin needle. Magnified about 430 times. *Courtesy R. L. Steer and R. C. Williams.*

TABLE 18.1. Some important diseases caused by viruses.

Disease	Pathogen	Hosts
Tobacco mosaic	Tobacco mosaic virus (TMV)	Pepper, tomato, tobacco, and others
Vein banding or streak	Potato virus Y (PVY)	Potato, pepper, tobacco
Tobacco vein mottle	Tobacco vein mottle virus (TVMV)	Tobacco
Tobacco etch	Tobacco etch virus (TEV)	Tobacco
Cucumber mosaic	Cucumber mosaic virus (CMV)	Cucumber, bean, tomato, tobacco, and others
Barley yellow dwarf	Barley yellow dwarf virus (BYDV)	Small grains
Maize dwarf mosaic	Maize dwarf mosaic virus (MDMV)	Corn
Maize chlorotic dwarf	Maize chlorotic dwarf (MCDV)	Corn
Tomato spotted wilt	Tomato spotted wilt virus (TSWV)	Tomato, tobacco, pepper, peanut, flowers

TOBACCO MOSAIC

Tobacco mosaic occurs all over the world. Although mosaic is so common, widespread epidemics seldom occur. However, if disease is severe within a field, an individual farmer can lose up to 90% of a tomato or tobacco crop. The biggest losses occur when the plants are infected while young. In such cases both yield and quality are seriously reduced.

Tobacco mosaic is the most important virus disease on tomatoes, especially those grown in greenhouses. Once a greenhouse crop becomes infected, the damage often is so great that the crop is of little value; moreover, it is expensive to clean up the greenhouse to get ready for the next crop. Therefore, every effort should be made to prevent mosaic from becoming established.

Much knowledge about viruses in general has been derived from research done on TMV. This virus has a wide host range. More than 350 species of plants are susceptible.

Symptoms

Mosaic symptoms vary with the plant species, age of the plant when affected, virus strain, nutrition, temperature and light conditions that occur after infection takes place. Tobacco plants infected with TMV have leaves mottled with light and dark green areas (Fig. 18.2). These areas are most easily seen on younger leaves and produce the mosaic pattern from which the disease gets its name. Some leaves are distorted with dark green blisters or swellings. Stunting of young plants also occurs.

On tomato, mottling occurs on older leaves; the leaflets become long and pointed rather like a shoestring. Fruit set is reduced, and those fruits that do form seldom reach full size. Parts of the pulp turn black; fruits from mosaic-infected plants have little market value.

Causal Agent and Disease Cycle

The TMV particle is rod-shaped, about 300 nm long × 15 nm in diameter. The nucleic acid in TMV is single-stranded RNA (ssRNA); the protein subunits are arranged in a helix around a spiral core of ribonucleic acid. The virus is one of the most stable viruses known and can remain infective for at least 50 years in dead, dried tissues and for many months on equipment, flats, greenhouse frames, sawdust, and tools and in the soil.

There are many strains of TMV. Some cause severe damage on a given host, whereas other strains cause such mild symptoms that they are barely visible.

Tobacco mosaic virus is transmitted mechanically and is only rarely transmitted by insects. Any means that permits the virus to touch an injured cell of a host plant many result in infection. One of the most common ways the virus is spread is by crop workers with TMV on their hands. This often happens during weeding, transplanting, cultivating, pruning, suckering, or tying operations. Tractors or equipment that brushes against a TMV-infected plant may spread the virus to other plants. Chewing insects such as grasshoppers have been shown to transmit the virus, but this means of spread is not very important.

Tobacco mosaic virus overwinters in a number of ways, which must be understood for successful management through sanitation practices. Infection of tobacco plants from overwintering sources of the virus is known as *primary infection.*

All forms of tobacco, including tobacco products used by humans, may carry TMV. The virus can survive for years in dried tobacco tissue, including trash contaminating field equipment and tobacco sheets used in marketing. The virus also overwinters in stalk and root tissue of mosaic-diseased plants from a previous crop. Transplants that come in contact with TMV-infested stalk and root tissue usually are infected.

FIGURE 18.2. Flue-cured tobacco infected with tobacco mosaic virus. A, B—Symptoms on young leaves. Leaf distortion (B) commonly occurs on plants infected at the four- to six-leaf stage. C, D—Symptoms on old leaves including line patterns along the veins (C) and mosaic burn (D). *Courtesy G. V. Gooding.*

Such infection occurs during transplanting when a plant is pushed against a piece of virus-infected tissue. The number of transplants that become infected in this way will depend on several factors, including the amount of mosaic present the previous year and the quantity of tissue surviving. The virus can overwinter in dead as well as living plant tissue, but dead tissue contains less active virus than living tissue.

Tomato, pepper, and eggplant also are hosts of TMV, and fruit from these crops may contain infective virus. There is, for example, enough active TMV in one infected tomato to infect every tobacco plant grown in North Carolina.

Weeds

A number of weeds are known to be hosts of TMV, including horsenettle (*Solanum carolinense*) (Fig. 18.3). The only way horsenettle can be eliminated as a source of TMV is to eradicate it by herbicides and/or cultivation.

Management

The most efficient way to control mosaic is to keep the crop free of TMV. This requires constant attention and an IPM program using every practical management measure available.

Only seed from reputable sources should be used. Tomato seed should have fermented or been treated with acid or bleach by the seedsman. Growers may treat their own seed by soaking it in a 1% water solution of trisodium orthophosphate for 15 min and then in 0.5% sodium hypochlorite (Clorox) for 30 min.

As mosaic infection often begins in seedbeds, growers should produce TMV-free plants and transplant them in noninfested soil. Seedbed soils should be fumigated with methyl bromide or other chemicals that kill weeds and inactivate TMV. Chemical weed management reduces the possibility of TMV spread by eliminating hand weeding. The seedbed area needs to be kept free of susceptible weeds, such as groundcherry or horsenettle, that can carry TMV from one year to the next.

Workers should not smoke or chew tobacco while working in and around seedbeds because their hands become contaminated by contact with tobacco or saliva. This TMV contamination can be removed by thorough washing with soap and water.

Spraying seedbeds with whole or skim milk (2 liters/10 m^2) 24 hr before pulling or handling plants helps to reduce the number of plants that become infected. Thorough coverage is important. Milk sprays should always be used if the use of tobacco by workers cannot be prevented. All workers should also wash their hands with soap or dip them in milk about every 20 min.

Other helpful management practices include removing diseased plants early in the season (do not touch healthy plants during this operation), destroying plants in seedbeds soon after transplanting is completed and in fields soon after harvest, and rotating susceptible crops (tobacco, tomato, and pepper) with nonhost crops (corn, pasture, or grains).

Finally, resistant cultivars should be planted whenever possible. Although mosaic-resistant tobacco cultivars historically have been lower in yield and quality than nonresistant ones, some of the newer releases compare favorably with nonresistant

FIGURE 18.3. Shoot from a horsenettle (*Solanum carolinense*) plant. This common weed frequently is infected with TMV and is suspected to be the source of TMV found in many tobacco fields. *Courtesy G. V. Gooding.*

cultivars. Even when resistant cultivars do not perform as well as susceptible ones on a given farm, there are two instances when they would be of benefit. The first is obviously on those farms where losses to mosaic exceed the differences in yield potential between a mosaic-resistant and a susceptible cultivar. Secondly, on farms where monoculture is practiced, a mosaic-resistant cultivar can be grown on a problem field to break the cycle of virus carryover.

VEIN BANDING OR STREAK

Potato virus Y (PVY), which causes vein banding, is so named because it was first found on Irish potato. The name is misleading in many areas because the virus may not be a problem on potato, but frequently causes mild to severe diseases on other solanaceous crops, including tobacco, tomato, and pepper. The virus is probably worldwide in distribution. It is a serious problem on both Irish potato and tobacco in Europe where tobacco and potato production areas overlap. Management of the disease and the insect vector on one crop influences management on the other. Thus, IPM methods for each crop need to be synchronized.

The diseases caused by PVY have not been thoroughly studied in regard to yield losses. Field observations, however, indicate that PVY is now a significant threat to the production of peppers, potatoes, tobacco, and tomato in most central European countries. Losses exceeding 50% occur in tobacco cultivars infected with a necrotic strain of the virus, and the severe strain on peppers probably shortens the productive life of these plants. Although not often seen, PVY in combination with potato virus X causes a severe disease of potato known as *rugose mosaic*. Because of the great variation in symptoms caused by different strains of PVY on different crops, field identifications always should be confirmed by clinical tests.

Symptoms

There are a number of strains of PVY, based on the symptoms they cause. One causes mild mosaic symptoms on all cultivars of tobacco. One causes vein and midrib *necrosis* (death) on leaves of root-knot-resistant tobacco and mild mosaic symptoms on root-knot-susceptible tobacco. A third strain causes vein and midrib necrosis on both root-knot-resistant and root-knot-susceptible cultivars of tobacco. Ironically, the mild strain on tobacco is severe on pepper; the two necrotic strains on tobacco are mild on pepper. All three strains cause mild mosaic symptoms on potato and tomato.

Causal Agent and Disease Cycle

Potato virus Y is classified with a group of approximately 16 related viruses, including tobacco etch and vein mottle, that are long flexuous rods. PVY is the type member for the potyvirus family. The individual particles look like short pieces of string (760 nm × 12 nm). The virus is an obligate parasite and cannot reproduce outside a living host.

PVY is carried in potato tubers. Plants grown from them or infected tomato, tobacco, or pepper transplants serve as inoculum. The virus overwinters in weeds or other solanaceous crops, especially in mild climates where annuals persist as semi-perennials. Spread of PVY by seed has *not* been demonstrated.

The perennial weeds ground cherry (*Physalis* sp.) and nightshade (*Solanum aculeatissimum*) may serve as overwintering hosts. Potato virus Y is not seed-transmitted, nor does it survive in dead host tissue.

The only vectors known to transmit PVY are several species of aphids. These sucking insects have mouth parts called *stylets* that are specialized for sucking in a liquid diet. When the aphids insert their stylets into the cells of an infected plant, some of the juice-containing PVY particles stick to the stylets. If the aphid then probes the cells of a healthy plant, the PVY particles are injected into the leaf through the tiny puncture; infection has begun. Aphids can pick up PVY after feeding for 5 seconds and transmit the virus within 10 seconds of feeding.

One of the most important vectors of PVY is the green peach aphid, which inhabits temperate, subtropical, and tropical areas in all parts of the world. It is host-alternating. The most important primary host plants are *Prunus* species, on which, in cold climates, the aphids overwinter as black shiny eggs (0.6 mm × 0.3 mm). The eggs hatch in early spring. Where winters are mild, the insects colonize winter (secondary) hosts, principally *Brassicae* crops, including more than 400 species, and breed by giving birth to living, wingless, pale yellow to green young females. Mature females (1.9 mm × 0.8 mm) produce more wingless young, and after about three generations winged migrant forms appear that fly to summer hosts, including tobacco and weeds. Reproduction continues into late summer when winged forms migrate to winter hosts and lay eggs.

The movement of insects from one host plant to another determines the transmission of plant-virus diseases by arthropods. This movement is conditioned by the life history of the insect, host range, host preference, availability, and condition of these hosts and their status as virus reservoirs. Superimposed on these biotic factors are physical factors of the environment. Production of winged aphids is influenced by parentage as well as light, temperature, crowding, starvation, host wilting, and lack of water. The time of aphid flight is more important than the number of aphids. Mild winters lead to early flight by permitting survival of aphids on alternate hosts.

M. persicae can survive when the mean temperatures remain above 4°C (39°F) during the three coldest months. Early-season arriving aphids are more likely to be carrying virus than late-season arrivals. Plants infected early are more important virus sources for secondary spread than plants infected later.

Winged aphids are present in the air during the entire growing season. Evidence indicates that aphids fly only in light, mostly between 15°C (59°F) and 32°C (90°F), in both calm and windy weather during high or low humidity. Younger aphids are more active than older ones. Strong winds may carry aphids over large areas.

Management

There are three general approaches to managing PVY: elimination of the virus source, reduction of vectors, and use of resistant or tolerant cultivars.

The PVY sources can be eliminated by plowing under the remains of the preceding crop as soon as possible after harvest so that aphids cannot feed on them. In addition, weed hosts should be removed in fields near Irish potato, pepper, or tobacco. Farmers should grow their own transplants or use only certified transplants or seed. Importation of plants from regions where PVY persists on growing plants year-round is unwise. Where PVY is found on only a few plants, they may be pulled and removed from the field, thus preventing the virus from being spread to other plants.

The use of insecticides to reduce the number of aphids available for virus transmission helps prevent PVY from getting started in seedbeds. Cultivars resistant or tolerant to PVY should be used wherever possible.

TOBACCO VEIN MOTTLE

Tobacco vein mottle is caused by tobacco vein mottle virus (TVMV), which is closely related to PVY. The TVMV is widespread in the Burley tobacco growing areas of the United States. The symptoms of tobacco vein mottle, often confused with vein banding, vary with age of the plant when infected and cultivars. The virus is a flexuous rod spread only by aphids; TVMV can be distinguished from PVY serologically. The virus overwinters in perennial weed hosts and living tobacco roots. Management measures are the same as for PVY.

TOBACCO ETCH

Tobacco etch is caused by tobacco etch virus (TEV), which is another aphid-transmitted, flexuous rod virus related to PVY. Etch occurs in tobacco-growing areas all over the world. It causes serious losses in some years on flue-cured tobacco in Canada. TEV also infects pepper and tomato. There are many strains of TEV, which cause symptoms that vary from a mild mosaic or etching to chlorotic tattered leaves and eventual death of the plant. Pepper and tomato plants may be stunted. On peppers, both leaves and fruit may be distorted. Management measures are the same as for PVY.

TOMATO SPOTTED WILT

Spotted wilt, sometimes called "vira-cabeca" and "kromnek," causes damage in Europe, North and South America, South Africa, and Australia. Apparently the virus is worldwide in distribution. In the United States the disease was relatively unknown in the 1970s and early 1980s. By 1989, however, the disease had become widespread in peanuts, tobacco, tomatoes, pepper, and several ornamental crops. Evidence indicates that TSWV occurrence and prevalence will increase in the United States in both field and greenhouse crops. TSWV has an extremely wide host range (about 200 species). It infects many crop plants (including peanuts, peppers, pineapple, tobacco, and tomatoes), ornamentals, and herbaceous weeds. Many of these are reservoirs for the virus from one season to the next.

Symptoms

Spotted wilt symptoms vary with the age of the plant, environmental conditions (especially temperature), level of infection, and number of virus strains infecting the

FIGURE 18.4. Symptoms of tomato spotted wilt on tomato plant (center). The diseased plant is stunted, and the leaves appear to droop or bend over. *Courtesy NCSU Plant Pathology Photo Collection.*

host. Infection occurs in epidermal cells through wounds made by the shallow probes of feeding thrips. Symptoms may appear within 2 to 4 days. Seedling plants may wither and die within a week from systemic infection. The virus causes concentric necrotic rings and zonate necrotic spots of tissue on the younger leaves. Frequently the spots coalesce. At first the spots are yellow, but later the dead areas turn red-brown. Young seedlings to mature plants may be attacked. Infected plants are stunted, and the apical bud droops or bends over (Fig. 18.4). Young leaves infected on one side frequently are distorted or puckered as a result of unequal growth especially in tomatoes. Plants severely attacked do not grow for several weeks; the leaves droop and finally die. Some affected plants appear to recover and produce secondary growth that later develops necrotic symptoms.

Causal Agent and Disease Cycle

TSWV belongs to the group of unstable "spherical" plant viruses (Fig. 18.5) with an average particle size of 75 to 100 nm. The nucleic acid component of TSWV is ssRNA. The virus is physically and chemically unstable at room temperature. There are several strains that produce mild to severe symptoms on indicator plants. Often more than one strain is present in a given area.

At least nine species of thrips can transmit TSWV. *Thrips tabaci* and several other species of thrips (*Frankliniella* spp.) are the principal vectors of TSWV. These tiny

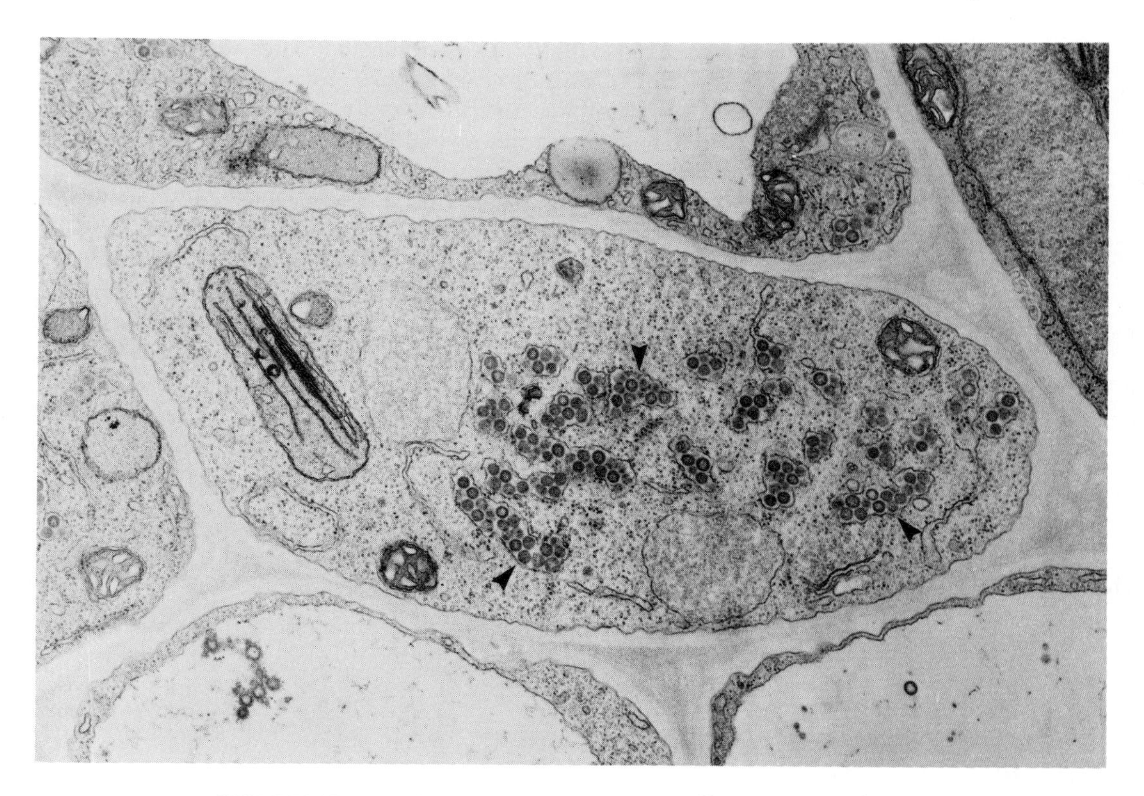

FIGURE 18.5. Clusters of particles (near arrows) of tomato spotted wilt virus (TSWV) in a host parenchyma cell. *Courtesy J. W. Moyer.*

insects, about 1 mm long, occur in huge numbers in the air up to high altitudes and may be blown many miles by the wind (like dust). *Thrips tabaci* can feed on at least 140 species of plants in over 40 plant families.

Thrips are minute, slender, and agile insects. They spring or fly readily when disturbed. The dark brown to black adults and straw-colored larvae overwinter on plants or rubbish. Females reproduce regularly without males. These insects squirm in between the young leaves and feed. They have rasping-sucking mouth parts that puncture the leaf surface and make shallow wounds. They swallow the sap with bits of leaf tissue. Only the larval stage is capable of acquiring the virus. The larvae acquire the virus after feeding about 15 to 30 min. After an incubation period in the insect of about 4 to 18 days, the larva and subsequently the adult arising from it become infective.

The white bean-shaped eggs are inserted into leaves and stems and hatch in 5 to 10 days. The virus is not passed through the egg. Eggs, larvae, and adults are found throughout the summer. There are five to six generations a year. Usually individual thrips live about 20 days. High temperature and low humidity are unfavorable for thrips. Heavy precipitation and freezing temperature in winter reduce populations considerably.

The long growing season in the southeastern United States, where many susceptible hosts are grown (often adjacent to each other), provides suitable conditions for perpetuation of TSWV and thrips vectors. In this region, where wild hosts may survive the mild winters, susceptible host crops are in the field most of the time. In addition, the interstate shipment of tomato, pepper, tobacco transplants, and ornamental greenhouse plants provides a continual transfer of virus-infected plants from one region to another.

As is true of other insect-transmitted viruses, disease severity is dependent upon interaction of the environment, vector, viruses, and host. Factors affecting vector activity in moving from plant to plant appear to be more important than the size of the insect population because one viruliferous thrip can inoculate a plant. However, late in the growing season more than 400 thrips have been counted on a single plant.

The disease is favored by warm temperatures and abundant moisture. At 20°C (68°F) symptoms may appear in 5 days, but they take 3 weeks to appear at 10 to 12°C (50-54°F). Incidence and severity also vary with rainfall, its pattern of occurrence, and the presence of preferred hosts attractive to the thrips.

Management

Efficient management of TSWV is difficult. In general, the same measures used to manage PVY are used to reduce losses to TSWV.

CUCUMBER MOSAIC

The virus that causes cucumber mosaic occurs worldwide and has a wide host range. Many vegetable and ornamental crops are affected in both the greenhouse and the field; the result is reduced yields of low quality. The disease occurs all over the world. The more important hosts include banana, bean, celery, crucifers, cucumbers, gladiolus, lilies, melons, petunias, spinach, squash, tomatoes, and zinnias. Many weeds also carry the virus.

In some areas up to 50% of the plants may be destroyed; susceptible crops such as cucumbers may have to be replaced or abandoned. In Japan, where many vegetable crops and tobacco are grown intensively in the same neighborhoods, cucumber mosaic is a constant threat. There, entire communities cooperate in conducting an IPM program for CMV.

Symptoms

Cucumber mosaic symptoms resemble those of tobacco mosaic. Usually first infections on cucumber appear 4 to 6 weeks after the crop is planted. Young developing leaves become mottled, distorted, and wrinkled. Their edges curl downward. Growth is reduced, the plants appear stunted, and few runners are produced with only a few flowers and fruit. Older leaves develop chlorotic areas that later kill the entire leaf. Infected cucumber fruits are distorted and misshapen with light and dark green mottling on the surface. Such fruits often have a bitter taste and become soggy when pickled.

On other hosts, symptoms may range from mild to severe, depending on the strain and host. Cucumber mosaic on tobacco is hard to distinguish from tobacco mosaic. Tomato leaves become long and stringy like shoestrings. Certain weeds show hardly any symptoms at all.

Causal Agent and Disease Cycle

Cucumber mosaic virus (CMV) is a polyhedral virus whose particles are about 30 nm in diameter. CMV is the type member for the cucumovirus family. There are numerous strains that vary in pathogenicity. The virus is easily transmitted by rubbing and other mechanical means. At least 60 species of aphids also transmit it. Aphids can suck up CMV in less than a minute and inoculate it into plants in less than a minute.

The virus can overwinter in many perennial weeds, flowers, and crop plants such as ground cherry, horsenettle, milk weed, nightshade, ragweed, and various mints. These hosts act as a continuous reservoir of CMV from year to year; the virus remains infective in their roots during the winter. When growth occurs in the spring, the virus is transported to the leaves, where it is picked up by feeding aphids and carried to nearby susceptible crops. Once a few cucumbers or squash become infected, the virus is spread during cultivation and harvesting. Sometimes all plants in a field become infected after the first picking. Another source of CMV is infected seed. Several weeds and flowers carry the virus in the seed; however, cucumber does not.

Management

Cucumber mosaic can be controlled, but it takes IPM. Resistant cultivars of cucumber, muskmelon, spinach, and tobacco have been developed, and should be used whenever the disease is a problem. Big losses often can be avoided by preventing or delaying infection in the crop and restricting aphid movement (which is hard to do). Aphids can be reduced by elimination of weed hosts, especially perennials in nearby fields, by disking or plowing under, or by spraying the weed source with an aphicide. Aphid transmission of the virus also is reduced by planting the crop during seasons of the year when aphid populations are low or before they build up. Cucumbers should not be grown near other host crops such as tobacco, tomatoes, or peppers, and weed hosts should be eliminated. Growers should plant a nonhost crop such as sunflowers between the source of inoculum and the cucumbers to act as a barrier against aphid flights, or interplant with barley; and should plow under or otherwise destroy the remains of the preceding crop as quickly as possible.

Greenhouse crops should be isolated from other crops susceptible to CMV. Before handling transplants, workers should dip their hands in milk. Insecticides should be used in seedbeds to prevent aphids from infecting transplants.

Virus-free seed should always be used for those crops where CMV is known to be seed-transmitted.

The fact that aphids are repelled by certain wavelengths of light and attracted by others offers another possible future method of management, by making plants invisible or unattractive to alighting aphids. Reflective surfaces such as aluminum strips placed between the rows have offered protection in some crops; as the aphids come in to land, the reflected ultraviolet light is thought to act as a repellent.

BARLEY YELLOW DWARF

Barley yellow dwarf (BYD) is widely distributed throughout the world. It is one of the most important diseases of small grains in the southeastern United States. It affects barley, oats, rye, and wheat; oats are the most seriously affected and suffer serious losses annually. The barley yellow dwarf virus (BYDV) also attacks many weed and pasture grasses.

The yield loss from BYD depends on the time of infection. Plants infected when young may die, or if they live, produce no grain; however, infection at heading time may reduce yields only slightly. In years of severe BYD epidemics, yields losses can be as high as 50% on oats and up to 30% on barley and wheat. Infection by BYDV also makes plants more susceptible to winter killing and to infection by root-rotting and leaf-spotting fungi.

Symptoms

Symptoms of BYD include yellowing and dwarfing of leaves, stunting of the plant, reduced tillering, sterility, and shrunken kernels; root systems also are reduced.

The leaves of infected oat plants usually turn red or bronze. The disease on oats has been called *red leaf*. The red color appears to be the result of reduced amounts of chlorophyll and increased amounts of various yellow and red pigments. In wheat, barley, and rye, infected leaves usually turn yellow, beginning at the tip and progressing toward the base. Leaf veins frequently remain green longer than the area between the veins and form green lines that project into the yellow area. The disease usually occurs in patches in the field and frequently is more prevalent along the edges of the fields.

Causal Agent and Disease Cycle

Barley yellow dwarf virus, a luteovirus, has a polyhedral particle about 25 nm in diameter. The virus concentration in infected plants is very low. It is not mechanically transmissible and is not seedborne. Barley yellow dwarf is difficult to diagnose because at present there is no satisfactory method of routinely detecting the presence of the virus in plants. The virus cannot survive from season to season in the soil or in dead straw or debris; it survives only in living plants or in living aphids. The spread of BYDV depends upon the spread of the insect vectors. At least nine species of aphids can transmit BYDV.

Some aphid species transmit BYDV more efficiently than others. There are a number of strains of the virus; certain strains can be transmitted by some species of aphids, but not by others. The most important transmitter of BYDV in the fall in the southeastern United States is the oat-bird cherry aphid, a small brown aphid that usually feeds on the lower parts of the plants. In the spring the most common transmitter is the English grain aphid, a larger green aphid that usually feeds on the lower portions of the plants.

The virus remains infective over winter in bromegrass, orchard grass, ryegrass, tall fescue, and other grasses that harbor the aphids. The aphids then crawl or fly from the infected grass to grain crops and carry the virus from grass to grain. The only way by which small grain plants can become infected with BYDV is from feeding by aphids carrying the virus.

When winters are mild, there is abundant survival of aphids in the southern United States. Warm south winds blow or carry the aphids northward in stages as spring advances. The aphids feed on the new tender leaves of wheat, oats, and grass hosts as they begin to grow. Mean daily temperatures of 10 to 18°C (50-64°F) are most favorable for succulent grass growth, aphid activity, and reproduction. Thus, epidemics of barley yellow dwarf do the most damage when the spring and early summer weather is cool and moist.

Symptoms usually appear 3 to 6 weeks after infective aphids feed on plants. The first leaves to show symptoms are best for diagnosis.

Management

As is the case with other viruses transmitted by aphids, management of BYD depends on managing the aphid vector. Attempts to control aphids by spraying in the field have not been satisfactory. Infective aphids that move into a field can transmit the virus before the insecticide kills them.

Several cultural practices help reduce losses: Do not plant small grains near large grass areas, such as pastures, that may act as a source of virus. Do plant oats or barley near the end of the normal seeding period, as BYD is generally more prevalent in early seeded grain.

All small grain cultivars are subject to damage by BYD, but some are more affected than others. Therefore, growers should use only those cultivars recommended for their area.

SOILBORNE WHEAT MOSAIC

Soilborne wheat mosaic (SBWM) occurs throughout the eastern and central United States and in Argentina, Brazil, Egypt, Italy, and Japan. Normally only fall-sown wheats develop symptoms of the disease although spring wheats are susceptible. Yield losses due to SBWM vary with the cultivar, virus strain, and environment. When the disease is severe, entire fields can be affected, and yields are reduced so much that harvest may not be profitable. In the central grain belt of the United States, SBWM annually results in losses and can be as important economically as barley yellow dwarf.

Symptoms

Symptoms range from light green to yellow leaf mosaic. Plants may be stunted slightly or severely. Certain strains of the virus cause rosetting, where the plants have a short, bunchy habit of growth. Fields may show a uniform chlorosis, or the disease may occur in patches. Symptoms are most prominent on early spring growth, and new leaves may appear mottled.

Causal Agent and Disease Cycle

The virus, soilborne wheat mosaic virus (SBWMV), survives in soil in a fungus, *Polymyxa graminis* Led., which is an obligate parasite in the roots of many plants. This fungus is in the division Myxomycota and thus produces a plasmodium instead of mycelium. *P. graminis* produces motile zoospores when soil is water-saturated and between 10°C and 20°C (50°F and 68°F). The zoospores enter roots hairs and epidermal cells of roots. The zoospores carry the virus (SBWMV) into the plant, where it continues to multiply and to spread throughout the plant. The fungus rapidly expands and replaces the contents of the infected root cells with plasmodial bodies. These bodies will either split into additional zoospores or will develop thick-walled resting spores within 2 to 4 weeks after infection. The fungus, infected by SBWMV, can survive in soil for many years. The fungus and the virus can be disseminated by any method that moves soil among fields.

Management

Management of this virus disease is really brought about by management of the fungal vector. Resistant cultivars offer the best management of soilborne wheat mosaic. Cultivars are resistant to the fungal vector, not to the virus. Late autumn planting may allow the wheat to grow at a time when the fungus is not very active in soil because of cold temperatures. Crop rotation with noncereal crops also may reduce the population level or inoculum density of *P. graminis*. Continuous wheat cultivation within a field should be avoided because the fungus will increase each year, and the disease caused by the virus will become increasingly severe.

MAIZE DWARF MOSAIC AND MAIZE CHLOROTIC DWARF

Two major virus diseases of corn are maize dwarf mosaic (MDM) and maize chlorotic dwarf (MCD). Before 1960, virus diseases of corn did not receive much attention or were overlooked or confused with other growth problems. However, we now know that these two diseases can cause losses of 5 to 90% in some fields. Maize dwarf mosaic is worldwide in distribution. Less is known about MCD; however, corn scientists are on the lookout for MCD, and reports are being made from new areas.

All types of corn—field, pop, and sweet—are hosts of the viruses that cause MDM and MCD. Johnsongrass is a principal host; other weeds, including barnyard grass, crabgrass, foxtail, and quack grass, also are hosts. Pearl millet, sorghum, Sudan grass, sugarcane, and teosinte also serve as hosts.

Symptoms

Many virus diseases are not easily diagnosed, as the symptoms are similar to those caused by other growth problems. Therefore, it is necessary to be sure that other factors including nutrition, herbicides, soil, or other pathogens are not causing the trouble.

Symptoms of MDM and MCD overlap and may be influenced by environmental conditions and the cultivar of corn grown.

Maize dwarf mosaic first produces a faint, pale green streaking, mottle, or mosaic in younger leaves along the veins. Spots of dark green tissue appear on a lighter green background. As infected plants mature, this discoloration increases, and the leaves become yellowish-green. Infected plants may be stunted and bunchy, tiller excessively, produce multiple ear shoots, but set few seed. Early infection may predispose the corn plant to root and stalk rots and sometimes premature death. Corn ears on infected plants remain small and do not fill completely.

Maize chlorotic dwarf, which used to be called *corn stunt*, causes more severe stunting than MDMV. Infected leaves turn yellow, but no mosaic pattern develops. As the plants mature, the yellow discoloration increases; the leaves become red. Infection with MCDV stunts or shortens plants, giving them a "bunchy" appearance. Infection often occurs late in the season. On such plants the lower portion grows normally with the upper portion red and stunted. If susceptible cultivars are infected early in the growing season, they produce only short, stunted ears (sometimes called "nubbins").

Causal Agents and Disease Cycle

There are several strains of MDMV, which belongs to the potyvirus group. The virus particles are flexuous rods, about 750 nm × 12 nm. Particles of MCDV are small spheres that average about 25 nm in diameter; they can persist in the leafhopper vector for several months. Both viruses are transmitted by insects: MDMV by aphids and MCDV by leafhoppers. Soil and seed spread have not been shown. Johnsongrass is the most important overwintering host. Aphids and leafhoppers feed on infected Johnsongrass and carry the viruses to corn.

Management

Management of MDM and MCD is based primarily on the use of resistant cultivars and elimination of Johnsongrass, the principal host. If these diseases have been identified in an area or have caused losses on a farm, resistant or tolerant cultivars should be grown. Resistant corn cultivars adapted to particular areas are generally available. The county extension agent has the latest information on hybrids suitable for a local area.

Particular attention must be paid to controlling Johnsongrass—in the field, fence rows, road sides, and other areas adjacent to corn fields. An efficient weed management program is essential for control of MDM and MCD.

A good corn production program must be followed: Prepare the land properly, use proper amounts of lime and fertilizer, plant corn at the correct time and depth, rotate it with other crops, and use herbicides if necessary.

As corn yields gradually creep upward, it may become feasible to use systemic insecticides on corn crops. Compounds that act both as nematicides and insecticides are now being applied to corn soils before planting. An unexpected benefit of such compounds may be the help they give in controlling the vectors of MDMV and MCDV.

VIROID DISEASES

For a hundred years or more, farmers have recognized that newly developed cultivars or introduced varieties of certain crops are subject to afflictions known as "running out," "decline," "reversion," or "senility." The cause of many of these *degenerative diseases* was unknown until 1972 when T. O. Diener showed that the potato spindle tuber disease was caused by a *viroid*. Since then, at least 13 plant diseases, including potato spindle tuber, citrus exocortis, chrysanthemum stunt, coconut cadang-cadang, and cucumber pale fruit have been shown to be caused by viroids.

Reports indicate that viroid diseases occur worldwide, and several are destructive and significant threats to crop production. So far, viroids have not been shown to cause animal or human diseases.

The symptoms caused by viroids in plants vary and are similar to those caused by virus infection. Symptoms include dwarfed or stunted plants, rolled or twisted leaves, and mottled or yellowed leaves. Potato tubers infected with potato spindle tuber characteristically are elongated with prominent bud scales and sometimes have severe growth cracks. Yields are reduced, often by 25% or more. Citrus exocortis may cause severe dwarfing, cracking, and scaling of the bark of the rootstocks.

Viroids are small, circular, single-stranded units of ribonucleic acid (RNA) that can infect plant cells, replicate in these cells, and cause disease. Viroids are only about one-tenth the size of viruses, and the "naked" viroid RNA, unlike that of plant viruses, is not enclosed in a protein shell. The way in which viroids multiply is not well understood. They seem to be associated with the cell nuclei.

Viroids are spread from diseased plants to healthy plants primarily by mechanical means. It is possible that viroids can be spread in plant sap carried on workers' hands and tools during propagation and other cultural practices and in propagation material such as cuttings or tubers.

There are severe and mild strains of viroids. Some viroids are transmitted through pollen and seed, but no insect vectors have been reported.

Viroid diseases can be managed by using viroid-free propagating materials, by eradicating viroid-infected plants, and by sterilizing tools to remove viroid-infected plant sap before they are used with healthy plants. Workers should wash their hands thoroughly after handling infected plants.

19

Diseases Caused by Parasitic Plants

Usually we think of parasites as lower forms of plant life such as bacteria, fungi or nematodes. However, a number of flowering or seed plants are parasites on other plants. Parasitic seed plants vary widely in their dependence on a host. The more independent ones, called half-parasites or hemiparasites, have chlorophyll and roots and can manufacture their own food; but they depend on their host for dissolved minerals and organic substances. Others, such as mistletoe, have chlorophyll but no roots and depend on their host for minerals and water. Some other parasitic seed plants, such as dodder, having neither chlorophyll nor true roots, depend entirely on their host for their existence.

Parasitic seed plants cause disease principally by depriving their hosts of water and nutrients and by disturbing the balance of growth regulators. The full spectrum of diseases caused by parasitic seed plants still is not known. Some of the important parasitic seed plants are discussed in this chapter and are summarized in Table 19.1.

TABLE 19.1. Diseases caused by parasitic plants.

Parasitic plants		
Common name	Scientific name	Host(s)
Broomrape	*Orobanche* spp.	Lettuce, tobacco, tomato, other vegetables, weeds
Witchweed	*Striga* spp.	Corn
Leafy mistletoe	*Phoradendron* spp.	Hardwood trees
Dwarf mistletoe	*Arceuthobium* spp.	Coniferous trees
Dodder	*Cuscuta* spp.	Many crops

BROOMRAPES

The broomrapes are leafless herbs that appear above ground as clumps of brown, purple, white, or yellow stems. Most broomrapes belong to the genus *Orobanche.* They are widely distributed.

Broomrape attacks lettuce, potato, tobacco, tomatoes, clover, and alfalfa and may live on weeds between crop plantings. The broomrape plant arises from the ground at or near the base of the host plant (Fig. 19.1). The stems are 15 to 50 cm high. Exudates from the growing roots of a host plant stimulate broomrape seeds to germinate. The slender root of the germinating seed attaches itself to the root of the host, forms a tuberous enlargement, and absorbs food from the host. As a result, the weakened host is stunted, does not grow properly, and has reduced yields.

A mature broomrape plant may produce up to one million dustlike seeds that are easily spread by soil, water, and wind. They may remain dormant in the soil for years.

Management

Broomrape is hard to manage. Hand weeding often is too expensive, the seed lives too long for rotations to be practical, and resistant cultivars have not been developed. Methyl bromide can be used to kill seeds in infested soil. In some developing

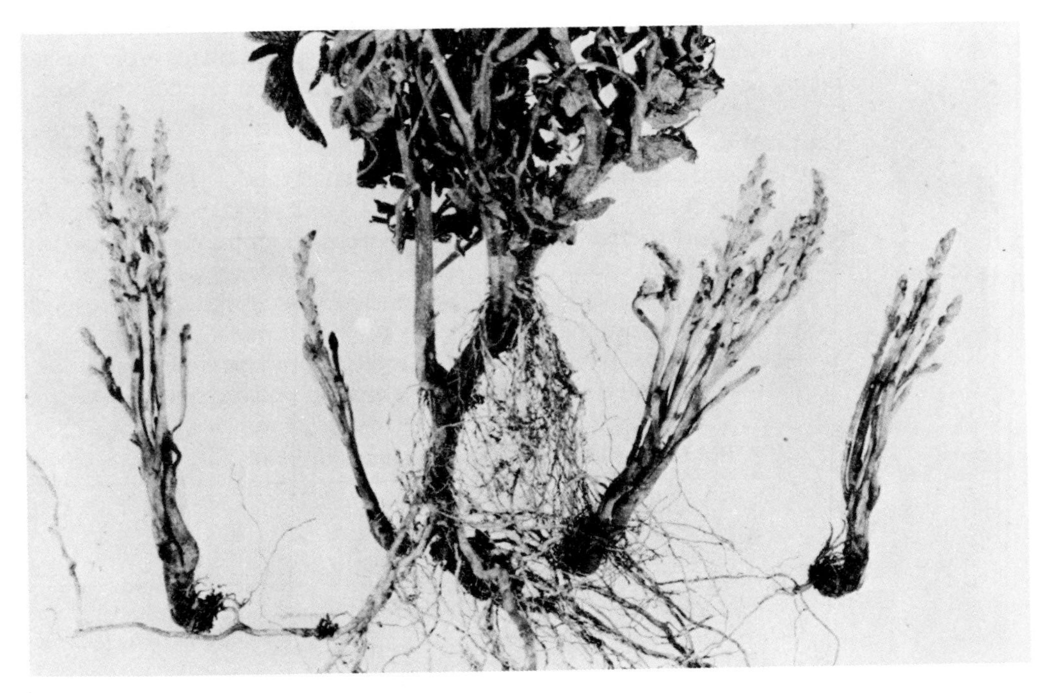

FIGURE 19.1. Four broomrape plants attached to roots of a tomato plant. The soil has been washed from the roots to show the attachment of the parasitic plant to the tomato roots. *Courtesy California Department of Agriculture.*

countries, geese are allowed to graze in infested fields. Two geese per hectare grazing on growing broomrape shoots are said to give good management.

The most promising management methods involve injecting the soil with chemicals that stimulate the seeds to germinate in the absence of the host, treating the soil with herbicides that kill the broomrape and not the host, or spraying the broomrape with a selective herbicide after it emerges from the ground.

WITCHWEED

Witchweed is a pretty plant with bright red flowers, which grows in clumps to about 50 cm in height. It looks innocent enough, but underground it is attached to the roots of a host plant that it robs of food and nutrients.

Witchweed belongs to the genus *Striga*. It is a serious pest of corn, rice, sorghum, sugarcane and tobacco; it also grows on crabgrass. It has been of serious concern in Africa, Asia, and Australia for many years. Witchweed was discovered for the first time in the United States in 1956 in North and South Carolina. As a result of effective quarantines, the parasite appears to have been limited to the area of original infestation. However, a constant watch is maintained for outbreaks in other states.

Plants attacked by witchweed look like they are suffering from drought. They are stunted, wilted, and yellowish. Heavily parasitized plants eventually die and produce little or no yield.

A single witchweed plant may produce half a million tiny brown seeds, which are easily spread by water and wind. Most seeds require a rest period before they will germinate. Chemicals that leak out of the host roots stimulate the seeds to germinate, and one of these stimulants is ethylene. Each seed produces a tiny rootlet that grows toward a host root, attaches itself, and draws sap from the host for its own nourishment. The witchweed plant then produces more rootlets, which attach themselves to other roots of the host. Sometimes several hundred witchweed plants may parasitize a single corn plant at the same time. Such infested plants often die prematurely, especially during dry weather.

It takes about 30 days after the seed germinates for the witchweed plant to emerge above ground, and another 30 days before it flowers. Seeds are produced within 5 to 6 months. Although the witchweed plant has leaves containing chlorophyll and probably can manufacture its own food, it depends upon the host for water, minerals, and organic substances as well.

Management

Several practices help growers to manage witchweed. Movement of seed to new areas on transplants, machinery, soil, or agricultural products should be prevented. Injecting soil with ethylene gas induces witchweed seed to germinate in the absence of a host plant. Ethylene also can be injected during the growing season in fields growing such nonhosts as cotton or soybeans.

Herbicides also help. They may be used in corn fields before witchweed emerges, or specific chemicals may be used after emergence. Seedbeds too can be treated with methyl bromide.

Trap crops of nonhost legumes also can be useful; the nonhost plants stimulate the germination of seeds. Because the witchweed cannot infect the nonhost plants, the seedlings literally starve to death.

MISTLETOES

Almost 2300 years ago, the Greek botanist Theophrastus wrote down observations that showed he recognized mistletoe as a parasitic plant. Mistletoe has been featured in European legend for centuries. It was considered to have unusual medical and magical value and frequently was used in religious rites. The early American settlers recognized mistletoes in the hardwood forests of the eastern United States and perpetuated the legends and symbolism that had developed in their homelands. Today, mistletoe is highly prized for Christmas decorations. It is often hung in doorways during the holiday period, and those who stand beneath it may receive an unexpected kiss or some other blessing.

There are two kinds of mistletoe, leafy and dwarf. In most parts of the United States the leafy mistletoes are scarce and regarded more as botanical curiosities than as serious pests. Most of the mistletoe for the Christmas trade is gathered in the southern United States, where it is abundant in some localities and provides off-season income for farm workers. Most leafy mistletoes found in the United States belong to the genus *Phoradendron.* They grow principally on hardwoods but also parasitize cedar, citrus, cypress, juniper, and pecans! The leafy mistletoe is spread from tree to tree by birds. *Phoradendron* spp. cannot stand cold winters and seldom occur in the forests of the northern United States (Fig. 19.2).

The dwarf mistletoes are a serious pest in pine forests of the western United States. Trees of any age may be dwarfed, deformed, or killed. Timber quality is reduced, and affected trees produce less seed and are subject to wind breakage, insects, and other diseases. Seedlings and young trees often are killed by dwarf-mistletoe infections.

Infected branches are easily recognized by the branched shoots of the mistletoe plants, which occur in tufts or distributed along the twigs of the host. Diseased limbs become swollen and branch excessively, forming witches' brooms. Heavily infected stands contain deformed, stunted, and dead trees or trees broken off at trunk cankers.

Dwarf mistletoes that parasitize conifers belong to the genus *Arceuthobium;* all species are regarded as hemiparasites. They produce clusters of green, perennial, jointed stems, up to 10 cm long, on the branches and trunks of the host. Usually the inconspicuous leaves are green, leathery, and scalelike. The stems are supplied with water and nutrients from a rootlike absorbing system, made up of haustoria, which developed in the bark and wood of the host. The flowers are either male or female. Pollinated female flowers produce shiny, white explosive berries in late fall or early winter. Within each berry is one seed coated with a sticky substance called *viscin.* When the fruits are ripe, they develop an internal pressure. If fruits are touched or disturbed, the ripe seed may be shot 12 m or more away. The sticky seeds adhere to anything they touch. Birds also feed on the pulp of the berries and carry sticky seeds from one place to another. The seeds may germinate within a few weeks or delay until the next spring.

FIGURE 19.2. Clusters of mistletoe plant growing on an oak tree. *Courtesy L. F. Grand.*

A mistletoe plant removes water and nutrients from its host and saps the vitality of the entire tree. It also upsets the hormonal balance, with resultant swellings and deformities of the limbs and excessive branching.

Management

Clear cutting (the practice of cutting every tree of a given kind or species in a selected area) is one method of reducing losses caused by dwarf mistletoe, followed by burning to eliminate any undetected infection. In the past, forest fires helped keep mistletoe in check, but with improved fire control mistletoe continues to spread. For large infection centers where burning all infected trees is not feasible, it is necessary to plant a nonsusceptible host between infected trees and newly cleared areas. In areas where dwarf mistletoe is a problem, trees should not be cut when berries are ripe and seed dispersal is in progress. Pruning of affected branches may be effective in high-value stands if infection is light.

Chemical management of dwarf mistletoe with herbicides is being investigated. Biological management using fungi that attack the mistletoe or insects that feed on it also show some promise. Resistant cultivars are not available.

Management of leafy mistletoe is similar to that for dwarf mistletoe. Individual, infected branches can be removed from valuable trees.

DODDER

Among the unique types of seed plants that parasitize green plants are the dodders, members of the genus *Cuscuta*. There are more than 100 species, which grow almost everywhere within the temperate and tropical zones. Dodder plants appear as conspicuous tangles of intertwined yellow threads on the aerial parts of host plants. Sometimes dodder is tinged with red or purple. Occasionally it is almost white. The

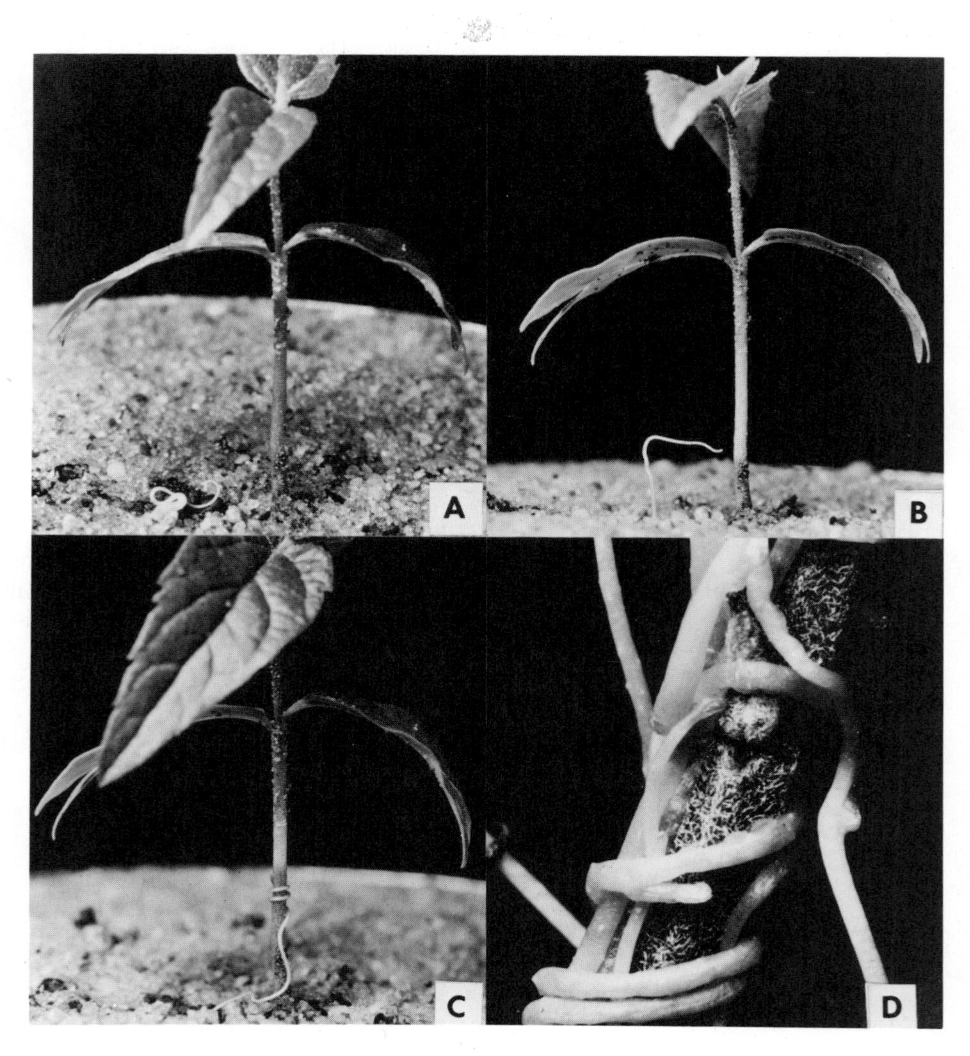

FIGURE 19.3. Growth of dodder on a host plant. (A) Sprouting seed with tendril-like filament. (B) 12 hr later, the filament has extended upward and outward. (C) 24 hr later, the dodder has twined itself around the host's stem. (D) Close-up of dodder on chrysanthemum, showing coils and haustoria. *Courtesy R. E. Hutchins.*

plant has a number of unusual and poetical names such as devil's hair, gold thread, hair weed, hell vine, love vine, pull down, and strangle weed.

Dodder is particularly troublesome in regions where clover and alfalfa are grown extensively. Dodder reduces the growth and the yield of infected plants and clogs harvesting machinery. In some European countries production of clover seed is difficult because of dodder. Other crops that suffer losses are flax, lespedeza, onions, potatoes, sugar beets, and several ornamentals. Cereal plants never are attacked.

Dodder seed are gray to red-brown and somewhat smaller than alfalfa seed. The seeds overwinter as contaminants among crop seeds or in the soil where they may survive for 5 years. When a seed germinates, it produces only a rudimentary root system and sends up a yellow shoot whose tip grows in a spiral fashion (Fig. 19.3). If the shoot touches the stem of a susceptible plant, it twines around the stem and quickly produces haustoria, which penetrate the stem and absorb juices from the plant. The dodder continues to grow and may cover many nearby plants with its colorful net of threadlike stems. If the dodder shoot finds no host, it may live for several weeks and then die of starvation.

Management

Using dodder-free seed and transplants and fumigating seedbed soils are important steps in managing dodder. So is thorough cleaning of equipment before moving it from dodder-infested to dodder-free soils. Animals should not be grazed in dodder fields and then moved to dodder-free soils, as the seed is spread in the manure. Spot infestations can by sprayed with contact herbicides. Heavily infested fields can be treated with soil herbicides.

20

Abiotic Agents

Some plant diseases are not caused by living agents, but are due to an upset in the environment—usually too much or too little of something necessary for growth—or the presence of some harmful substance in contact with the plant. We call these diseases *abiotic diseases* or *noninfectious diseases*. Noninfectious diseases occur in the absence of pathogens and thus cannot be transmitted from one plant to another. These diseases can occur at any stage from seed to mature plant; they may cause damage in the field, during storage or transit, or at the market. Symptoms may range from slight to severe, and affected plants may be killed.

Like every other living thing, plants require specific environmental conditions to prosper. Some plants thrive under moist conditions; others do well in dry atmospheres. Some plants grow fastest in bright sunshine; others need shade. Some require fertile soils; others do not. Whenever the proper balance of temperature, soil moisture, light, purity of air, nutrients, and pH is upset to a smaller or a larger degree, this imbalance is reflected in the growth of plants. Moreover, cultivated plants frequently are grown in artificial environments or subjected to cultural practices (e.g., fertilization, irrigation, pesticide, or hormone treatments) that may affect their growth considerably.

DEFICIENCIES AND TOXICITIES

Nutrient Deficiencies

Plants require mineral elements for normal growth. Nitrogen, phosphorus, potassium, calcium, magnesium, and sulfur are needed in large amounts and are called *major elements*. Iron, boron, manganese, zinc, copper, molybdenum, and chlorine, needed in very small amounts, are called *trace* or *minor elements.* When one or more of these nutrients are lacking or deficient, plants become diseased and exhibit various symptoms that may appear on all organs, including flowers, fruit, leaves, roots, seed, and stems.

Certain symptoms are the same regardless of the missing element, but other diagnostic features usually accompany a deficiency of a particular nutrient. The lack of normal amounts of most essential elements usually results in a reduction of growth and yield. As the deficiency becomes greater, symptoms become more pronounced, and affected plants may even die. There are characteristic symptoms of specific deficiencies: nitrogen-deficient plants grow poorly and are light green in color; calcium-deficient plants have young leaves that are distorted with hooked tips and curled margins; boron-deficient plants often show brown or dead patches of tissue in roots or fleshy organs.

Nutrient deficiencies can be corrected by fertilization with the missing elements. When a deficiency is suspected, samples of soil and plant tissues should be sent to a soil analysis laboratory, which can determine what elements are lacking. Based on the test results, growers can devise fertilization programs to supply the missing nutrients.

Nutrient Toxicities

One of the biggest problems of the home gardener or owner of house plants is overfertilization (Fig. 20.1). Fertilizer, when applied in excessive amounts or placed incorrectly, will "burn" roots. Root burning occurs when the concentration of salts or fertilizer in the soil water becomes so high that water is drawn from inside the roots (*reverse osmosis*), which turn brown (burn) and die. Evaporation can concentrate salts in the root zone and also burn roots. This often happens with a potted plant. Also, salt applied to icy roads in winter can damage roadside plants the following spring when the dissolved salt washes off the road.

In addition, there should be a proper balance of nutrients in the soil. If certain nutrients are present in excess, they may interfere with the uptake of others.

Soil analyses help determine if certain elements are present in excess. Also, the level of soluble salts can be determined, which indicates if too much fertilizer has been applied. Homeowners should be careful not to overfertilize potted plants, shrubbery, or lawns.

Acid and Alkali Soils

Most plants will not grow in extremely acid (sour) or extremely alkaline (sweet) soils. The acidity or alkalinity of soil is measured by a scale of pH units on which 7 is neutral, 0 if the most acid, and 14 is the most alkaline. Most soils range between 4 and 8 on the pH scale. In acid soils, mineral salts become more soluble and may become so concentrated that they are toxic to plants or interfere with the absorption of other necessary elements. Symptoms of mineral deficiency then develop.

Excessive amounts of certain salts raise the soil pH and cause alkali injury. This often happens when desert soils are irrigated.

FIGURE 20.1. The poinsettia plant is dying from too much fertilizer. The high level of soluble salts in the soil causes water to be drawn from the plant roots (reverse osmosis), and the plant wilts and dies. *Courtesy N. C. Cooperative Extension Service.*

It is relatively easy to determine the pH of soil with instruments called pH meters. Most soil laboratories have them. All farmers and gardeners should know the pH of their soil. Once the pH is known, steps can be taken to change it to a range favorable to plant growth. Acid soils can be corrected by spreading lime on the land.

WEATHER EXTREMES

Frost and Cold

Low temperatures injure plants primarily by causing ice formation, between and/or within cells, that injures membranes and other cell components. Low temperatures, even if above freezing, may damage warm-weather crops such as corn and beans. Plants suffer a variety of injuries from temperatures below freezing. Late frosts kill tender plants, buds, leaves, or flowers. Fleshy tissues such as potato tubers become black and blotchy. Low winter temperature may kill roots of trees and also cause bark-splitting and cankers.

Rapid decreases in temperature cause more damage than slow decreases. Rapid increases in thawing have the same effect. Frost-injured plants wilt quickly as soon as the sun hits them, and in a few days they turn brown.

Valuable plants can be protected from extremes of temperature by planting them at the proper time. Proper shading, covering, mulching, irrigating, fertilization and heating (during cold snaps) will also help. Also, some bacteria serve as nuclei for ice crystals or frost. Through biotechnology scientists have created strains of these bacteria that will not serve as ice nuclei. Application to plants of the ice-minus bacteria may help reduce the likelihood of frost damage in the future.

Heat

Leaf scald ("sunburn" or "sun scald") occurs during hot weather after a plant has made rapid growth, especially after periods of rainy, cloudy weather. Large, irregular, water-soaked or dead areas may form on the sun-exposed sides of fleshy fruits and vegetables, such as apples, tomatoes, onions, and potatoes, Succulent leaves also may be injured.

Hot-weather injury is increased by the excessive light, drought, high winds, and low RH that usually accompany high temperatures.

Lightning

Lightning frequently strikes in cultivated fields. Growers often mistake lightning injury for an outbreak of some new and destructive disease, because the area has the appearance of a spot in which a disease is spreading (Fig. 20.2). The affected area in a

FIGURE 20.2. Lightning injury on cotton may cover an area 15 m in diameter. Plants near the point of the lightning strike usually wilt and die; plant damage decreases near the outer edge of the strike zone. Symptoms can easily be confused with high application rates or carryover of herbicides. *Courtesy N. C. Cooperative Extension Service.*

field usually is circular and may be 15 m or more in diameter. The plants in the center of the area may be killed to the ground line, the injury being less severe toward the borders. Roots of injured plants show a broad brown streak from the soil level upward along one side of the stem into the midribs of the leaves. The injured midribs and veins on the lower leaf surface turn black and collapse. Some leaves may be brown and dry at the tips along the midveins. The uneven growth of the injured midrib and the normal leaf blade results in leaves becoming puckered and distorted. Also the continued growth of the uninjured side of the whole plant and the simultaneous failure of the injured side to elongate may cause the stem tip to turn and twist. The leaves may be green for several days after the stalk and midribs are shriveled and dead. When the stalk of an affected plant is split open lengthwise, the pith is found to be separated into disks, giving a peculiar ladder-like effect.

ABNORMAL WATER CONDITIONS

All living things are composed mostly of water—often 80 to 95% by weight—and require water to live. It acts as a solvent for and transports food and food materials from soil into a plant and from cell to cell throughout a plant. Also water serves as one of the crude materials in the formation of sugars by photosynthesis. It preserves turgidity within cells. It permits transpiration and thus promotes and regulates growth.

Plants are very sensitive to moisture; too much water or too little will cause injury. Water must be available at the correct time in the correct amounts. The health of a plant is affected by the moisture supply of the soil and air, with the amount of water needed for proper growth influenced by temperature, sunshine, soil structure, and wind.

Drought

The first symptom of water shortage in plants is wilting, which results in drooping of succulent shoots and rolling of the leaves. This is a common sight on a hot, summer day. Plants may recover during the night, but if the lack of water continues, they become dwarfed and stunted. The leaves yellow or redden, begin to die along the margins and tips, and finally drop off. Without sufficient water, root crops remain small; cereals produce shriveled grains; fruits become spotted, deformed, or stunted and may shrivel and fall prematurely. Prolonged drought leads to plant death; also drought may increase the damage from certain root diseases.

Obviously, the solution to a shortage of water is to supply more of it. Nature, however, does not always cooperate; farmers gamble each year that there will be enough rain at the right times. In other areas, irrigation is widely used to supply water at the proper time. However, irrigation is expensive and can be used profitably only on high-value crops.

Sometimes house plants, greenhouse plants, or seedbed crops are neglected, and the plants become dry and wilted. If so, they should be adequately watered as soon as possible.

Drowning

Most cultivated plants cannot obtain enough oxygen from water; hence, a soil saturated with water will not have enough oxygen for the root system—and oxygen is necessary for life. When soils are flooded, poorly drained, or water-logged, many plants that cannot stand "wet feet" will grow poorly, drown, or die.

However, lack of oxygen is not the entire explanation of injuries due to too much water. Roots injured from oxygen deficiency often are attacked by bacteria. The dying root cells and bacteria produce toxins that injure the plant and prevent proper absorption of water; the plants dry out, and the leaves wilt and droop. Injury is more rapid at higher temperatures, probably because roots require more oxygen and bacteria grow faster at higher temperatures. All degrees of injury may result, including destruction of the entire root system and complete wilting to death of only a few of the deeper roots—in which case only temporary wilting occurs.

Drowning can be prevented by good drainage; thus, management of erosion and drowning are closely connected. For row crops, the rows should be located to decrease runoff and soil loss, low areas should be tiled or drained, and terraces should be maintained properly with adequate ditches provided.

Seeds sown too deeply or plants set too deeply in wet soil often suffer from drowning injury. Adequate drainage should always be provided.

Homeowners should avoid overwatering and possibly drowning their plants.

PHYTOTOXIC CHEMICALS

Pesticides

Herbicides have come into wide use in the last 50 years. These chemicals are used to kill weeds in many different crops, and are highly specialized: chemicals suitable for one type of crop may not be applied to another. The rapid introduction and the increasing use of chemical weed management in crop production have led to dramatic increases in crop yields, but also have resulted in wide misuse and carelessness by some growers and homeowners. Plant injury (Fig. 20.3) may occur.

Occasionally, the misuse of pesticides also results in plant injury. Many insecticides and fungicides are now mixed together in the same tank and applied simultaneously. This step saves time and money, but some pesticides are incompatible; that is, when they are mixed, reactions occur that reduce their effectiveness and also may cause injury to plants. Often, the injury may be caused in part by the diluents, solvents, or carriers used to dilute the insecticide or fungicide.

Symptoms of Pesticide Injury

Symptoms of herbicide injury vary widely (Figs. 20.4 through 20.6). Malformation, stunting, yellowing and/or death of leaves and shoots quickly develop, often overnight or in a few hours. Although herbicide injury may be confused with virus symptoms, it usually is possible to determine whether herbicides might have been used nearby and caused the damage.

FIGURE 20.3. Excessive rates herbicide application the previous year along the edge of this field caused severe stunting and purpling of the corn plants, a condition similar to phosphorus deficiency. Injured plants have a greatly reduced root system that cannot absorb enough phosphorus to supply the plant's needs. *Courtesy N. C. Cooperative Extension Service.*

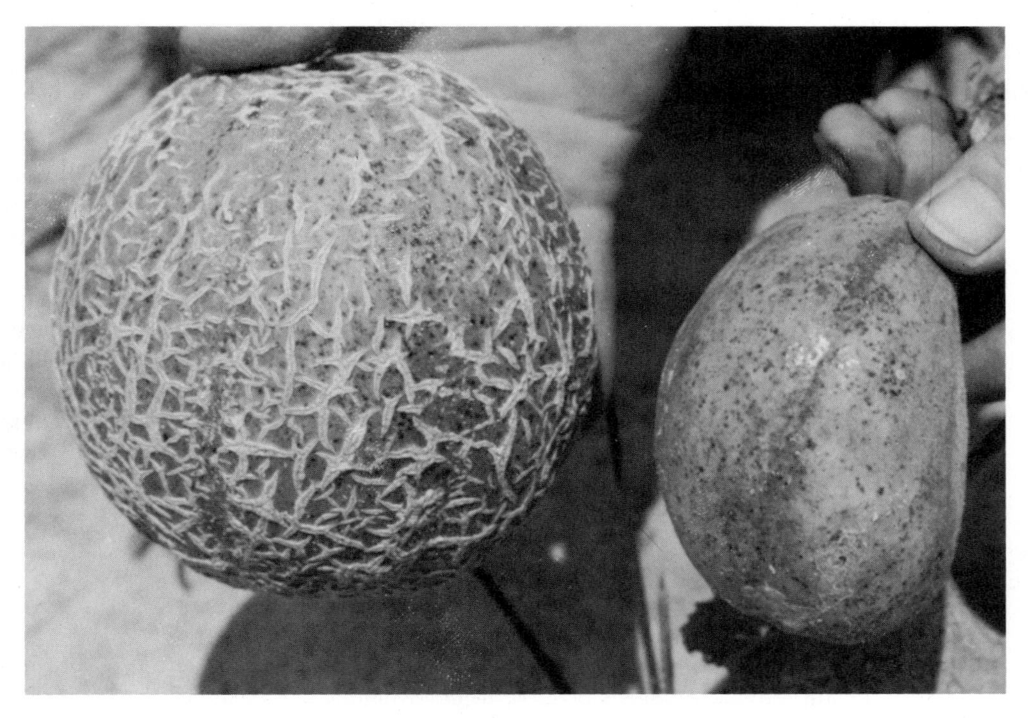

FIGURE 20.4. Malformation of the cantaloupe on the right was caused by an over-the-top postemergence application of a herbicide. *Courtesy N. C. Cooperative Extension Service.*

FIGURE 20.5. Veinal yellowing of grape leaves can be caused by an overdose of some herbicides. This injury usually occurs during a wet season. *Courtesy N. C. Cooperative Extension Service.*

Other types of injury include dead, burned, or scorched spots on or at the tip of leaves; russeting of fruit; misshapen fruit, leaves, or plants; off-color and/or yellowing or delayed development; poor germination; and complete death of plants. Unfortunately, these symptoms of chemical injury often do not appear until several days after exposure to a herbicide. If they do appear, they are not always clear-cut and often may be confused with other problems. It may be necessary to call on experienced individuals to help identify the cause of the symptoms.

Symptoms of insecticide and fungicide injury to plants are generally similar to those of herbicide injury. They include stunting, yellowing, leaf burn, malformation, and death of certain tissues. The degree of injury often is influenced by temperature and RH.

FIGURE 20.6. Spray drift or vapor drift from the growth regulator herbicide 2,4-D caused this injury on tomato leaflets. Notice the curved feather-like leaves. *Courtesy N. C. Cooperative Extension Service.*

Management of Pesticide Injury

There are three principal causes for chemical injury to crop plants: misuse and misapplication of pesticides, drifting of pesticide vapors beyond the target area, and pesticide contamination of soil and water. The most effective management of chemical injury is to avoid it by applying appropriate pesticides at the proper rates and in such a way as to reduce drift and contamination of soil and water. See Chapter 7 for a detailed discussion of the proper use of pesticides and the problems associated with their misuse.

AIR POLLUTION

Air pollution is one of the most serious environmental problem facing the United States and the world. As the world's population has grown and urbanization and industrialization have increased with the mounting energy demands of transportation, industry, and electricity, air pollution also has increased; and it is now one of the critical problems in the survival of humankind.

FIGURE 20.7. A power plant in the southwest corner of Poland near the German and Czechoslovakian border emits a high concentration of sulfur dioxide in smoke, from the burning of high-sulfur (up to 18%) coal mined in that region. Clean air laws in the United States make it impractical to legally burn coal containing more than 1% sulfur. Thus, it is easy to understand why air pollution in eastern Europe is much greater than it is in the United States. *Courtesy H. Morton.*

Many health and agricultural problems are due to dirty air. Air pollution causes acid rain, ozone injury, smog, haze, lung disease, and the greenhouse effect. Dirty air damages trees and crops, obscures scenic views, pollutes streams and lakes, corrodes buildings and statues, and contributes to global environmental problems.

Air pollution is a community problem, particularly in large cities. As long as we have automobiles and smokestacks (Fig. 20.7) producing an abundance of smog and other toxic substances, plants will become injured and die prematurely. The only sensible management seems to be to reduce the amount of poisonous gases released into the air. This feat will be difficult. It will cost much money. It will take community, national, and international action. Meanwhile, people and plants will suffer.

Pollutant Damage to Plants

Leaves of growing plants are subject to a great variety of diseases that are nonparasitic. Symptoms vary, and diagnosis sometimes is difficult because the symptoms may be mistaken for and confused with those of parasitic leaf spots or leaf spots that are the result of unbalanced nutrition, such as potassium or phosphorus deficiency.

FIGURE 20.8. Ozone injury on a corn leaf. The symptoms can be confused with those of damage caused by postemergence application of some herbicides. *Courtesy N. C. Cooperative Extension Service.*

Probably all types of plants are susceptible to ozone (O_3) injury, also known as weather fleck (Fig. 20.8). Ozone injury has been shown to occur on 28 plant species. Alfalfa, bean, corn, ornamentals, grapes, spinach, small grains, soybeans, and tobacco are sensitive. Annual economic losses for the United States are estimated at $2 to $5 billion for major economic crops. Comparable or greater losses probably occur in Europe.

Weather fleck appears on leaves as numerous, irregular, water-soaked, small to large spots, flecking, stippling, mottling, or chlorosis. On soybean leaves, the spots may be purple, on clover the leaves turn yellow, grapes leaves become bronzed or

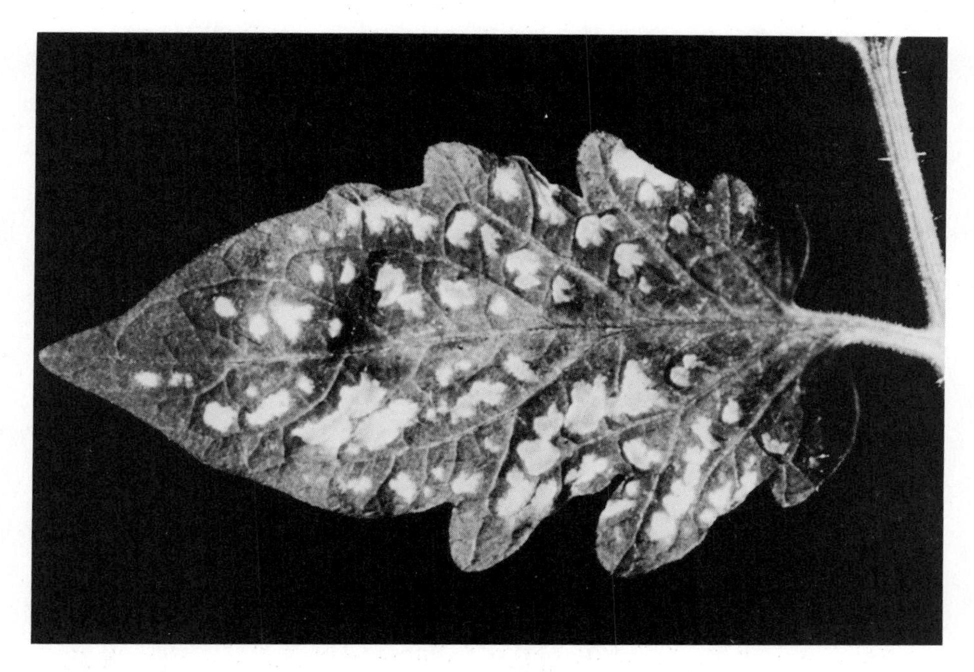

FIGURE 20.9. Injury of a tomato leaflet by exposure to sulfur dioxide. The injured areas are light tan in color. *Courtesy R. A. Reinert.*

yellowed, and on pine the needles drop prematurely. In 48 hours or less after exposure to ozone the lesions change from brown to white and become necrotic. When lesions coalesce, leaf tissues fall out, producing a ragged shot hole effect, or, in severe cases, only the midrib and veins remain. Foliar symptoms on crops often appear as early senescence and may be difficult to assess. The response of many crops to ozone is affected by the presence of other pollutants.

Weather fleck is caused by phytotoxicants in the air, with ozone being the principal agent. Ozone is a natural component of the atmosphere, and is a powerful oxidant. Much of its injurious effect may be due to this property. Ozone is generated by the photochemical action of solar ultraviolet radiation, and nitrous oxides and hydrocarbons when irradiated by sunlight also react to form ozone. However, other pollutants are involved as well. The synergistic action between ozone and sulfur dioxide (Fig. 20.9) enables a mixture of the two gases to cause symptoms although the critical concentration of each is below the individual threshold levels. The effluent ("smog") from automobile exhausts and industrial and domestic furnaces injects large amounts of hydrocarbons, nitrogen dioxide, sulfur dioxide, and dust into the atmosphere, where they become a source of photochemical air pollution. Smog can be blown many kilometers and injure the tender tissue of rapidly growing plants by causing leaf burning, leaf spots, silvering, specking, flecking, or rusting. Trees, flowers, and many other plants are very susceptible to smog. Affected plants age early, and their foliage become discolored. Sometimes they are killed (Fig. 20.10).

FIGURE 20.10. Trees in Eastern Europe killed by air pollution and acid rain. *Courtesy H. Morton.*

The polluted air must enter the leaf through open stomates. This is more likely on turgid leaves exposed to sunlight. Once inside the leaf, the air pollutants disrupt the carbon cycle of the plant, increase respiration, reduce photosynthesis, alter sugar metabolism, interfere with the distribution of metabolic products, and reduce starch reserves that supply energy for normal growth.

The severity of weather fleck depends upon the interaction of several factors: (1) the frequency and magnitude of ozone concentration and duration; (2) moisture, nutrient level, and plant density; (3) light intensity, RH, air and soil temperature, moisture, and rainfall immediately preceding, during, and after ozone exposure; (4) plant health, vigor, age, and stage of leaf development; (5) the cultivar; and (6) the interaction of air pollutants with other pathogens.

Management is difficult. The best management strategy is to avoid the use of cultivars sensitive to ozone injury. Also, sensitive crops should not be grown near cities where smog and polluted air are prevalent. One long-term solution that has been suggested for polluted air is to substitute nuclear power plants for coal and oil to generate electricity, as nuclear plants do not emit significant amounts of air pollutants. However, radioactive waste must be disposed of safely, and this, too is a cumbersome, expensive task. Therefore, other means of generating and conserving energy must be implemented. These include solar, wind, hydroelectric, and other renewable energy sources. An equally important solution is to implement and enforce programs that conserve energy resources and environmental quality. Automobiles that get better gas mileage and more efficient machinery are needed. People also need to conserve energy whenever possible.

The attempt to solve the needs of booming population growth and search for improved living standards around the world while keeping contamination to a minimum is a formidable undertaking.

Acid Rain

When pollutants in the air such as sulfur and nitrogen oxide combine with moisture in the upper atmosphere, chemical reactions occur, and acid rain forms. In this way the pH of the rain is lowered. The lower the pH of the rainwater, the more acidic the rain. Rain with a pH of 1.5 has been reported in West Virginia, which is more acidic than vinegar with its pH of 3.0. Acid rain is a by-product of other pollutants, a result of the combustion of coal, oil, and natural gas and the smelting of sulfide ores. It is created when oxides of sulfur and nitrogen react with water to form acid. Sulfur dioxide and nitrogen oxide are the primary causes of acid rain.

Acid rain contributes substantially to water pollution. Many pollutants eventually fall from the atmosphere and end up in streams, lakes, swamps, or coastal waters where they kill microorganisms, plants, and fishes. These poisonous fumes are discharged into the air from smokestacks and automobile exhausts.

The occurrence of acid rain is becoming more frequent throughout the world as the air becomes more polluted. Scientists believe that acid rain may affect plants and plant diseases. In some instances, plants may be stressed by acid rain, and the yield may be reduced. Acid rain may interact with plant pathogens and either reduce or increase the amount of damage from diseases.

The effects of acid rain are subtle and may not be immediately noticed. Although many scientists believe that acid rain damages some crops, it is difficult to separate the damage caused by acid rain from other factors (insects, nutrition, pathogens) that may be contributing to the problem.

Global Climate Change

Ozone in the stratosphere (above 8 miles) protects the earth from harmful solar radiation, principally ultraviolet radiation. On the other hand, too much ozone in the troposphere (everything below 8 miles) acts as a serious pollutant to plants and animals.

The use of chlorofluorocarbons (CFCs), such as the Freon used in refrigerators and air conditioners, solvents, and foam insulation, has increased. These stable chemicals have extremely long lives. They leak out of air conditioners and drift up into the upper atmosphere, where intense solar radiation breaks apart the CFC molecules and releases chlorine atoms that destroy the ozone at a rapid rate. International agreements are now in place to drastically reduce the use of CFCs, but studies have shown that stratospheric ozone levels have been decreasing since 1969, particularly over the North and South Poles, in some years by as much as 3%. A serious hole in the ozone layer over the antarctic region has been observed annually since the late 1980s. The diminished ozone layer allows more of the sun's ultraviolet radiation to penetrate to the earth's surface, where it may be harmful to animals and plants. Increased ultraviolet radiation would increase risks of cancer and eye damage in humans and could increase the susceptibility of plants to disease.

Another aspect of global climate change is the potential for global warming. Many scientists believe such warming or a "greenhouse effect" will result from the burning of fossil fuels (coal, natural gas) and deforestation, which add carbon dioxide and other gases (methane, nitrous oxides) to the atmosphere every year. The gases trap the sun's energy, much like the glass in a greenhouse, raising the earth's temperature.

The consequences of global climate change may be staggering. If the buildup of heat-absorbing gases continues, as it will with increasing population and energy demands, the global climate will change. Global precipitation will increase, but the effect will not be uniform.

Growing seasons in the higher latitudes would become longer and warmer, a change that would shift the suitable ranges of temperature and humidity for rice, corn, wheat, and other crops away from the tropics.

Increased temperatures would increase evaporation and the loss of water from soils and plants. Semiarid or desert areas would increase. More irrigation water would be needed. In some areas, water supplies are already overdrawn.

Increasingly violent storms would occur, especially in deforested areas, causing increased runoff, erosion, and loss of valuable land. If deforestation continues unabated, the land will become so degraded through soil erosion and watershed damage that it will no longer be able to support even a low standard of living. Such land would be largely worthless. Disease incidence, in both plants and animals, would change. The range of tropical plants, diseases, and insects would be extended.

The warming would cause large sea level rises (maybe as much as 3 feet in the twenty-first century). Many coastal areas would be flooded, and hurricanes would become stronger and more devastating. The rising sea level would push salt water farther up into freshwater streams and into underground aquifers. Salt marshes would disappear or retreat inward. Rising seas would devour beaches, moving the water line several hundred feet inland in Maryland, Virginia, North Carolina, South Carolina, Georgia, and Texas, and several miles inland in Florida, Alabama, and Louisiana. The loss of wetlands would diminish flood control, water conservation, and fish and wildlife preservations. Flooding of beachfront property would increase. The economic losses that could occur are difficult to calculate.

In the realization that increased ultraviolet radiation due to depletion of the ozone layer and accelerated global warming are likely to cause severe and unacceptable costs to human societies and the planetary environment, strategies and ways to pay the costs to ameliorate these trends should be devised.

Global warming is an international problem that will require the mutual efforts of all nations to solve it adequately. Strategies to prevent global warming require cooperation of the scientific, business, and political communities worldwide. Such strategies include taxation, regulation, lending and investment, guidance of government-owned and private companies, and funding of research and development. Strategies to limit the buildup of greenhouse gases must be integrated with clean air, flood control, deforestation, shore protection, environmental protection, manufacturing, employment, and overall economic policies, development programs (particularly third-world), and geopolitical planning. Failure to integrate these strategies will diminish or cancel their effectiveness.

Prospects for the Future

Air pollution, acid rain, and global climate change are popular topics of discussion in the press, on television, or in government and industry. The accumulation of contaminates in our environment since the beginning of the industrial revolution occurred gradually for a long time, but now is accelerating rapidly. The range of toxic compounds released into the environment also is increasing.

More frequent and more severe damage to plants will occur in the future if steps are not taken to reduce pollutants and acid rain, to slow the addition to the atmosphere of gases associated with global warming, and to preserve the stratospheric ozone layer. The increase in damage to plants will occur directly from exposure to pollutants or as a result of the interactions of pathogens and pollutants. Also, if the climate changes within an area, different plants and pathogens then may exist in that area. Research will be needed so that planners can understand and predict the problems that will arise as a result of environmental changes.

The preservation of the environment and of the quality of life we have come to expect is possible and essential. Every farmer, nursery operator, commercial grower, and backyard gardener can play a significant role in reducing the pollution of the air, soil, and water. The responsible use of agricultural chemicals, such as pesticides and fertilizer, the conservation of energy, and the realization of the importance of the environment are the responsibility of each person. The challenge, as René Dubos stated it, is to "Think globally and act locally."

Glossary

The terms included here usually appear in text discussions. Some specialized terms not listed here can be found in the Index.

Abscission. The formation of a layer of cells that results in fruit, leaf, or stem being separated from a plant.
Acervulus (pl. acervuli). An asexual, saucer-shaped fruiting body in which conidia are produced on short conidiophores.
Acid rain. Precipitation with low pH due to the presence of sulfuric and nitric acids; formed by the combination of air pollutants and water.
Acre. An area of land containing 43,560 ft^2 equal to 0.40 hectare.
Actinomycetes. A group of microorganisms similar to bacteria that produce long filaments.
Adjuvant. In relation to pesticides, any substance added to a formulation to improve its physical or chemical properties.
Adventitious roots. Roots that appear in unusual places, e.g., on the stem.
Aeciospore. A dikaryotic spore of a rust fungus produced in an aecium.
Aecium (pl. aecia). A fruiting body of the rust fungi in which aeciospores are produced.
Aerobic. Relating to the growth of organisms in the presence of molecular oxygen.
Alternate host. One of two different types of plants on which some pathogens may develop to complete their life cycle or to survive between seasons.
Amino acids. The building blocks of proteins. There are 20 common amino acids; they are joined together in a strict order that determines the character of each protein.
Anaerobic. Relating to biological processes that occur in the absence of molecular oxygen.
Anion. An ion having a negative charge.
Annual. A plant that completes its life cycle from seed to seed in 1 year.
Antagonism. The phenomenon that results in a depression of activity of one organism when two or more organisms occur in close proximity.

Anthracnose. A leaf- or fruit-spot disease caused by fungi that produce spores in an acervulus.
Antibiotic. A substance produced by one microorganism that inhibits or kills other microorganisms.
Antibody. A protein produced by the immune system in a warm-blooded animal in reaction to a foreign antigen, such as a pathogen; antibodies react specifically with the antigen.
Antidote. A substance used to counteract the effects of a poison.
Antigen. A protein that induces the production of antibodies by the immune system of warm-blooded animals.
Antiserum. The blood serum of warm-blooded animals that contain antibodies.
Apothecium (pl. apothecia). An open cup- or saucer-shaped structure in which some fungi in the Ascomycotina (ascomyceti) produce asci and ascospores.
Appressorium (pl. appressoria). The swollen tip of a fungus hypha that attaches to and leads to the penetration of the host.
Aseptic. Free of microorganisms.
Ascocarp. The fruiting body of some fungi (Ascomycetes) that contains asci.
Ascomycotina (Ascomycetes). A group of fungi that produce sexual spores called ascospores.
Ascospore. A sexual fungal spore produced by ascomycetes.
Ascus (pl. asci). A microscopic structure containing ascospores (usually eight).
Asexual. A type of reproduction that does not involve the union of gametes; not sexual.
Auxin. A plant growth-regulating substance that controls cell length.

Bacillus. A short rod-shaped bacterium.
Bacteriocide. A chemical that kills bacteria; usually an antibiotic.
Bacterium (pl. bacteria). A prokaryotic microorganism that is usually single-celled and increases by binary fission.
Basidiomycotina (Basidiomycetes). A group of fungi that produce sexual spores called basidiospores.
Basidiospore. A sexual fungal spore produced by basidiomycetes.
Basidium (pl. basidia). A fungal structure on which basidiospores are produced.
Biennial. A plant that completes its life cycle in 2 years. The first year it produces leaves and store food; the second year it produces fruits and seeds.
Bioassay. Determination of the relative strength of a substance (such as a drug) by comparing its effect on a test organism with that of a standard preparation.
Biocide. A chemical that is toxic to many organisms, including both plants and animals.
Biological management (control). Techniques for managing pests by use of predators, parasites, and disease-producing organisms.
Biosynthesis. The production of a chemical compound by a living organism.
Biotechnology. The use of biological systems and processes in industrial production; genetically engineered microorganisms, plants, or animals.
Blight. A disease that causes rapid killing of many leaves, flowers, and stems.
Blotch. A disease with irregularly shaped spots on leaves, fruits, and stems.
Broad-spectrum pesticide. A chemical that controls a wide range of pests.
Budding. A method of vegetative propagation of plants by implantation of buds from the mother plant onto a rootstock.

Callus. A mass of undifferentiated cells produced on wounds or in tissue culture.
Canker. A necrotic (dead) lesion on a stem or branch.
Capsid. The protein coat of viruses forming a shell around the nucleic acid.
Carbohydrate. A type of foodstuff composed of carbon, hydrogen, and oxygen; includes sugars, starches, and celluloses.

Carcinogen. A substance or agent capable of producing cancer.
Carrier. A liquid or solid material added to a chemical compound to prepare a proper formulation.
Cation. An ion having a positive charge.
Causal agent. The organism or abiotic factor that produces a given disease.
Cellulose. A polysaccharide composed of many glucose molecules and found in plant cell walls.
Center of origin. The geographical area where a plant or animal originated.
Centigrade (C). A thermometer scale in which the freezing point of water is set at 0° and the boiling point at 100°. To change to degrees Fahrenheit, multiply degrees centigrade by nine-fifths and add 32. See also *Fahrenheit*.
Chemotherapy. Control of a plant disease with chemicals that are translocated internally.
Chlamydospore. An asexual fungal spore formed from a cell of a fungal hypha.
Chloroplast. A structure in plant cells that contains chlorophyll and is the site of photosynthesis.
Chlorosis. Yellowing of normally green plant tissue resulting from lack of chlorophyll.
Chromosomes. Threadlike components of a cell that are composed of DNA and contain genetic information.
Circulative viruses. Viruses that are acquired by a vector, accumulated internally, passed through vector tissues, and introduced into other plants.
Cleistothecium (pl. cleistothecia). An entirely closed fruiting body produced by certain fungi in the Ascomycotina (ascomycetes) that contains asci and ascospores.
Clone. Many genetically identical individual organisms produced asexually from one individual (e.g., rooted cuttings).
Coccus. A spherical or round bacterial cell.
Compatible. Two compounds that can be mixed without affecting each other's properties.
Concentration. The amount of a chemical compound in a given volume or weight of diluent.
Conidiophore. A specialized fungal hypha on which asexual spores (conidia) develop.
Conidium (pl. conidia). An asexual fungal spore formed on a conidiophore.
Contact pesticide. A compound that kills by contact with plant tissue or pest rather than as a result of translocation.
Cotyledon. The seed leaf. Monocotyledons have one seed leaf; dicotyledons have two.
Crop rotation. The alternation of different crops in the same field or area, designed to increase soil fertility and reduce disease and pest problems.
Crop tolerance. The ability of a crop to endure treatment with a chemical without adverse effect; also, the amount of a pesticide that is legally allowable on or in a crop at harvest time.
Cultivar. Short for cultivated variety. See *Variety*.
Culture. To grow microorganisms on a prepared nutrient medium.
Culture medium. Prepared nutrient medium on which microorganisms are grown.
Cuticle. A thin waxy layer on the outer wall of epidermal cells.
Cyst. In fungi, an encysted zoospore; in nematodes, structure that may contain eggs.
Cytoplasm. Living substances in a cell except for the nucleus.

Damping-off. Rapid dying of seedlings near or below the soil surface, resulting in the seedlings falling over or failing to emerge, a disease caused by soilborne fungi.
Deciduous plant. A perennial plant that loses its leaves during the winter.
Dermal. Pertaining to the skin.
Dermal toxicity. Property of a compound to be absorbed through the skin of animals to produce symptoms of poisoning.
Detoxification. The inactivation or destruction of a toxin by alteration, binding, or breakdown of the toxic molecule.

Deuteromycotina. A group of fungi (also known as deuteromycetes) for which no sexual stage was originally known or has yet been found.

Dieback. Death of shoots, branches, and roots, starting at the tip.

Dikaryotic. The nuclear condition of some fungi (particularly rust and smut fungi) in which each cell contains two different haploid nuclei; usually designated as n+n.

Diluent. Any liquid or solid material used to dilute or carry an active ingredient.

Diploid. An individual with double the number of chromosomes (2n) per cell, as in plants, animals, and fungi.

Disease. Abnormal development of a living organism as a result of association with a pathogen, that interferes with its normal growth and structure and reduces its economic or aesthetic value.

Disease cycle. Stages involved in the development of a disease; it includes the development of the pathogen and the effects of disease on the host.

Disease incidence. The proportion of plants in a population that have disease symptoms.

Disease intensity. The amount of disease present in a population of plants or on a particular plant; a measure of disease incidence or disease severity.

Disease severity. The amount or proportion of diseased tissue present on a plant or in a population of plants.

Disease tolerance. In plants, the ability of a host to survive and produce satisfactory yields at a level of infection that causes economic loss to other cultivars or types of the same host plant.

Disease triangle. A concept describing the simultaneous occurrence of a pathogen, a susceptible host, and a favorable environment such that a disease may develop.

Disinfectant. An agent that can be used to free plant tissue from infection.

Disinfestant. An agent that kills pathogens on a plant or plant part before infection occurs or kills or inactivates a pathogen in the environment.

Dispersal. See *Dissemination.*

Dissemination. In relation to plant diseases, the transfer of inoculum to healthy plant tissue.

DNA (deoxyribonucleic acid). The genetic material found in all living organisms. Every inherited characteristic has its origin in the code of each individual's complement of DNA.

Dormant. Not actively growing (as seeds or other plant organs) owing to internal causes.

Dose, Dosage. Quantity of a substance applied per unit of plant, soil, or other surfaces; also called *rate.*

Downy mildew. A plant disease in which the fungus appears as a downy growth on the host surface, usually the lower surface of a leaf.

Drift. In relation to pesticides, the movement of a portion of the airborne particles of a spray or a dust away from the target area.

Economic poison. As defined under the Federal Insecticide, Fungicide and Rodenticide Act (FIFRA), an economic poison is "any substance or mixture of substances intended for preventing, destroying, repelling, or mitigating any insects, rodents, nematodes, fungi, or weeds, or any other form of life declared to be pests." As so defined, economic poisons now are generally known as *pesticides.*

Economic threshold. The pathogen density or disease intensity at or above which the value of crop losses—in the absence of management efforts—would exceed the cost of management practices (especially, the use of pesticides).

Ectoparasite. A parasite that feeds from the outside of a host.

Egg. In nematodes, the first stage of the life cycle.

ELISA (enzyme-linked immunosorbent assay). A serological method in which one antibody is linked to an enzyme such that with the addition of a substrate a colored compound is formed to indicate the presence of an antigen.

Emulsion. A mixture in which one liquid is suspended as small globules in another liquid (e.g., oil in water).
Encyst. To form a cyst.
Endemic. Naturally occurring in geographical region.
Endodermis. A layer of cells that surrounds the vascular tissues of plant roots.
Endoparasite. A parasite that enters and feeds inside a host.
Enzyme. A protein that catalyzes a specific biochemical reaction.
Epidemic. A rapidly developing and widespread occurrence of a disease in a population.
Epidemiology. The study of epidemics.
Epidermis. The outer layer of cells on all plant parts.
Epiphytotic. Relating to a widespread and destructive disease of plants (epidemic).
Eradication. Management of plant disease by attempting to eliminate the pathogen from a large area.
Ergotism. A disease in animals and humans caused by eating grain contaminated with alkaloids of the ergot fungi; also known as holy fire or St. Anthony's fire.
Etiolation. Elongation of stems caused by reduced light intensities.
Eukaryote. An organism with a membrane-bound nucleus and other organelles. Fungi, higher plants, and animals are eukaryotic organisms.
Exclusion. Management of plant disease by preventing the pathogen or infected plant material from being in crop production areas.
Exotic. An organism that is not native to an area; an introduced organism.
Exudates. Liquid discharge from plant tissue.

Facultative parasite. An organism that is usually saprophytic but may become parasitic. See also *Obligate parasite.*
Facultative saprophyte. An organism that is usually parasitic but may live as a saprophyte.
Fahrenheit (F). A thermometer scale in which the freezing point of water is set at 32° and the boiling point is 212°. See also *Centigrade.* To change to degrees Centigrade, subtract 32 and multiply by five-ninths.
Fermentation. Enzyme-catalyzed oxidation of organic substances in the absence of free oxygen.
Filamentous. Threadlike; usually refers to fungi.
Fission. A form of asexual reproduction in which a bacterial cell splits into two cells.
Flagellum (pl. flagella). A whiplike structure on bacterial cells or zoospores that functions in movement.
Flagging. The drooping of leaves and shoots before wilting of a plant.
Fleck. White to tan necrotic lesions up to a few millimeters in diameter, usually confined to the upper surface of leaves; a characteristic response of tobacco to ozone.
Formulation. In relation to pesticides, a specific mixture containing the active ingredient, the carrier, and other additives (adjuvants) required to make it ready for sale.
Fruiting body. A fungal structure in which spores are produced.
Fumigant. A toxic volatile substance that is used to manage various pests.
Fumigation. The application of a fumigant to an area.
Fungicide. A toxic substance to fungi; a type of pesticide.
Fungus. A eukaryotic microorganism composed of hyphae and lacking chlorophyll that is saprophytic and/or parasitic.

Gall. A tumorlike overgrowth produced on a plant.
Gametangium. A male or female sexual organ in some fungi.
Gene. The hereditary unit; a linear segment of DNA coding for a specific protein.
Gene expression. The expression of the genetic material of an organism as specific traits.

Genetic code. The biochemical basis of heredity, consisting of codons (base triplets along the DNA or RNA sequence) that determine the specific amino acid sequence in proteins and that are the same for all forms of life studied so far.

Genetic diversity. See *Genetic variability.*

Genetic engineering. Alteration of the hereditary apparatus of a living cell using enzymes and laboratory techniques rather than biological hybridization so that the cell can produce more or different chemicals, or perform completely new functions.

Genetic variability. The property or ability of a species to change its inherited characteristics from one generation to succeeding ones.

Genome. The basic chromosome set of an organism—the sum total of its genetic information.

Genotype. The genetic constitution of an individual or a group.

Germ tube. The growth of hyphae from a germinating fungus spore.

Germicide. A substance that kills microorganisms.

Germination. The beginning of growth of a spore, or seed sprouting.

Germplasm. The total genetic variability available to an organism, represented by the pool of germ cells or seed.

Global climate change (global warming). The changes expected to occur in the world climate over a very long period of time due to changes (increased levels of carbon dioxide and pollutants) brought about in the global environment by human activities.

Greenhouse effect. A misnamed theory that temperatures of the earth's atmosphere will increase with the increased concentration of certain gases which trap solar radiation. Also known as global warming or global climate change.

Haploid. Having a single complete set of chromosomes (the nuclear state is written as 1n or n).

Haustorium (pl. haustoria). A structure of specialized fungal hyphae in host cells that absorbs nutrients.

Hazard. In relation to pesticides, the probability that injury will result from the use of a particular compound; the sum of the toxicity plus the exposure to a pesticide.

Hectare (ha). An area of land in the metric system equal to 2.47 acres.

Herbicide. A chemical used for killing or inhibiting plant growth; a weed or grass killer; a type of pesticide.

Hormone. A growth regulator (e.g., auxins and gibberellins in plants). See also *Plant growth regulator.*

Host. A living organism from which parasites or pathogens obtain nutrients.

Host range. The hosts that may be attacked by a pathogen.

Hyaline. Transparent; without color.

Hydathodes. Structures in which water moves from the interior to the surface of the leaf.

Hydrogen ion concentration. A measure of acidity. See *pH.*

Hypersensitivity. A quick response in which plant cells die in response to certain pathogens and thus block the advance of the pathogen.

Immune. Cannot be infected by a given pathogen.

Imperfect stage. The portion of a fungal life cycle in which no sexual spores are produced; the anamorph stage.

Incubation period. The time between penetration of a host by a pathogen and the appearance of symptoms.

Inert ingredient. Ingredient in a product (pesticide) that does not contribute to the activity of the active ingredient.

Infection. The early stage of disease caused by a pathogen.

Infectious disease. A disease that is caused by a pathogen that can be spread from a diseased to a healthy plant.

Infectious period. The time during which inoculum is produced in or on infected plant tissues.

Infested. Containing or covered by great numbers of pathogens (insects, nematodes, bacteria, or fungi). Refers to the presence of pathogens in the environment, not to their infection of hosts.

Inoculation. The arrival of a pathogen on a host.

Inoculum (pl. inocula). Part of a pathogen that comes in contact with a host and can cause disease.

Insecticide. A substance or mixture of substances that kills, repels, or mitigates insects; a type of pesticide.

Intercellular. Between cells; between cell walls in plants.

Intracellular. Within cells; inside cell membranes.

In vitro. Outside the living body (e.g., a host) and in an artificial environment; in culture.

In vivo. In the living body of a plant or animal.

Ion. An electrically charged particle, atom, molecule, or radical in which the charge is due to the gain or loss of one or more electrons.

Koch's postulates. A set of rules or procedure for establishing the pathogenicity of an organism.

LD_{50}. A measure of toxicity; the dose of a compound that causes death in 50% of the test animals so treated. If a compound has an LD_{50} of 10 mg/kg, it is more toxic than one having an LD_{50} of 100 mg/kg.

Larva. The life stage of certain animals between the embryo and the adult (e.g., an immature nematode).

Latent period. The time between infection of a plant by a pathogen and the production of inoculum by that pathogen.

Leaching. Movement of a substance downward or out of the soil as the result of water movement.

Leaf spot. A lesion (chlorotic or necrotic spot) on a leaf.

Lesion. A localized area of diseased tissue.

Life cycle. The stages in the growth and development of an organism.

Local lesion. A small localized area of diseased tissue produced on a leaf upon mechanical inoculation of the leaf.

Mastigomycotina. A classification (subdivision) or fungi that generally produce zoospores and have mycelium without cross walls (aseptate); includes the oomycetes.

Meristem. Formative plant tissue made of cells that can divide continually.

Metabolism. The processes by which organisms utilize nutrients to build structural components, and break down cellular material to obtain energy and simple substances for special functions.

Micrometer (μm). A unit of length equal to 1/1000 of a millimeter; also called *micron.*

Micron. See *Micrometer.*

Mildew. A disease of plants in which the fungus produces a whitish growth on the host surface; also the growth of a fungus on a household surface.

Millimeter (mm). A unit of length equal to 1/10 of a centimeter (cm) or 0.039 of an inch.

Miscible. Capable of being mixed and remaining mixed under normal conditions.

Mode of action. The way in which a chemical compound, such as a pesticide, functions or acts.

Mold. Fungus growth on decaying matter or on the surface of plant tissues.

Monoclonal antibody. An antibody that is produced by a laboratory-grown cell clone and is more abundant and uniform than a natural antibody; shows specificity to a single antigenic fitting site.

Monoculture. The production of successive crops of the same plant species.

Mosaic. Symptom of some viral diseases of plants characterized by normal and light green or yellowish patches in the foliage.

Mottle, mottling. An irregular pattern of light and dark areas.

Multiline. A mechanical mixture of cultivars or varieties that vary primarily in their genes for resistance to diseases or insects.

Mummy. A dried, shriveled fruit.

Mycelium (pl. mycelia). Mass of fungal hyphae that may be visible to the unaided eye.

Mycoplasma-like organism. A prokaryote that lacks a cell wall and causes diseases in plants with typical symptoms of yellowing, stunting, and witches' brooms.

Mycorrhiza (pl. mycorrhizae). A fungus in a mutualistic relationship with the roots of a plant.

Mycotoxin. A poisonous compound produced by a fungus.

Nanometer (nm). A unit of length equal to one millionth of a millimeter.

Necrosis. Death of tissue.

Necrotic. Exhibiting varying degrees of dead areas or spots.

Nematicide. A chemical compound that kills nematodes.

Nematode. Any of a group of generally microscopic, wormlike animals that live as saprophytes in water or soil, or as parasites of plants and animals.

Noninfectious disease. A disease that is caused by an abiotic agent that cannot be transmitted from one plant to another.

Nonionic. Not having a positive or a negative charge; electrically neutral.

Nonseptate. Without cross walls.

Nucleic acid. Any of various acids composed of sugar or a derivation thereof, phosphoric acid, and a base involved in transferring and expressing the genetic properties of organisms. See also *DNA; RNA.*

Obligate parasite. An organism that can grow and multiply only on living organisms. See also *Facultative parasite.*

Organ. A differentiated structure in an animal or plant made up of various cells and tissues and adapted for the performance of specific functions.

Organic compounds. Chemical compounds that contain carbon.

Ozone. A highly reactive form of oxygen that may injure plants; a component of smog formed from oxygen in the presence of sunlight.

Pandemic. A widespread and destructive outbreak of disease simultaneously in several countries.

Parasite. An organism that lives in or on another organism and obtains food from that organism (host).

Parts per billion (ppb). A method of expressing the concentration of chemicals or other proportions. One pound is 1 ppb in 500,000 tons or in 1,000,000,000 (1 billion) pounds.

Parts per million (ppm). A method of expressing the concentration of chemicals or other proportions. One inch is 1 ppm in 16 miles; 1 pound is 1 ppm in 500 tons or 1,000,000 (1 million) pounds.

Pasteurization. The destruction, by heat or chemicals, of selected, harmful organisms (e.g., of plant pests) in soil or other propagating media.

Pathogen. A living agent that can cause disease.

Pathogenicity. The capability of a pathogen to cause disease.

Pathovar. In bacteria, a subspecies or group of strains that can caused disease only on plants within a particular genus or species.

Penetration. In relation to diseases, the initial invasion of a host by a pathogen.

Perennial. A plant that lives for more than 2 years.

Perfect stage. The sexual stage in the life cycle of a fungus; the teleomorph stage.

Perithecium (pl. perithecia). The flask-shaped fruiting structure of some fungi in the Ascomycotina (ascomycetes); it has an opening at maturity and contains asci.

Pest. Any life form that under some circumstances is detrimental to humans or to things they value (e.g., crops, stored food, domesticated animals).

Pesticide. A chemical substance used to kill pests (weeds, insects, rats and mice, algae, nematodes, etc.). See also *Economic poison.*

pH. A measure of acidity and alkalinity within the range of 0 to 14. The midpoint value (7) represents neutrality (neither acid nor alkaline), values below 7 indicate increasing acidity or hydrogen ion concentrations, and values above 7 indicate increasing alkalinity or lower hydrogen ion concentrations.

Photoperiod. The optimum duration of light and darkness for the normal growth of a plant.

Photosynthesis. The process by which plants combine carbon dioxide and water in the presence of light and chlorophyll to form carbohydrate.

Physiologic race. A group of microorganisms similar to others in morphology but having distinct cultural, physiological, pathological, or other characteristics.

Phytopathogenic. Capable of causing disease in plants.

Phytotoxic. Poisonous, injurious, or otherwise harmful to plants.

Pith. The parenchymatous tissue in the central area of a plant stem.

Plant–disease interaction. The concurrent parasitism of a host by more than one pathogen, in which the symptoms or other effects produced are of greater magnitude than the sum of the effects of each pathogen acting alone; an example of synergism.

Plant growth regulator. A substance that affects plants or plant parts through physiological rather than physical action by accelerating or retarding growth, prolonging or breaking a dormant condition, promoting decay, or inducing other physiological changes.

Plant pathology. The study of plant diseases; phytopathology.

Plasmid. A spherical, self-replicating hereditary element that is not part of a chromosome, found in certain bacteria and fungi. Because plasmids are generally small and relatively simple, they are used in recombinant DNA experiments as acceptors of foreign DNA.

Poison. A substance that, when absorbed or taken orally can cause illness, death, or retardation of growth or shortening of life; also called *biocide.*

Polysaccharide. A large organic molecule consisting of many units of a simple sugar.

Postemergence. The period after the appearance of a specified weed or crop.

Powdery mildew. A white, powdery growth produced on the surface of leaves and stems by certain fungi that are obligates parasites in the Ascomycotina (ascomycetes).

ppb. See *Parts per billion.*

ppm. See *Parts per million.*

Predisposition. The tendency of nongenetic conditions, acting before infection, to increase the susceptibility of a plant to disease.

Preemergence. The period before the emergence of a specified weed or crop.
Primary infection. The first infection of a plant by a pathogen.
Primary inoculum. Survival structures of a pathogen that cause primary infections.
Prokaryote. An organism, often unicellular, in which the cells lack a membrane-bound nucleus or organelles; bacteria are prokaryotes.
Protectant. With pesticides, a substance that prevents infection by a pathogen.
Protective fungicide. A pesticide that protects an uninfected plant or seed by killing or inhibiting the development of fungus spores or mycelia on the surface.
Protein. A compound consisting of amino acids.
Protoplast. Part of a single cell surrounded by a membrane.
Protoplast fusion. The joining of two protoplasts or the joining of a protoplast with any of the components of another cell such that genetic transfer may occur.
Pseudothecium (pl. pseudothecia). A fruiting body produced by certain fungi in the Ascomycotina (ascomycetes) that has an opening through which ascospores are discharged; similar to a perithecium in appearance.
psi. Pounds per square inch.
Pustule. A small blisterlike structure in which some fungi, particularly rusts, produce spores.
Pycnidium (pl. pycnidia). An asexual, spherical or flask-shaped fruiting body produced by certain fungi in the Deuteromycotina (imperfect fungi) that contains conidiophores and conidia.

Quarantine. A legal restriction placed on the movement of plants or animals from one area to another to prevent the spread of diseases.

Range. The geographic region throughout which a kind of organism lives. Not to be confused with *host range.*
Rate. The amount of pesticidal chemical (formulated product) applied to a given surface area; also called *dose, dosage.*
Recombinant DNA. The hybrid DNA produced by joining pieces of DNA from different sources through genetic engineering.
Relative humidity (RH). The ratio of the quantity of water vapor present in the atmosphere to the quantity that would saturate the air at the same temperature.
Residue. In relation to pesticides, the amount of chemical that remains on or in a harvested crop.
Resistance. The ability of an organism to overcome a pathogen or other damaging factor; ability to resist disease in plants.
Resistant. Possessing qualities that hinder the development of a given pathogen or the effects of other damaging factors.
Resting spore. A fungal spore that is resistant to extremes in temperature and moisture and often germinates only after a period of time.
Restriction enzyme. An enzyme within a cell that recognizes and degrades DNA from foreign organisms, thereby preserving the genetic integrity of the cell. In recombinant DNA experiments, restriction enzymes are used as tiny biological scissors to cut up foreign DNA before it is recombined with a vector.
Rhizome. A modified underground rootlike stem that produces roots and leafy shoots (e.g., the white underground parts of Johnsongrass).
Rhizosphere. The area near the surface of a living root.
Ringspot. A circular area of chlorosis which is characteristic of many virus diseases.
RNA (ribonucleic acid). A nucleic acid involved in protein synthesis; also, the genetic material of most plant viruses.

Rogueing. The practice of removing diseased, inferior, or abnormal plants from a population.
Rosette. Short, bunchy type of plant growth.
Rot. The softening and disintegration of plant tissue by enzymes produced by fungal or bacterial infection.
rpm. Revolutions per minute.
Rust. A disease caused by fungi in the Basidiomycotina (basidiomycetes), often characterized by orange-red or orange spores produced in pustules or on galls.

Sanitation. The destruction of infected or infested plants or plant parts in order to eliminate the inoculum of pathogens and insect vectors.
Saprophyte. An organism that can obtain nutrients from dead organic material.
Scab. A crustlike diseased area on the surface of a plant organ.
Sclerotium (pl. sclerotia). A compact mass of fungal hyphae, capable of surviving under unfavorable environmental conditions; usually with a dark-colored rind that resists desiccation.
Scorch. "Burning" of leaf margins from disease or environmental conditions.
Secondary infection. Any infection from inoculum produced from primary or subsequent infections. See also *Primary infection.*
Selective pesticide. A chemical that is more toxic to some specific biological species than to other species.
Septate. In fungi, hyphae with cross walls.
Serological specificity. The pattern of reactions of a given antigen (e.g., protein from cell wall of a bacterium or from the protein coat of a virus) and various antibodies, used in the detection and identification of antigenic substances.
Serology. A method that uses antibodies to detect antigens.
Seta. A bristle-like protuberance from a spore or on some fungal fruiting bodies.
Sexual reproduction. Any type of reproduction that involves a union of male and female gametes.
Sign. Parts of a pathogen or its products seen on a diseased host plant.
Silvering, silverleaf. A symptom of leaves or fleshy tissues characterized by grayish or shiny lesions; often induced by exposure to certain air pollutants.
Silviculture. The science of developing, caring for, or cultivating forests.
Soil sterilant. A chemical that, when present in the soil, prevents the growth of plants, microorganisms, etc.
Somatic cell. Cells other than sex or germ cells.
Sooty mold. A sooty fungal growth on foliage and fruit in the honeydew secreted by insects such as aphids.
Species. A class of related organisms having common characteristics and capable of interbreeding.
Sporangiophore. A specialized fungal hypha on which sporangia are produced.
Sporangium (pl. sporangia). Type of asexual reproductive structures in some fungi, particularly in the Mastigomycotina (oomycetes).
Spore. In fungi, the sexual or asexual reproductive unit consisting of one or more cells.
Sporodochium. A fruiting structure of fungi made up of a cluster of conidiophores.
Sporulate. Production of spores.
Stem pitting. A symptom of some viral diseases with depressions on the stem.
Sterile fungi. Fungi that are not known to produce spores.
Sterilization. The elimination of living organisms including pathogens (e.g., from containers, soil, or other rooting media) with heat or chemicals.
Stoma (pl. stomata or stomates). Minute, organized openings on the surface of leaves or stems through which gases pass.

Stylet. A long, needle-like feeding structure of nematodes and some insects.

Succulent. Tender, young, or watery plant tissues.

Surfactant. A material that increases the emulsifying, dispersing, spreading, wetting, or other surface-modifying properties of a pesticide formulation.

Susceptibility. The inability of a plant to overcome disease.

Susceptible. Lacking the ability to resist the effect of a given pathogen, or other damaging factor.

Symbiosis. A mutually beneficial relationship between two different organisms.

Symptom. The internal or external reactions or alterations of an organism due to a disease.

Systemic. Spreading internally throughout the plant.

Systemic pesticide. A pesticide that is translocated within the plant.

Teliospore. Type of spore produced by rust and smut fungi.

Tip burn, tip necrosis. Necrosis of apical tissues of leaves affecting only a small percentage of the entire leaf.

Tissue culture. An *in vitro* method of propagating heathy cells from plant tissues.

Tolerance. The ability of a plant to withstand the effects of a disease; also, the amount of toxic residue allowable on edible plant parts.

Tolerant. Having the ability to withstand to a certain degree the effects of disease or other unfavorable circumstances.

Toxic. Poisonous, injurious, or harmful to life.

Toxicant. An agent capable of being toxic; a poison.

Toxicity. The relative capacity of a compound to cause injury. See also LD_{50}.

Toxin. A compound produced by an organism that is toxic.

Tracheid. In plants, a xylem element that functions in conducting water and minerals.

Transformation. The transfer of genetic material from one organism to another by means of genetic engineering.

Transgenic. An organism that possesses a gene from another species as a result of genetic engineering.

Transpiration. The loss of water vapor from plant surfaces.

Tumor. An uncontrolled growth of tissue.

Urediniospore. A dikaryotic (n+n), repeating spore produced by rust fungi.

Variability See *Genetic variability.*

Variety. A subdivision of a species where individuals within a species are distinct in form and function from other similar arrays of individuals; in crops, uniformly growing plants with certain characteristics.

Vascular. Relating to a plant tissue consisting of conductive tissue.

Vector. A living agent able to transmit a pathogen.

Vein banding. Retention of bands of green tissue along the veins while the tissue between the veins has become chlorotic.

Vein clearing. Destruction of chlorophyll adjacent to or in the vein tissue as a result of infection by a virus or other pathogen.

Vessel. In plants, a xylem element that functions in conducting water and minerals.

Viricide. A substance capable of inactivating or suppressing the multiplication of a virus completely and permanently.

Virion. A virus particle.

Viroid. A small RNA unit that can infect plant cells, replicate itself, and cause disease.

Virulence. The degree of pathogenicity of a pathogen.

Virulent. Capable of causing a disease.
Virus. A submicroscopic obligate parasite composed of nucleic acid and protein that causes disease and multiplies only in living cells.
Volatile. Able to vaporize (changes from a solid or liquid to a gas) at ordinary temperatures on exposure to air.

Wetting agent. A compound that reduces surface tension and causes a liquid to contact plant surfaces more thoroughly, usually a surfactant.
Wilt. Loss of rigidity of plant parts due to insufficient water in plant tissues.
Witches' broom. Broomlike growth caused by the dense clustering of branches on woody plants.

Xylem. A complex tissue (tracheids, vessels, parenchyma cells, fibers) that typically constitutes the woody element in plants and functions in conduction of water and nutrients, storage, and support.

Yellows. A plant disease characterized by yellowing and stunting of plants.

Zoospore. An asexual spore capable of moving in water by means of flagella.

Suggested Readings

Agrios, G. N. 1988. *Plant Pathology,* 3rd ed. Academic Press, New York.

Ainsworth, G. C. 1981. *Introduction to the History of Plant Pathology.* Cambridge University Press, New York.

Alexopoulus, C. J. and Mims, C. W. 1979. *Introductory Mycology,* 3rd ed. Wiley-Interscience, Somerset, NJ.

Anon. 1972. *Genetic Vulnerability of Major Crops.* National Academy of Sciences, Washington, DC.

Barker, K. R., Carter, C. C., and Sasser, J. N., eds. 1985. *An Advanced Treatise on Meloidogyne,* Vol. 2. North Carolina State University Graphics, Raleigh, NC.

Beckman, C. 1987. *The Nature of Wilt Diseases of Plants.* APS Press, St. Paul, MN.

Billing, E. 1987. *Bacteria as Plant Pathogens. Aspects of Microbiology 14.* American Society for Microbiology, Washington, DC.

Bruehl, G. W. 1987. *Soilborne Plant Pathogens.* Macmillan, New York.

Burdon, J. J. 1987. *Diseases and Plant Population Biology.* Cambridge University Press, Cambridge, UK.

Burdon, J. J. and Leather, S. R., eds. 1990. *Pests, Pathogens and Plant Communities.* Blackwell, Oxford, UK.

Campbell, C. L. and Madden, L. V. 1990. *Introduction to Plant Disease Epidemiology.* John Wiley, New York.

Campbell, R. 1989. *Biological Control of Microbial Plant Pathogens.* Cambridge University Press, Cambridge, UK.

Carefoot, G. L. and Sprott, E. R. 1967. *Famine on the Wind.* Rand McNally, Chicago.

Chet, I., ed. 1987. *Innovative Approaches to Plant Disease Control.* Wiley-Interscience, New York.

Christensen, C. M. 1966. *Molds and Man.* University of Minnesota Press, St. Paul, MN.

Clifford, B. C. and Lester, E., eds. 1988. *Control of Plant Diseases: Costs and Benefits.* Blackwell, Oxford, UK.

Cook, R. J. and Baker, K. F. 1983. *The Nature and Practice of Biological Control of Plant Pathogens.* APS Press, St. Paul, MN.
Delp, C. J., ed. 1988. *Fungicide Resistance in North America.* APS Press, St. Paul, MN.
Dropkin, V. H. 1989. *Nematology,* 2nd ed. John Wiley, New York.
Fry, W. E. 1982. *Principles of Plant Disease Management.* Academic Press, New York.
Fulton, R. H., ed. 1984. *Coffee Rust in the Americas.* APS Press, St. Paul, MN.
Hampton, R., Ball, E., and DeBoer, S. 1990. *Serological Methods for Detection and Identification of Viral and Bacterial Plant Pathogens: A Laboratory Manual.* APS Press, St. Paul, MN.
Heitefuss, R. 1989. *Crop and Plant Protection: The Practical Foundation.* Ellis Horwood, Ltd., Chichester, UK.
Hickey, K. D., ed. 1986. *Methods for Evaluating Pesticides for Control of Plant Pathogens.* APS Press, St. Paul, MN.
Horst, R. K. 1990. *Westcott's Plant Disease Handbook,* 5th ed. Van Nostrand Reinhold, New York.
Kranz, J., ed. 1991. *Epidemics of Plant Diseases: Mathematical Analysis and Modelling,* 2nd ed. Springer, Berlin.
Large, E. C. 1940. *The Advance of the Fungi.* Dover, New York.
Littlefield, L., ed. 1981. *Biology of the Plant Rusts: An Introduction.* Iowa State University Press, Ames, IA.
Lucas, G. B. 1975. *Diseases of Tobacco,* 3rd ed. Biological Consulting Associates, Raleigh, NC.
MacKenzie, J. J. and El-Ashry, M. T. 1989. *Air Pollution's Toll on Forests and Crops.* Yale University Press, New Haven, CT.
Manion, P. D. 1991. *Tree Disease Concepts,* 2nd ed. Prentice-Hall, Englewood Cliffs, NJ.
Marasas, W. F. O. and Nelson, P. E. 1987. *Mycotoxicology.* The Pennsylvania State University Press, University Park, PA.
Marco, G. J., Hollingworth, R. M., and Durbam, W., eds. 1987. *Silent Spring Revisited.* American Chemical Society, Washington, DC.
Matossian, M. K. 1989. *Poisons of the Past: Molds, Epidemics, and History.* Yale University Press, New Haven, CT.
Matthews, R. E. F. 1991. *Plant Virology,* 3rd ed., Academic Press, San Diego, CA.
National Research Council. 1991. *Managing Global Genetic Resources: The U.S. National Plant Germplasm System.* National Academy Press, Washington, DC.
Nelson, R. R., ed. 1973. *Breeding Plants for Disease Resistance: Concepts and Applications.* The Pennsylvania State University Press, University Park, PA.
Nyvall. R. F. 1989. *Field Crop Diseases Handbook,* 2nd ed. Van Nostrand Reinhold, New York.
Palti, J. 1981. *Cultural Practices and Infectious Crop Diseases.* Springer-Verlag, Berlin.
Parry, D. 1990. *Plant Pathology in Agriculture.* Cambridge University Press, Cambridge, UK.
Saettler, A. W., Schaad, N. W., and Roth, D. A., eds. 1989. *Detection of Bacteria in Seed and Other Planting Material.* APS Press, St. Paul, MN.
Schaad, N. W. 1988. *Laboratory Guide for Identification of Plant Pathogenic Bacteria,* 2nd ed. APS Press, St. Paul, MN.
Schumann, G. L. 1991. *Plant Diseases: Their Biology and Social Impact.* APS Press, St. Paul, MN.
Sinclair, W. A., Lyon, H. H., and Johnson, W. T. 1987. *Diseases of Trees and Shrubs.* Cornell University Press, Ithaca, NY.
Spencer, D. M., ed. 1981. *The Downy Mildews.* Academic Press, London.
Tattar, T. A. 1989. *Diseases of Shade Trees,* rev. ed. Academic Press, San Diego, CA.
Teng, P. S., ed. 1987. *Crop Loss Assessment and Pest Management.* APS Press, St. Paul, MN.
Thurston, H. D. 1984. *Tropical Plant Diseases.* APS Press, St. Paul, MN.

Vanderplank, J. E. 1963. *Plant Diseases: Epidemics and Control.* Academic Press, New York.
Walkey, D. G. A. 1985. *Applied Plant Virology.* John Wiley, New York.
Ware, G. W. 1989. *The Pesticide Book,* 3rd ed. Thomson Publications, Fresno, CA.

The following publications of the American Phytopathological Society (APS Press, 3340 Pilot Knob Road, St. Paul, MN 55141, 612/454-7250) also provide valuable information on plant diseases and their management:

Compendium of Alfalfa Diseases, 2nd ed. (1990)
Compendium of Apple and Pear Diseases (1990)
Compendium of Barley Diseases (1982)
Compendium of Bean Diseases (1991)
Compendium of Beet Diseases and Insects (1986)
Compendium of Citrus Diseases (1988)
Compendium of Corn Diseases, 2nd ed. (1980)
Compendium of Cotton Diseases (1981)
Compendium of Elm Diseases (1981)
Compendium of Grape Diseases (1988)
Compendium of Ornamental and Foliage Plant Diseases (1987)
Compendium of Pea Diseases (1984)
Compendium of Peanut Diseases (1983)
Compendium of Potato Diseases (1981)
Compendium of Raspberry and Blackberry Diseases (1991)
Compendium of Rhododendron and Azalea Diseases (1986)
Compendium of Rose Diseases (1983)
Compendium of Sorghum Diseases (1986)
Compendium of Soybean Diseases, 3rd ed. (1989)
Compendium of Strawberry Diseases (1984)
Compendium of Sweet Potato Diseases (1988)
Compendium of Tobacco Diseases (1991)
Compendium of Tomato Diseases (1991)
Compendium of Turfgrass Diseases (1983)
Compendium of Wheat Diseases (1987)

Index

Abiotic agents, 6, 316-331
  acid rain, 12, 329
  acid soils, 317-318
  air pollution, 12, 324-331
  alkaline soils, 317-318
  chemicals, 317
  cold, 318-319
  drought, 320
  drowning, 321
  frost, 318-319
  fertilizer, 317
  heat, 319
  herbicides, 321-324
  lightning, 319-320
  noninfectious diseases, 275
  nutrient deficiencies, 11, 316-317
  nutrient toxicities, 317
  pesticides, 321-324
  sun scald, 319
Abiotic diseases (*see* Abiotic agents)
Abnormal water conditions, 320-321
  drought, 320
  drowning, 321
Acid rain, 12, 329
Acervuli, 160, 234
Active ingredient, 73
Adjuvant (additives), 75
Aecia, 195
Aeciospores, 194, 195
Aflatoxins, 53, 54, 155
Agitators (for spray tanks)
  by-pass, 90
  hydraulic, 90
  jet, 90
  mechanical, 90
Agonomycete, 158, 160
Agricultural consultants, 36
Agrobacterium tumefaciens, 124-125, 266, 279-280
  crown gall, 124-125, 278-280
  plasmid, 124-125, 278-280
Air pollution, 12, 324-331
Alfalfa, 65
  anthracnose, 233
  broom rape, 310-311
  crown rot, 186
  dodder, 314-315
  lesion nematode, 147
  Phytophthora root rot, 165
  southern blight, 183
  stem and bulb nematode, 150
Alternaria spp., 49, 193, 230-233
Alternate hosts
  for cedar-apple rust, 198-200
  for wheat stem rust, 194
Annosus root and butt rot, 244, 249-250

Anthracnose, 233-235, 244
Antibiotic, 11, 68
Antibody
  monoclonal, 112
  polyclonal, 112
Aphids, 23, 303, 307
Apothecia, 159, 187, 188
Apples, 54, 198
  Alternaria leaf spot, 230
  apple scab, 5, 82, 113, 236-240
  cedar apple rust, 198-200
  fire blight, 250, 269-272
  Gymnosporangium juniperi-virginianae, 183, 198
  powdery mildew, 212
  Venturia inaequalis, 193, 236
Appressoria, 199
Arceuthobium spp., 309, 312
Aristotle, 15
Arthur, J. C., 18
Ascomycetes, 158, 159, 160
Ascomycotina, 158, 159, 160, 178, 187
Ascospores, 159, 188, 190, 213, 237, 238
Ascus, 159, 213, 232
Aseptate, 143
Aspergillus flavus, 53, 155-156
Aster yellows, 23, 280
Azaleas
  gray mold, 219
  Phytophthora root rot, 167-170
  Sclerotinia spp., 187
    petal blight, 187

Bacillus, 55
Backcrossing, 61
Bacteria, 9, 11
  dispersal, 264
  distribution, 264
  growth, 263-264
  identifying, 260-262
  reproduction, 263-264
  size, 260
  structure, 260
  type, 260-261
Bacterial diseases
  bacterial canker of tomato, 274-276
  bacterial spot of peach and plum, 276-278
  bacterial wilt, 3, 13, 25, 267-269, 279
    of cucumber, 274
  crown gall, 278-280
  fire blight of apple and pear, 269-272
  soft rots, 272-273
Bactericides, 11
Bacteriocin, 55
Bacteriology, 3
Bacteriophages, 260
Balance of nature, 107
Banana, 2
  bacterial wilt, 267
  cucumber mosaic, 302
  gray mold, 219
Bark beetles, 23, 244, 245, 247
Barley, 2, 53
  Barley yellow dwarf, 292, 304-305
  Barley yellow dwarf virus, 292, 304-305
  bunt, 203
  covered smut, 203
  stinking smut, 203
Barn rot, 266
Basidiomycetes, 158, 159, 160
Basidiomycotina, 158, 159, 173, 183
Basidiospores, 143, 173, 176, 178, 179, 180
Bawden, F. C., 18, 284
Bean, 2, 179
  anthracnose, 234
  crown and stem rot, 186
  cucumber mosaic, 302
  Fusarium rots, 179, 181
  gray mold, 219
  hypocotyl rot, 179
  Rhizoctonia solani, 173, 174
  root rot, 179
  southern blight, 183
  stem and bulb nematodes, 150-151
  stubby root nematode, 139
  web blight, 173-174
Beet (sugar), 2
  Cercospora leaf spot, 221
  cyst nematode, 146
  dodder on, 315
  Rhizoctonia solani, 173
  Sclerotium rolfsii, 183
  southern blight, 183
  stem and bulb nematode, 150-151
Beet curly-top virus, 23
Beijerinck, M. W., 18
Belonalaimus spp., 139
Bengal famine, 229
Berkeley, M. J., 17
Binary fission, 263
Biochemistry, 3
Biological control, 54-58
Biotechnology, 19
  bacterial vectors of DNA, 124-125
  citizen concern, 125-126
  gene cloning, 123
  gene splicing, 121-123
  organ culture, 115
  plants from cells, 118-120
  protoplasts, 120-121
  recombinant DNA, 121
  single cell cultures, 117

tissue culture, 117
vegetative propagation, 115-118
viral vectors of DNA, 125
Biotic agents, 9
Bipolaris spp., 59
maydis, 119, 193, 226-228
oryzae, 193, 229
Black root rot, 115
Black shank, 166-167
Bordeaux mixture, 18, 32
Botany, 4
Botrytis spp., 193, 220-221
Botrytis blight, 219-221
Broom rape, 309, 310-311
Buildings, wood decay, 255-258
Bunt (*see also* Covered smut), 203
Burrill, T. J., 18, 259, 269

Calibration of pesticide equipment
application equipment, 98-105
factors affecting, 98-99
fumigators, 100-104
granular applicators, 104-105
sprayers, 98-100
Callus, 116
Calonectria crotalariae, 189-191
Camellia japonica, 168-169
Camellia sasanqua, 168-169
Camellias
petal blight, 187
Phytophthora root rot, 163, 168-169
Sclerotinia spp., 187
Cantaloupe
cucumber mosaic virus, 115
herbicide injury, 322
powdery mildew, 211
Carbamate fungicides, 18
Carnation, Alternaria leaf spot, 230
Carson, Rachel, 67
Cassava, 2
Cedar, 198
Cedar-apple rust, 40, 198-200
Celery
Cercospora leaf spot, 221
Rhizoctonia solani, 173
Center(s) of origin, 61
Centrifugal pump, 90
Ceratocystis
fagacearum, 244, 248
ulmi, 244, 245, 246, 247
Cercospora leaf spots, 221-225
Cercospora, 49, 193, 222, 224
Certified seed, 63-64
Chemical management, 67
Chemistry, 4
Chemotherapy, 38
Cherry
lesion nematode, 147
Rhizopus soft rot, 240
Chestnut blight, 23, 244-245
Chisel-injection method, 77-78
Chitwood, B. G., 18
Chlamydospore, 160, 180
Chlorofluorocarbon, 329-330
Chrysanthemum
dodder, 314
stunt, 308
Chytridiomycete, 158
Citrus, 2, 134
bacterial canker, 40
stubborn, 13
Citrus nematode, 131
Citrus tristeza virus, 288
Claviceps purpurea, 155, 193, 216
Cloning, 115, 123
Clover, 65
anthracnose, 233
black root rot, 27
broom rape, 310-311
crown rot, 186
dodder, 314
Clover yellow vein virus, 286
Club root or cabbage, 27
Cobb, N. A., 18
Coccus, 260, 291
Cochliobolus spp.
heterostrophus, 226
miyabeanus, 229
Coconut, 2
Coconut cadang-cadang, 308
Coconut palm, lethal yellowing, 280
Coffea arabica, 200
Coffee, 54
Coffee rust, 3, 200-201
Colchicine, 119
Coleomycete, 158, 159
Colletotrichum, 49, 193, 234, 235
Colonization, 24
Common names
of diseases, 14
of pesticides, 69
Conidia, 156, 160, 211, 213, 214, 217-218, 221, 222, 224, 226, 228, 229, 231-233, 235
Conidiophore, 217, 218, 220, 224, 226, 228, 229, 231
Conducive soils, 54
Control of plant diseases (*see* Management of plant diseases)
Controlled droplet application (CDA), 83
Copper sulfate, 17
Coremia, 246

Corn, 2, 25, 48, 53, 54, 62, 65
  anthracnose, 233
  Bipolaris spp., 226-229
  ear and kernel rot, 179
  fertilizer injury, 322
  Fusarium rot, 179
  lesion nematode, 147
  maize chlorotic dwarf, 306-307
  maize dwarf mosaic, 306-307
  ozone damage, 326
  Rhizopus soft rot, 240
  smut, 201-203
  southern leaf blight, 225-229
  stalk rot, 179
  stem and bulb nematode, 150-151
  stubby root nematode, 151-153
  stunt, 282
  witchweed, 311-312
Cornus florida, 253
Corynebacterium michiganense, 266, 275, 276
Cotton
  leaf curl virus, 23
  lightning injury, 319
  wilt, 23
Cotton seed, 54
Cover crops, 50, 52
Covered smut, 203
Crabgrass, witchweed, 311-312
Crick, F. H., 284
Criconemella, 139
Cronartium quercuum f. sp. *fusiforme,* 244, 252
Crop loss, 32
  nematode, 128
Crop rotation, 33, 136, 149
Crop yields, US, 114
Cropping sequence, 48
Cropping systems, 48
Cross protection, 57
Crown gall, 124, 226, 278-280
Crown rust, 59
Cryphonectria parasitica, 244-245
Cucumber
  Alternaria leaf spot, 230
  Anthracnose, 237
  bacterial soft rot, 272
  bacterial wilt, 274
  belly rot, 175
  cucumber mosaic, 302-303
  downy mildew, 211
  pale fruit, 308
  Rhizoctonia solani, 175
Cucumber beetle, 23
Cucumber mosaic, 23, 292, 302-303
  causal symptoms, 303
  disease cycle, 303
  management, 303
  symptoms, 302-303
Cucumber mosaic virus, 14, 40, 115, 292, 302-303
Cucumovirus, 303
Cucurbits
  anthracnose, 233
  bacterial wilt of cucumber, 274
  cucumber mosaic, 302-303
  powdery mildew
  Rhizopus soft rot, 240
Curative, 68
Cuscuta spp., 309, 314
Cycle, 22
Cylindrocladium crotalariae, 162, 189-190
Cyst nematode, 145-147
  causal agent, 145
  disease cycle, 145
  Globodera, spp., 139, 145
  Heterodera, spp., 139, 145
  management, 146
  symptoms, 145
Cytoplasmic genes, 113

Dagger, 139
Damping-off, 162, 166, 170-173
Damping-off diseases, 26, 52
Darwin, Charles, 17
Dates, 54
DeBary, H. A., 17
Debilitating disease, 5
Decay, 244
Deficiencies, nutrient, 316-317
Deoxyribonucleic acid (DNA), 113, 122, 284, 287
  movement of
    bacterial vectors, 124
    gene cloning, 128
    microprojectile bombardment, 125
    viral vectors, 125
Deuteromycotina, 158, 159, 173, 178
Devastating diseases, 5
Diaphragm, pump, 88
Diener, T. O., 19, 285, 308
Diploid cells, 119
  growth hormones, 119
  regeneration of haploid cells, 119
Discomycete, 158, 159
Discula spp., 244, 253
Disease complexes, 12
Disease cycle, 21, 25
Disease focus, 29
Disease gradient, 29
Disease progress curve, 27-28
Disease resistance, 5, 6, 32
  developing resistant cultivars, 61-62

genetic uniformity, 59-60
kinds of, 60-61
Ditylenchus spp., 139, 150
dipsaci, 150
Dodder, 304, 314-315
Dogwood anthracnose, 253-254
Doi, Y., 18
Downy mildews, 3, 25, 59, 204-211
of cucurbits, 185-186, 211
of grape, 18, 209-211
of potato, 205, 209
Dubos, René, 331
Dutch elm disease, 23, 244, 245-248
Dwarf mistletoe, 309

Early blight, 230-231
Economic thresholds, 35
Ectomycorrhizae, 255
Ectoparasites, 9
Ectoparasitic nematodes, 9, 132, 151-152
dagger nematode, 152
lance nematode, 152
ring nematode, 152
sting nematode, 152
stubby root nematode, 151-153
stunt nematode, 152
Edaphology (soils), 4
Eggplant, tobacco mosaic, 295
Electron microscope, 18
Elm bark beetle, 247
Endomycorrhizae, 155, 255
Endoparasites, 10
Endoparasitic nematodes, 9, 132, 151
lesion nematode, 147-150
root knot nematode, 140-145
Endothia parasitica (see *Cryphonectria parasitica*)
Engineering, 4
English grain aphid, 304
Entomology, 4
Environment, 5
effects on disease development, 25-27
cultural practices, 27
light, 26
moisture, 25
nutrition, 26
soil type and pH, 26-27
temperature, 25
wind, 26
protection of, 107-111
Environmental pollution, 5
Enzyme-linked immunosorbent assay (ELISA), 288, 289
Enzymes, 24
Epidemic, 18, 27
Epidemiology, 21
Epinasty, 267
Eradicant, 68
Ergot, 15, 23, 215-217
Ergotism, 215
Erwinia spp.
amylovora, 266
carotovora, 266, 273
herbicola, 57
tracheiphila, 266, 274
Erysiphe spp., 193, 213
Erysiphe cichoracearum, 214
Evasion, 38-39
Exclusion, 38-39
Explants, 116
Exudates, 56

Figs, 54
Fire blight, 18, 23, 266, 269-272
Fleming, Alexander, 54
Flexible impeller pump, 88
Flower crops
cucumber mosaic, 302
Rhizopus soft rot, 240
stem and bulb nematode, 150
Forage crops
anthracnose, 233
Cercospora leaf spot, 221-225
crown and stem rots, 186-189
dodder, 314-315
Helminthosporium leaf spots, 225
powdery mildew, 212
Sclerotinia spp., 186
southern blight, 183
stem and bulb nematode, 150-151
stubby root nematode, 151-153
Forecasting plant diseases, 82, 105-107
Forest trees
annosus root and butt rot, 249-250
chestnut blight, 144-145
dogwood anthracnose, 253-254
Dutch elm disease, 245-248
dwarf mistletoe, 312-313
fusiform rust, 250-253
leafy mistletoe, 312-313
little leaf of pine, 254-255
oak wilt, 248-249
Foundation seed, 63-64
Frogeye leaf spot, 222-223
Fumigants, 76-79
application, 76-79
factors affecting performance 74, 76-77
soil treatment, 75
Fungi, 9-10
classification, 157-160
diseases caused by airborne fungi, 192-242

Fungi *(continued)*
diseases caused by soilborne fungi, 162-191
diseases of shade and forest trees and decay in wood, 255-258
distribution and dispersal, 160-161
growth and reproduction, 157-160
hyphae, 154
importance, 154-155
mycelium, 154
reproduction, 157-160
structure, 157-160
Fungi imperfecti, 158-159
Fungicides, 11
Fungistasis, 56
Fusarium, 49, 57, 145, 162, 178
oxysporum, 14
solani, 180
Fusarium wilts, 181-182
Fusiform rust, 244, 150-253

Galls, 140-141, 278-280
Gear pump, 90
Gene, 112
Gene cloning, 123
Gene splicing, 121-123
Gene transformation, 115
General resistance, 60
Generic name of pesticide, 69
Genetic diversity, 6
Genetic engineering, 58, 114
Genetic resistance, 38
Genetic uniformity, 59, 225
Genetics and plant breeding, 4
Genus, 13-14
Germ theory of disease, 17
Germplasm, 58
Germplasm preservation, 118
Giant cells, 143
Gibberella (Fusarium) zeae, 25
Global climate change, 329-331
Global warming, 329-331
Globodera spp., 139, 145
rostochiensis, 145
Golden Age of Biology, 16
Gram stain, 260, 262
Granular pesticide applicators, 80-81, 105
calibration, 105
Granville wilt, 12, 266, 267
Grape, 54, 209
Botrytis blight, 220
downy mildew, 209-211
powdery mildew, 213
stubby root nematode, 151-153
Gray mold, 219-221
Green peach aphid, 298
Greenhouse management, 50-52
pesticide application, 96
Ground cherry, 298, 303
Gutenberg, John, 16
Gymnosporangium juniperi-virginianae, 193, 198

Haploid cells, 119-120
Hartig, Robert, 17
Hasimoto, H., 18
Haustoria, 207, 214
Helicotylenchus, 139
Helminthosporium (*Bipolaris*), 59
maydis, 225-229
oryzae, 229-230
Hemiascomycete, 158, 159
Hemibasidiomycete, 158, 159
Hemileia vastatrix, 193, 200
Hemiparasites, 309
Hemoglobin, 10
Heterobasidion annosum, 55, 244, 249, 250
Heterodera, spp., 139, 145
glycine, 145
schachtii, 145
High volatile fumigants, 79
Holy fire, 215
Horizontal resistance, 60
Horsenettle, 293, 296
Hydathodes, 24
Hyperparasites, 56
Hymenomycete, 158, 159
Hyphomycete, 158, 160
Hypovirulent, 245

Immunity, 60
Imperfect fungi, 158, 159, 160
Incubation period, 24
Indicator plants, 288
Infection, 24
Infection court, 52, 60
Infectious plant diseases, 9-12
Inoculation, 24
Inoculum potential, 107
Insect vectors, 24
Integrated pest management, 30-37
economic threshold, 35
evolution, 30-33
IPM concept, 34-35
management principles, 35
IPM (*see* Integrated pest management)
Irish famine, 205
Ivanovski, D. I., 18

Jet agitator, 90
Johnsongrass, 307

Jones, L. R., 18
Juveniles, 9, 146

Kausche, G. A., 18
Koch, Robert, 17, 259
Koch's Postulates, 7, 17
Knots (*see* Galls)
Kühn, J. G., 17

Lance nematode, 152
Larvae, 9, 132-133
  Ditylenchus spp., 150
  Meloidogyne, spp., 142-143
  Pratylenchus spp., 150
Late blight, 17, 205-209
Latin binomial, 14
Leaf blight, 225-229
Leafhopper, 11, 18, 19, 23, 282, 307
Leafy mistletoe, 13, 309
Leeuwenhoek, Anton von, 16, 259
Lesion nematode, 139, 147-150
  disease cycle, 148-149
  Pratylenchus spp., 139, 149
  symptoms, 147-148
Lespedeza
  dodder, 314-315
  Rhizoctonia solani, 174
Lethal yellowing, 280
Life cycle, 21
Limiting diseases, 5
Linneaus, 14
Little leaf of pine, 244, 254-255
Loculoascomycete, 158, 159
Loose smut, 203-204
Low-volatile fumigants, 77
Luteovirus, 304

Macroconidia, 180
Magnaporthe grisea, 193, 217, 218
Maize, 2
Maize chlorotic dwarf, 292
Maize chlorotic dwarf virus, 292, 306-307
Maize dwarf mosaic, 292, 306-307
Maize dwarf mosaic virus, 286, 292, 306-307
Male sterility
  southern leaf blight of corn, 225-226
  Texas cytoplasm, 225-226
Management of plant diseases, 7, 38-66
  avoidance of pathogen, 39
  bacteria, 264-265
  certified seed, 62-66
  chemicals, 67-111
  cropping systems, 47-50, 136-137
  disease resistant cultivars, 58-62
  economic threshold, 35
  environment, adjustment of, 52-54
  eradication, 38, 39-41
  evasion, 38-39
  exclusion, 38-39
  greenhouse management, 50-52
  inoculum reduction, 41-52
  insect management, 52
  nematodes, 136-138
  physical methods, 52-54
  postharvest treatment, 81
  rotations, 47-50
  sanitation, 41-47
  seed storage, 64-66
  seed treatment, 81
  viruses, 290
  weed management, 52
Mastigomycotina, 158-159
Material Safety Data Sheets (MSDS), 69-71
Matthews, R. E. F., 287
Mechanical agitator, 90
Meloidogyne spp., 13, 139
  arenaria, 142
  hapla, 142
  incognita, 13, 14, 142
  javanica, 142
Meristem culture, 117
Meteorology, 4
Micheli, P. A., 17
Microconidia, 180
Microsclerotia, 157, 190
Milk weed, 303
Millardet, P. M. A., 18, 204
Mistletoe, 12, 309, 312-313
Monceau, Duhamel de, 16
Monitoring pests, 34
Monoculture, 6, 47
Monocyclic diseases, 25
Moyer, Adolph, 18
Multipurpose fumigants, 76
Mycelia sterilia, 160
Mycelium, 10
Mycology, 3
Mycoplasma, 9, 11, 19, 283
  diseases, 28-282
Mycorrhizae, 56, 123, 155, 255
Mycotoxins, 53, 155
Myxomycota, 158

Needham, J. T., 17
Nematicides, 10, 18
Nematodes, 4, 9-10, 25-26, 182
  anatomy, 129-130
  assays, 138
  disease complexes, 135-136
  dispersal, 132-133

Nematodes *(continued)*
- distribution, 129-130
- ectoparasites, 132
- endoparasites, 132, 151
- enemies, 133-134
- feeding behavior, 130-131
- importance, 127-128
- lesion, 147-150
- management, 136-138
- plant injury, 130-131
- reproduction, 132-133
- root knot, 140-145
- size, 129-130
- soil assays, 138
- stem and bulb nematodes, 139, 150-151
- stubby root, 139, 151-153
- stylet, 130-131, 151
- survival, 134-135

Nematology, 18
Nicotiana rustica, 121
Nicotiana tabacum, 121
Nightshade, 298, 303
Nonfumigants, 76
- granular application, 80-81
- liquid application, 81

Noninfectious disease, 12, 316
Nonpersistent, 49, 108-109
- pathogens, 49
- pesticides, 108-109

Nonselective pesticide, 68
Nonspecific resistance, 60
Nozzles, 92-94
- boomless, 92-93
- disk-core cone, 93-94
- even fan, 92
- flat fan, 92
- flooding, 92
- hollow cone, 93
- solid cone, 94
- whirl chamber, 92

Nucleic acid, 9, 11
Nucleoprotein, 18, 284
Nucleotides, 121
Nutritional deficiencies, 12

Oak, 21
- mistletoe, 313
- wilt, 23, 244, 248-249

Oat-bird cherry aphid, 304
Oats, 2, 48, 59
- barley yellow dwarf, 304
- Obligate parasites, 155

Old white hose, 13
Oomycetes, 158, 160, 171
Oospores, 163
Orobanche spp., 309, 310
Organ culture, 115
Ornamentals
- anthracnose, 232-236
- Cercospora leaf spot, 221-225
- crown gall, 278-280
- dodder on, 314-315
- gray mold, 219-221
- petal blight of azaleas, 187
- Phytophthora root rot, 167-170
- powdery mildew, 212
- *Sclerotinia* spp., 187

Ozone, 325-329

Papaya ringspot virus, 286
Parasitic plants, 12
Paratrichodorus spp., 139, 152-153
Parthenogenesis, 133
Pasteur, Louis, 17, 259
Pasteurization, 43-46, 51
Pathogens, 6, 22-24
- dispersal, 22
  - plant material, 23
  - vectors, 23-24
  - water, 23
- wind, 22

Peach, 54
- anthracnose, 233-234
- bacterial spot, 276-278
- brown rot, 23
- crown gall, 278-280
- lesion nematode, 147-150
- powdery mildew, 212
- Rhizopus soft rot, 240-242
- stubby root nematode, 151-153

Peanut stunt virus, 23
Peanuts, 2, 53, 54, 65
- bacterial wilt, 234
- Cercospora leaf spot, 82, 222-225
- crown rot, 186-187
- Cylindrocladium black rot, 189-191
- Rhizopus soft rot, 241
- root knot, 140-145
- *Sclerotinia* spp., 186
- *Sclerotium rolfsii,* 183, 222
- southern blight, 183-186
- stem rot, 183-186, 333
- tomato spotted wilt virus, 299-302

Pear, fire blight, 260, 269-272
Peas, 2
Pectin, 118, 241, 273
Penicillin, 54, 55
Penicillium spp., 54, 55
Peniophora gigantea, 55
Pepper, 47, 51
- potato virus Y, 297
- tobacco etch virus, 299

tobacco mosaic, 295
tomato spotted wilt, 299-302
Pepper mosaic virus, 286
Perithecia, 159, 160
Persistent, 49, 109
pathogens, 49
pesticides, 109
Pest control (*see* Pest management)
Pest management principles, 32-33
Pesticide, 67-111
carryover, 99
chemical name, 69
combination, 74
common name, 69
drift, 110-111
label, 69, 71
misuse, 5
naming, 69
nematicides, 137
particle size, 110
protecting the environment, 107-111
protective equipment, 70, 85
residues, 72
safety, 69-72
signal words, 69-71
tolerance, 72
trade name, 69
Pesticide application
applying fumigants, 77-81
checklist, 70-72, 95-97
controlled droplet application, 83
electrostatic application, 82
equipment, 83-94
guidelines, 95-97
nonfumigant application, 80-81
on growing plants, 82-83
postharvest treatment, 81-82
seed treatment, 81
soil treatment, 75-81
Pesticide application equipment, 83-105
agitators, 87
calibration, 98-105
cleaning, 97-98
hoses, 91
nozzles, 92-94
pressure gauges, 84
pumps, 88-90
sprayer parts, 81
storage, 97-98
tanks, 87-88
Pesticide drift, 110-111
nozzle orifice, 110
particle size, 110
pressure, 110
vapor, 111
weather, 110
Pesticide dusters, 82
Pesticide formulation, 72-75
adjuvants, 75
combinations, 74-75
dispersal granules, 74
dusts, 73
emulsifiable concentrates, 73
flowables, 73
granules, 74
soluble powders, 74
solutions, 74
wettable granules, 74
wettable powders, 74
Pesticide safety, 69-71
label, 69
precaution, 70-72
residues, 72
tolerances, 73
Pesticide sprayers, 82
air blast sprayers, 86
general use, 85-86
hand sprayers, 84
high pressure sprayers, 86
parts, 87-94
small sprayers, 85
Pesticide types
antibiotic, 68
curative, 68
eradicant, 68
nonselective, 68
selective, 68
therapeutic, 68
Pesticide use records, 97
Pest management, 7, 34
Petri, R. J., 17
pH, 27-53, 57
Phaseolin, 124
Phoradendron spp., 13, 309, 312-313
Photosynthesis, 24
Phyllosphere, 24
Physalis spp., 298
Physics, 4
Phytopathology, 3
Phytophthora spp., 49, 145, 162, 164
cinnamoni, 163, 168, 244, 254, 255
fragariae, 163
infestans, 28, 119, 193, 207, 208, 230
megasperma, 163, 165, 167, 168
nicotianae, 167
Phytophthora root rots, 3, 163-170
black shank, 166-267
management, 165, 167, 169-170
of ornamentals, 167-170
of soybeans, 163-165
Phytotoxic chemicals
acid rain, 329

Phytotoxic chemicals *(continued)*
- air pollution, 324-331
- pesticides, 321-324
- smog, 325, 327
- smoke, 325

Pine
- annosus root and butt rot, 249-250
- fusiform rust, 250-253
- little leaf, 254-255

Pine wood nematode, 128
Pineapple, 2
Pinus palustris, brown spot needle blight, 41
Pirie, N. W., 18, 284
Piston pump, 88
Plant cells
- haploid cells, 18
- protoplast fusion, 120-121
- protoplasts, 118
- regeneration, 118

Plant disease, 3-7, 9-14
- causes of, 9-14
- definition, 6
- disease cycle, 21-25
  - pathogen dispersal, 22-24
  - pathogen growth, 24-25
  - pathogen survival, 21
- development of, 20-29
- epidemiology, 27-29
- forecasting, 83, 105-107
- losses, 4-5
- management, 30-111
- noninfectious, 12
- physiology, 18
- recognizing, 6-7

Plant disease epidemiology, 27
Plant disease interaction (*see* Disease complexes)
Plant pathology, 3
- history, 15-19

Plant pathosystem, 27
Plant viruses, 12
Plasmid, DNA, 124-125
- *Agrobacterium tumefaciens,* 124-125
- crown gall, 124-125

Plasmopara viticola, 193, 209
Plato, 15
Plectomycete, 158, 159
Plum, 54
- bacterial spot, 276-278

Plum curculio, 23
Podosphaera, 193, 213
Poinsettia
- fertilizer damager, 318

Pollutants, 12, 32
Pollution, 33
- air, 1, 41
- water, 1

Polycyclic diseases, 25
Polymyxa graminis, 306
Poria spp., 255
Poria incrassata, 255
Postemergence damping-off, 170
Potato, 2, 61
- Alternaria leaf spot, 230-233
- bacterial soft rot, 272-273
- bacterial wilt (Granville wilt), 267-269
- broomrape, 310-311
- cyst nematode, 145
- dodder on, 314-315
- early blight, 230-231
- late blight, 3, 24, 29, 41, 47, 82, 106, 199, 205-209
- leaf roll virus, 23
- lesion nematode, 147-150
- potato spindle tuber, 308
- *Rhizoctonia solani,* 173-178
- rugose mosaic, 297
- soft rot, 273
- southern blight, 183-189
- vein banding, 297-299

Potato murrain, 17
Potato virus X, 288
Potato virus Y, 23, 60, 115, 286, 292, 297-299
Potyviruses, 286, 297
Powdery mildew, 68, 212-215
Pratylenchus spp., 139, 149
- *brachyurus,* 149
- *scribneri,* 149
- *zeae,* 149

Preemergence damping-off, 170, 173
Prevost, Isaac Benedict, 17
Protection, 38
Protection with chemicals, 67-111
Protein, 9
Primary infection, 293
Protoplasts, 118-1290
- fusion, 120
- transplanting cell organelles, 121

Pseudomonas spp., 13
- *solanacearum,* 13, 266-268
- *syringae,* 115

Pseudosperonospora cubensis, 193, 211
Puccinia spp.
- *graminis,* 13, 28, 193-196
- *recondita,* 197-198

Pumps for sprayers, 88-90
- centrifugal, 90
- diaphragm, 88
- flexible-impeller, 88-90
- piston, 88
- roller-impeller, 88

Pustules, 194
Pycnia, 195

Pycnidia, 159
Pycniospores, 195
Pyrenomycete, 158, 159
Pyricularia grisea, 217
Pythium spp., 149, 162, 171-172, 175, 181

Quarantines, 39

Race specific, 60
Races
  Bipolaris maydis, 226
  of rust fungi, 195-196
Radopholus, 139
Ragweed, 303
Rate of disease increase, 27
Rate of epidemic increase, 27
Receptive hyphae, 195
Recombinant DNA, 121, 123
Red leaf, 305
Reduction of inoculum,
  cropping systems on rotations, 47-50
  greenhouse management, 50-52
  insect management, 52
  sanitation, 41-47
  weed management, 52
Registered seed, 63-64
Relative humidity, 25
Repeating spores (*see* Urediniospores)
Residue, pesticide, 72
Resistant cultivars, 50, 58
Respiration rate, 24
Restriction enzyme, 121
Rhizoctonia spp., 16, 49, 181
  cerealis, 173
  solani, 14, 56, 159, 162, 171, 173-178
    belly rot, 175
    brown patch, 174, 175
    crown rot, 173
    damping off, 173
    leaf blight, 173
    on potatoes, 175
    root rot, 173
    sore shin, 173
    target spot, 174
    thread blight, 173
    web blight, 173
    wire stem, 173
Rhizoids, 241
Rhizomorph, 255
Rhizopus spp., 193, 240-242
Rhizosphere, 24, 56
Rhododendron, Phytophthora root rot, 166-167
Ribonucleic acid (RNA), 11, 18
  double stranded, 287
  single stranded, 287, 293
Rice, 2, 53
  blast, 3, 217-219
  brown spot, 229-230
  Rhizoctonia solani, 176
  sesame leaf blight, 229
  witchweed, 311-312
Ring nematode, 139
Ring rot, 47
Rishbeth, J., 55
RNA, 11, 18, 287
Robigalia, 15
Robigus, 15
Roguing, 40
Roller-impeller pump, 88
Root inhabitors, 49
Root knot nematodes, 3, 12, 13, 17, 131, 139, 140-145, 167, 181, 268, 269
  development, 143
  disease cycle, 142-144
  environmental effects, 144
  Meloidogyne spp., 142
  other diseases, 145
  reproduction, 143-144
  symptoms, 140-141
Root-rotting fungi, 53
Rosa alba L., 12, 13, 14
Rose, 14
  gray mold, 219-221
  powdery mildew, 212-215
Rotation, 48
Rotation crops, 34
Rots, 173-178
  crown and stem, 186-189
  Cylindrocladium spp., 189-191
  Fusarium root rot, 178-181
  Phytophthora spp., 163-170
  Rhizoctonia spp., 173-178
Rotylenchulus, 139
Rugose mosaic, 297
Rusts, 25, 52, 192-201
  cedar apple, 198-200
  fusiform, 250-253
  leaf, 197-198
  stem, 193-196
Rye, 2, 65
  ergot, 3, 15, 23, 215-217

Sanitation, 41-47, 50
Scientific names, 14
Sclerotia, 25
Sclerotinia spp., 56, 162, 186-189
Sclerotium rolfsii, 4, 162, 183-185, 222
Seed
  certified seed, 62-64
  quality, 65
  storage of, 64-65

Seed corn maggot, 23
Seedling diseases
  damping-off, 166, 170-173
  Fusarium rots, 178
  Pythium spp., 170-173
  Rhizoctonia solani, 173
Seed treatment, 81
Selective pesticide, 68
Small grains
  anthracnose, 233
  barley yellow dwarf, 304-305
  smut diseases, 203-204
  stem and bulb nematodes, 150-151
Smith, E. F., 18
Smog, 325, 327
Smuts, 59, 201-204
  corn, 201-203
  covered, 203
  loose, 203-204
Soft rot, 23, 266, 272-273
Soil inhabitors, 49
Soil invaders, 49
Soilborne pathogens, 25
Soilborne wheat mosaic virus, 305-306
Solanum aculeatissimum, 298
Solanum carolinese, 296
Sorghum, 2, 53, 65
Southern blight, 42, 162, 183-186
Southern corn leaf blight, 5, 119, 225-229
Southern stem rot (*see* Southern blight)
Soybeans, 2, 48, 52, 53, 65
  anthracnose, 233-236
  crown and stem rot, 186-189
  Cylindrocladium black rot, 163-164
  Cyst nematode, 145-146
  Frogeye (Cercospora) leaf spot, 222-225
  Phytophthora root rot, 163-164
  Sclerotinia spp., 186-189
  southern blight, 183-186
  witchweed, 311-312
Spatial pattern, 29
Species, 13-14
Spermosphere, 24
Sphaerotheca spp., 193, 213
Spiral, 139
Spiroplasma, 13, 282
Spontaneous generation, 16
Sporangia, 160, 163, 164, 171, 207, 208, 209
Sporangiophore, 163, 210
Spore, 157-161
Sporodesmium spp., 56
Stakman, E. C., 18
Stanley, W. M., 18, 284
St. Anthony's fire, 215
Statistics, 4
Steiner, G., 18
Stem and bulb nematodes, 139, 150-151
  disease cycle, 150
  Ditylenchus spp., 150
  management, 151
  symptoms, 150
Sterol inhibitor fungicides, 18
Sting nematodes, 132, 139
Stinking smut, 203
Stomates, 24
Storage rot, 266
Strawberry, 57
  anthracnose, 234, 236
  gray mold, 219-221
  Phytophthora root rot, 163
  stem and bulb nematodes, 150-151
Streptomyces spp., 265
Streptomycin, 55
Striga spp., 309, 311
Stromata, 234
Stubby-root, 139, 151-153
Stylet, 9
Sugar beet (*see* beet [sugar])
Sugarcane, 2
  witchweed, 311-312
Sugarcane mosaic virus, 23
Summer spores (*see* Urediniospores)
Suppressive soils, 54
Sustainable agriculture, 33, 41, 50
Sweet corn, 48
Sweet potato, 2
  bacterial soft rot, 272-273
  Rhizopus soft rot, 240-242
Swimming spores (*see* Zoospores)
Symptoms, 6
Systemic pesticides, 68

Target spot, 174
Teliospores, 159, 160, 194, 195, 198, 199, 200
Thanetephorus cucumeris, 159, 173, 174
Theophrastus, 15, 312
Therapeutic, 68
Thielaviopsis basicola, 115
Thrips, 23
Thrips tabaci, 300
Tilletia spp., 193, 203
Tissue culture, 115, 117-118
Tobacco, 2, 47, 54, 59, 60, 115
  anthracnose, 233-236
  bacterial barn rot, 272-274
  bacterial wilt (Granville wilt), 267-269
  black root rot, 115
  black shank, 166-167
  blue mold, 59, 60
  broomrape, 310-311

Cercospora leaf spot, 221-225
cucumber mosaic, 302-303
etch, 299
gray mold, 219-221
lesion nematode, 147-150
ozone damage, 327
powdery mildew, 212-215
Rhizoctonia solani, 174
Rhizopus soft rot, 240-242
stubby root nematode, 151, 153
tobacco etch, 299
tobacco mosaic, 286, 292-297
tobacco vein mottle, 211
tomato spotted wilt, 299
vine banding, 297-299
vein mottle, 299
weather fleck (ozone injury), 327
wildfire, 115
witchweed, 311-312
Tobacco etch, 292
Tobacco etch virus (TEV), 292, 299
Tobacco leaf-curl virus, 23
Tobacco necrosis satellite virus, 288
Tomato, 47, 48, 51, 57, 115
Alternaria leaf spot, 230-233
bacterial canker, 274-276
bacterial wilt (Granville wilt), 267-269
broomrape, 310-311
cucumber mosaic, 302-303
early blight, 230
gray mold, 219-221
herbicide injury, 324
lesion nematode, 147-150
potato virus Y, 297
Rhizoctonia solani, 175
Southern blight, 183
spotted wilt, 299-302
stubby root nematode, 151-153
sulfur dioxide injury, 327
tobacco mosaic, 115, 295
vein banding, 247-291
Tomato spotted wilt virus, 299-302
Toxicities
acid soil, 317-318
alkali soil, 317-318
nutrient, 317
Toxins, 5, 25, 58, 60
Trade name, 69
Transgenic plants, 58, 114-115
Transport systems, 24
Trichoderma spp., 55
Trichodorus spp., 139, 152-153
Tumor-induced (Ti) plasmid, 124, 278, 280
Turfgrasses
anthracnose, 233
brown patch, 5, 174-178
dollar spot, 187
Helminthosporium leaf spots, 225
lesion nematode, 147-150
powdery mildew, 212
Rhizoctonia solani, 174-178
Sclerotinia spp., 187
stubby root nematode, 151-153
Tylenchulus spp., 139

Ultra low volume, 82, 83
Ultraviolet (UV) light, 54
Ultraviolet (UV) radiation, 330
Urediniospores (Uredospores), 28, 160, 194, 197, 200
Ustilago, spp.
tritici, 193, 203
zeae, 193, 202

Vanderplank, J. E., 18
Vectors, 11
Vegetables
Alternaria leaf spot, 230-233
anthracnose, 233-236
bacterial soft rot, 272-279
Cercospora leaf spot, 211-225
crown and stem rot, 186-189
cucumber mosaic, 302-303
dodder on, 314-315
gray mold, 219-221
lesion nematode, 147-150
powdery mildew, 212-215
Rhizoctonia solani, 173-178
Rhizopus soft rot, 240-242
Sclerotinia spp., 186-189
southern blight, 183-185
tobacco mosaic, 293-297
vein banding, 297-298
Vegetative propagation of plants, 115-118
adventitious shoots, 116-117
callus tissue, 116-117
meristem culture, 117
organ culture, 115-116
protoplasts, 118-119
single-cell culture, 117-118
tissue culture, 117-118
Vein banding, 60, 292, 297-298
Venturia inaequalis, 193, 236
Vertical resistance, 60
Verticillium spp.
albo-atrum, 182
dahliae, 182
Verticillium wilts, 181-183
Victoria oats, 59
blight, 59

Viral inclusions, 288
Viroid, 9, 12, 19, 285, 308
Viroid diseases
  management, 308
  symptoms, 308
Virology, 3
Virus, 9, 11, 18
Virus diseases
  detection, 288-289
  identification, 286, 288-289
  management, 286
  morphology, 287-288
  naming, 285-286
  properties, 287-288
  symptoms, 286
  transmission, 289-290
  vectors, 290
Virus indexing, 288
Virus serology, 288

Waite, M. B., 18
Wallace, G. W., 17
Watermelon, downy mildew, 241
Watson, J. P., 284
Weather extremes, 318-320
  cold, 318-319
  frost, 318-319
  heat, 319
  lightning, 319-320
Web blight, 173
Weed science, 4
Weindling, R., 54
Wheat, 2, 25, 53, 61, 64
  barley yellow dwarf, 304-305
  bunt, 203
  covered smut, 203
  loose smut, 203-204
  powdery mildew, 212-215
  Puccinia graminis, 193-196
  Puccinia recondita, 197-198
  rust, 5, 13, 28, 193-198
  smut, 203-204
  soilborne wheat mosaic, 305-306
  Tilletia spp., 203
  Ustilago tritici, 203-204
Wheat gall nematodes, 128
Whetzel, H. H., 18
White flies, 23
Witchweed, 309, 311-312
Wood decay, 255-258
  causal agents, 255
  management, 256-258
  prevention, 256-258
  symptoms, 255
Wood preservation, 258
Wound tumor virus, 288

***X**anthomonas campestris*, 266, 277-278
Xiphinema spp., 139

Yam, 2
Yellows diseases, 11
Yield losses, 8

Zoosporangium, 157, 160, 171
Zoospores, 23, 164, 165, 171
Zygomycete, 158, 160
Zygomycotina, 158, 159, 241
Zygospores, 160, 241